Applied Graph Theory

Applied Graph Theory

CLIFFORD W. MARSHALL

Professor of Mathematics
Polytechnic Institute of Brooklyn

WILEY–INTERSCIENCE

A Division of John Wiley and Sons, Inc.
New York · London · Sydney · Toronto

To

RONALD M. FOSTER

Our dreams have wings that falter,
　Our hearts bear hopes that die:
For thee no dream could better
A life no fears may fetter,
A pride no care can alter,
　That wots not whence or why
Our dreams have wings that falter,
　Our hearts bear hopes that die.

To a Seamew—A. C. SWINBURNE

Preface

This book is an introduction to graph theory and its applications. An understanding of the material presented here should allow the reader to continue his own study in more advanced or special aspects of graph theory. In addition he should be able to appreciate applications of graph theory in any of the variety of areas to which it may be applied. I do not attempt to present a complete coverage of the theory, since the subject, already vast, is increasing at a rapid rate. Present applications of the theory include many subjects, and the potential for even wider application seems almost without limit. Some typical applications are considered, both because they are interesting and also because they indicate the kind of applications possible in the theory of graphs. I attempt to give enough applications to demonstrate the potential of graph theory as an important area of applied mathematics while touching on topics interesting to many readers. Many applications are not included; the choice is based on the historical importance of some applications, the degree to which they typified utilization of graph theory, and of course my interests and experience.

Though the main intent of the book is to direct the reader's interest to applications of graph theory, the theory is purely mathematical. It can be studied without reference to applications outside of mathematics. As a mathematics subject, graph theory has much in common with both logic and topology, as well as being closely related to combinatorial theory. Except for discussion of certain combinatorial aspects of graph theory this book does not develop the connections between graph theory and other areas of abstract mathematical research. The mathematical content of graph theory is presented on its own terms in an elementary way. Thus except for certain aspects of the subject, for example, random graphs, the reader does not need much formal mathematical knowledge.

The book can be read by undergraduate or graduate students who wish to learn what graph theory is and how it may be applied. Much of the material presented does not require any prerequisite other than a degree of mature reasoning ability. Some topics that do require background knowledge are included, to give the book reasonable scope, on the assumption that most readers have a sufficient amount of such knowledge. Those readers who lack the background for a particular section or result will find guidance in the Appendix where some of the necessary material is summarized. Others can skip such sections and return to them after acquiring some background knowledge. The Appendix covers topics from linear and integer programming. Though some knowledge of probability and linear algebra is required in certain sections of the book, those subjects are not treated at all in the Appendix, because they are too elaborate for brief summation.

The first six chapters of the book present major topics and concepts of graph theory. Probably anyone interested in graph theory would know or want to know that body of material. Some applications are indicated within these chapters by illustration or explanation of concepts. Chapter 7 discusses combinatorial theory in relation to graph theory and presents some important results that are part of both theories. No attempt is made even to survey the extremely broad field of combinatorial theory in graphs; rather, the reader is shown what is contained in that theory, and he is made familiar with useful and basic results. The last three chapters present aspects of applications of graph theory to operations research, sociology, and physics. Again, the presentation is not intended to give surveys of many applications but to develop a few indicative utilizations of graph theory in sufficient detail to give the reader a feeling of when and how graph theory may be useful. The applications chapters include additional graph theory concepts that are introduced and developed as needed for the applications under discussion. It should be noted that one is often tempted to speak of a simple linear graph representation of a problem as an application of graph theory, for example, a network of roads as an application of graph theory, in traffic analysis. Such a representation may indeed provide a first step in some meaningful graph theory application but is hardly an application on its own. One must employ the detailed concepts and deductive results of graph theory in resolving a problem if one is to speak of applying graph theory. With the possible exception of a few simple examples and illustrations, these criteria have guided the selection and presentation of the applications of graph theory given here.

Concepts and theorems are illustrated in each chapter as they occur. Some of the more advanced or involved applications are illustrated by simplified results. Each chapter contains exercises for the reader to help him to understand and fix the material of the chapter in his mind. References are listed, in order of their occurrence in the text, at the end of the book. They are indicated in the text by a number in brackets. I think that the references given are the best (or not less than best) for each subject in which they occur and that they provide an efficient direction for further study. The Bibliography also cites general reference works, including extensive bibliographical material. Journals in which most of the work on graph theory and its applications is reported are also listed in the Bibliography. An index of concepts and names is provided which should prove helpful in using the book as a reference work and as a textbook.

The completion of each proof is indicated by a triangle ▲. This concise method of separating proof from the subsequent text was used in mathematical literature by P. Halmos in 1950.

Notation is not uniform in graph theory because of the continuing development of the subject by different groups of people. In each case the term here is either the most commonly used in mathematical literature or has the sanction of usage by a leading authority in graph theory. In no case have I imposed my own term among the standard definitions (some terms in random graphs are mine). An extremely helpful reference to the extensive notation can be found in Essam and Fisher[1].

I wish to emphasize my dependence on works of many authors in the fields of combinatorial theory and graph theory. I have selected what I feel to be an interesting and instructive subset of the extant literature. Beyond that selection, the method of presentation, and the illustration of ideas, there is little in the book that is mine originally. All material used directly is referenced and listed in the Bibliography. The general dependence on other works is too extensive for complete citation. However, special mention must certainly be made of the importance of the writings of F. Harary and W. T. Tutte in graph theory and of H. J. Ryser and J. Riordan in combinatorial theory in preparing this work.

I wish to thank Miss Beatrice Shube, Editor, John Wiley and Sons, Inc., without whose help and encouragement this book would not have been written.

The book has benefited from the comments of three reviewers, and I thank them for their consideration and constructive advice, which has been followed so far as it was practical. Particular thanks go to one of the publisher's reviewers for his extensive substantive suggestions and his additions to the exercises in Chapters 4 and 6.

Several of my students at the Polytechnic Institute of Brooklyn provided references or insights that contribute to the book. Special appreciation is due to R. Flynn, J. Parsons, and E. Schroeppel.

Creation of the book was considerably assisted by the encouragement and support of Professor Harry Hochstadt, Head of the Institute's Mathematics

Department, and by the attention given to the effort by Mrs. Helen Warren of the Institute's Long Island Graduate Center.

Special thanks are due to Mrs. Rose DiGiacinto for her considerate and expert efforts in typing the manuscript.

It is with particular pleasure that I acknowledge the friendship and instruction of Professor Ronald Foster over a period of many years.

CLIFFORD W. MARSHALL

Brooklyn, New York
April 1971

Contents

Chapter 1

Basic Concepts

BASIC DEFINITIONS OF LINEAR GRAPHS

The idea of a set of elements is one of the most basic building blocks in mathematics. We assume here that the reader is to some extent familiar with elementary aspects of set theory. Specifically the equality of two sets, set inclusion, subsets, intersection, union, complementation, the null set denoted by Φ, and basic set algebra (such as De Morgan's formulas $\overline{\cup_\alpha A_\alpha} = \cap_\alpha \overline{A_\alpha}$ and $\overline{\cap_\alpha A_\alpha} = \cup_\alpha \overline{A_\alpha}$) are used without particular notice being given. On the other hand, set theory can be contemplated from an extremely sophisticated point of view (see, e.g., Lipschutz [2] for an elementary presentation and Stoll [3] for a more advanced treatment), for which we have no need here. Between the normal elementary preparation in set theory assumed for the reader and the set specialists detailed studies there are a few facts that will be very useful to have clearly in mind throughout the book.

For one thing the definition of a set is so elementary as to sometimes be given too little importance. Thus we emphasize that when defining a set as a collection of elements one must have a method of determination, applicable to an arbitrary element belonging to the totality of all such elements, which declares the element to be either a member of the set or unequivocally not a

1

member of the set. In addition we must clearly understand that the elements are distinct. This means that a rule exists that can be applied to an arbitrary element to determine whether the specific element has been previously identified, selected, or otherwise described.

Consider a set of elements $S = \{a_1, a_2, \ldots, a_n\}$, which we take as finite for illustrative purposes though it could just as well contain an infinite (countable or not) number of elements. It will be useful for us to have a concept for a collection of some of the elements of S including the possibility of repeated occurrence of elements. Such a collection cannot be a subset of S since we have observed that sets must have distinct elements. The required concept is given by a *k-sample of S*. A k-sample of S is a collection of k elements $\{b_1, b_2, \ldots, b_k\}$ where b_i is some element of S, thus for $i = l. \ldots, k$ there is an element a_{j_i} of s such that b_i is a_{j_i}. Of course this allows a single element of S to occur up to k times in a k-sample which is just what we require. If k is not specified we may speak of a sample of S and remark that a sample may contain more elements than S. When the elements in a k-sample are in fact distinct we may speak of a distinct k-sample or distinct sample. Of course distinct samples are subsets.

Consider the set (a, b, c, d, e) of five elements. The collection (a, b, c) is a subset or distinct 3-sample whereas the collection (a, a, b) is a 3-sample whose elements are not all distinct.

Another important concept is that of a mapping or function. We use these words synonymously to indicate a correspondence between two sets S_1 and S_2 as is usual. As in set theory we assume the reader has the basic ideas of a mapping in hand, and we do not need extremely sophisticated notions. It should be pointed out, however, that at times we may use a symbol such as F to stand for a function and the notation $F(a)$ to stand for the image element corresponding to the element a under the mapping F. At other times it may be convenient to use an explicit (though symbolic) formula to express the function value such as in a mapping of the real line into the real line given by $G(x) = 3x^2$. When this is done we may also refer to the function $G(x)$ or the function $3x^2$ particularly when dealing with examples and applications. Though this is not according to modern usage it is felt that having pointed out such usage here may avoid any misunderstanding by the reader and allow the term function to be rather freely utilized throughout the text.

A linear graph, as other mathematical objects, may be defined in several ways. Once a definition is selected, its alternatives become representations of the same object and often play an equal or even more dominant role in the development of the subject than is occupied by the definition itself. Thus the choice of definition becomes a matter of choice or chance, often being one of the more abstract available possibilities or one of the most succinct. Following this custom we choose to make the following abstract definition.

A linear graph $G(N, C)$ is a set of elements N and a collection C of un-ordered pairs (a, b) of elements of N. It is sometimes useful to indicate the dependence of N and C on the graph G by the notation $N(G)$ and $C(G)$. The elements a and b need not be distinct and a given pair (a, b) can occur more than once in C (it can in fact occur an infinite number of times). Thus C is a sample from the set of all distinct unordered pairs of elements of N. The terms linear graph and graph may be used interchangeably provided there is no chance of confusion with some other object, for example, the " graph " of a function. The correct term is linear graph which indicates that the subject deals with incidence relations between one simplices (line segments) and zero simplices (points). Though more general incidence relations can be studied they do not form part of (basic) graph theory.

For purpose of discussion and explanation it is convenient to introduce a representation of a linear graph G. This representation makes use of pictorial language and allows pictorial presentation of linear graphs. The elements of N are represented by points and the pairs (a, b) of elements of N are repre-sented by lines (which need not be straight lines), and drawn in a plane. A line is conceived for each pair (a, b) so that when a pair occurs several times several lines must also be formed, one for each occurrence of the pair (a, b). The pictorial representation is clear, we set out points corresponding to the elements of N and for every pair (a, b) of such points in the collection C we draw a line (of arbitrary length and shape). Such a representation leads to various names for the elements of N and the elements of C. In formulating the representation we have used points and lines for these quantities. The terms nodes and arcs are widely used as well as other alternatives. For most of our presentation we shall use the term *vertices* for elements of N and the term *edges* for elements of C. This terminology is reasonably descriptive and seems to be the most widely used at present. When using the terms vertices and edges one must remember that they stand for very general concepts here and are not to be confused with the same terms in the geometry of simple Euclidean figures. When we have occasion to deal with the latter the context will make the usage clear. In this way we utilize a unified and widely accepted terminology with only a slight danger of misunderstanding. Occasionally the terms point and line or the terms node and arc will be used, particularly in certain applications where such terminology seems to have become either standardized or is classical.

To make our discussion of linear graphs less formal and abstract and also to provide for the representation of particular graphs we introduce labels for the vertices. Such labels may be letters, numbers, or both, according to con-venience. For general discussion we denote the vertices by letters so that N becomes a set of letters (if more than 26 vertices are required we can use subscripts etc.). The edges are then pairs of letters such as (a, b) representing the edge associated with vertices a and b where a, $b \in N$. Thus the elements of the collection C are pairs (x, y) with x, $y \in N$ but not necessarily every possible pair is in C; in fact the general case has only some of the possible pairs comprising C, where some pairs may be repeated and pairs like (a, a) (self loops) are allowed. For many studies it is desirable to exclude self loops and multiple edges. Linear graphs resulting from such restrictions are called *simple graphs*. In general the term simple is used in graph theory to mean without repetition.

If the number of elements in N is k then we can call any graph based on these k vertices a *k*-graph. As we have seen above a convenient notation for a graph is $G = (N, C)$ or if the number of vertices is to be more explicitly put into evidence $G = (N(k), C)$. When both N and C are finite G is called a *finite graph*, otherwise G is an infinite graph. In general we will assume G is finite unless otherwise specified.

Though the elements of C, that is, pairs (a, b) of elements of N are introduced as being unordered it is sometimes useful to treat them as ordered pairs. If all pairs are ordered one calls G a *directed graph* or *digraph*. (This term is attributed to G. Polya by F. Harary, see [4], p. 2). Thus when we speak of digraphs the pairs of vertices are ordered and conversely. In the pictorial representation of digraphs one places an arrowhead on each edge pointing away from the vertex which is the first element of the ordered pair and pointing toward the vertex which is the second element. It is not important where the arrow head is placed on the edge. It is common to call the edges of a digraph arcs and the vertices nodes though such usage is not universal.

As an example consider two 5-graphs based on the vertex set $N = (a, b, c, d, e)$. One such graph has the edge collection $C = [(a, b), (b, b), (b, c), (c, d), (c, d), (d, e), (e, a), (e, a)]$ and the other 5-graph has the edge collection $C = [(a, e), (b, c), (c, d), (d, b)]$. A directed graph based on the same vertex set N has collection $C = [(a, e), (b, a), (c, b), (d, c), (e, d), (d, e)]$. In the last collection C we note that (e, d) and (d, e) both occur and since a digraph is being represented these correspond to two different edges; the first is from vertex e to vertex d whereas the second is from vertex d to vertex e. When an arc is directed from a vertex r to a vertex s it is said to originate at r and to terminate at s.

The pictorial representation is an obvious and simple one which is nevertheless extremely useful in guiding our reasoning, explaining concepts and results and serving as a direct model for structural situation over a wide range of applications. The placement of the points and the shape of the lines are of no importance in the general study of graphs, though they may be of value in specific types of theoretical studies and in model formulations. Thus the lines are most often shown as straight but only for convenience. Lines may cross each other and such crossover points are not to be considered as vertices. In general such crossover of lines is of no interest but for some studies we will be particularly concerned with the occurrence of such situations. Unless specified as being important we shall not take notice of lines crossing.

The pictorial representation of Figure 1 yields the three representations for the three graphs given above.

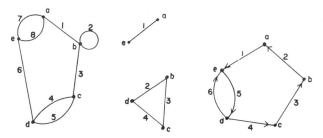

Figure 1.   Linear graphs.

Note that the lines have been drawn in various ways to emphasize the fact that their *form* is of no importance (in general studies). The pictorial representation does not consider detailed concepts that arise when one considers the representation as a point set in the plane. Such considerations lead into various topological studies most of which lie beyond the scope of this book. Some of the concepts are discussed in Chapter 3.

In many studies and applications of graph theory it is necessary to associate numerical values with each edge (arc) or vertex (node) of a linear graph (digraph). When numbers are associated with a graph in this way the result may be called a network and the numbers are given names relating to the intended usage or application. We believe this distinction between the underlying linear graph and the network concept can be a useful distinction in many applications. For example Figure 2 shows a network representation of the roads connecting several cities; the assigned numbers are the lengths in

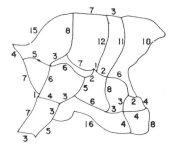

Figure 2.   A network of roads.

miles for each road. As a convenience numbers assigned to edges (arcs) are often given the general name of capacities and in applications to flow problems do in fact represent the capacity to pass material through each edge.

In [5] Fraley et al. have shown how linear graphs may be used to solve so-called crossing puzzles. Such puzzles are based on the classical puzzle of the cannibals and missionaries: "A group of three cannibals and three missionaries seeks to cross a river. The only available boat holds at most two people. If the cannibals outnumber the missionaries at any time, on either side of the river, they will do away with the missionaries. How can the group divide so as to move across the river?"

A graph formulation of the problem can be made by letting each vertex correspond to the number of cannibals c and missionaries m on the first bank of the river (when the boat is not in midstream). Such a graph can be represented with its vertices on appropriate lattice points of the plane, each vertex is determined by an allowable pair of values (c, m). Some consideration will show that: $0 \leq c \leq 3$, $0 \leq m \leq 3$, and $c = m$, or $c = 0$ or $m = 3$. Since the boat holds two people, the total decrease in c and m is either one or two when the boat goes from the first to the second bank. In addition the rules of passage require that $c \leq m$ in each case (in addition to the above restrictions). Similarly, one can stipulate the rules in moving from the second bank to the first. Fraley et al. employ a reachable points diagram to help formulate the graph of the problem. Vertices are connected by edges only when a transformation from one vertex (c, m) pair to the adjacent vertex pair can be achieved by a crossing. Once the corresponding graph has been drawn a solution is obtained by finding a directed path from (3, 3) to (0, 0) that alternates between a passage from the first bank to the second bank and a passage from the second bank to the first bank. One solution, given in [5] is the following vertex sequence: (3, 3), (1, 3), (2, 3), (0, 3), (1, 3), (1, 1), (2, 2), (2, 0), (3, 0), (1, 0), (2, 0), (0, 0). The reader is invited to formulate the graph representation of this problem and look for other solutions. If he does so he will find that some solutions require more steps than other solutions. Fraley et al. point out this aspect of the problem and discuss the possibility of finding minimal step solutions to crossing puzzles. In addition they study more general crossing puzzles by means of the corresponding transition graphs. Generalization takes place in both the number of cannibals (or missionaries) and in the capacity of the boat. It is found that four cannibals and four missionaries can not be taken across in a two person capacity boat. If the boat capacity is increased to three, up to five cannibals and five missionaries can be safely transported across; however six cannibals and six missionaries cannot be transported. If the boat capacity is increased to four any number of cannibals and missionaries can be taken across. An alternating directed path along the diagonal vertices is always a solution when the boat capacity is at least four. Such a solution is not always the minimum number of crossings path. It is a minimal crossings path if the boat capacity is even, and is less than the number of cannibals (or missionaries). An example in which the diagonal solution is not a minimal crossings path is given by boat capacity of five with six cannibals and six missionaries. These results and various other aspects of such crossing problems may be found in [5].

EDGE SEQUENCES AND CONNECTEDNESS

We now introduce a number of concepts which can be utilized either in undirected or in directed graphs. The literature of graph theory contains several different names for some of these and often uses a different name when the object occurs in a directed graph from that used in undirected cases. This situation is treated here by defining concepts in undirected graphs then using the prefix directed when the concept is used in a directed graph. Moreover, the names and definitions which seem to be most widely used or most clearly expressed shall be given.

Suppose a graph has labelled vertices so that the vertices can be denoted by $a_1, a_2, \ldots, a_n$ for an n-graph. An edge determined by two vertices a_i and a_j is written (a_i, a_j) and an *edge sequence* is a sequence of such pairs in which a vertex occurs in a pair and in the next pair of the sequence (note that for undirected graphs the order of the vertices within a pair is of no consequence). It may be that two vertices occur in only one edge each but all other vertices occur in two edges. The special vertices occur in the first and last edges of a finite edge sequence, they are called *initial* and *terminal* vertices respectively, other vertices are called *inner vertices* (of the edge sequence). Symbolically one can express an edge sequence in terms of its constituent edges as (a_0, a_1) (a_1, a_2) $(a_2, a_3) \ldots (a_{k+1}, a_k)$ where there are k edges, a_0 is the initial vertex and a_k the terminal vertex. Edge sequences are also called *chains* but we shall not use this term. It is possible for the same edge to occur more than once in an edge sequence. When a vertex occurs in only two edges we say the vertex occurs once in the edge sequence, a vertex may occur more than once in an edge sequence. Some edge sequences based on the graph of Figure 3 follow.

The (a_1, a_2) (a_2, a_3) (a_3, a_4) (a_4, a_6) is an edge sequence from a_1 to a_6 with no repeated edges or vertices, (a_1, a_5) (a_5, a_3) (a_3, a_4) (a_4, a_5) (a_5, a_1) is an edge sequence using edge (a_1, a_5) twice and vertex a_5 twice, (a_5, a_4) (a_4, a_6) (a_6, a_5) is an edge sequence in which the initial and terminal vertices are the same, namely vertex a_5. In this last sequence a_4 and a_6 are examples of inner vertices.

In an edge sequence the number of edges (counting multiplicities when they occur) in the sequence is called the length of the sequence, the three sequences above are of lengths 4, 5, and 3 respectively.

If the edges are directed we speak of a directed edge sequence. In this case the order of vertices within a pair is important and a directed sequence requires the first vertex of one pair to be the last vertex of the preceding pair except for the initial and terminal vertices. Of course, one may also form general edge sequences on the edges of a digraph by ignoring the directed character of the edges, such sequences are not to be considered as directed sequences.

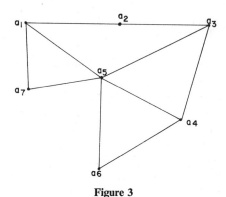

Figure 3

An edge sequence may contain an infinite number of edges. There are two types of infinite sequences. A sequence may have an initial or a terminal vertex but not both; such a sequence is called a one-way infinite sequence. If a sequence has neither an initial nor a terminal vertex it is called a two-way infinite sequence. We may observe that infinite edge sequences can occur in finite graphs. An example will be given after a few more concepts have been introduced.

We shall now introduce some special types of edge sequences which are particularly useful in graph theory. An edge sequence in which no edge is used more than once is called a _path_ or a simple chain. It may also be called a simple edge sequence, the term simple being widely used to denote lack of multiplicity in graph theory concepts. A path can utilize a single vertex several times as in the path from a_1 to a_6 of Figure 3 specified by: (a_1, a_5) $(a_5, a_7), (a_7, a_1) (a_1, a_2) (a_2, a_3) (a_3, a_5) (a_5, a_6)$. No edge is used more than once but vertex a_5 occurs twice.

A _way_ is a path in which no vertex occurs more than once (the term seems to be due to Nash-Williams). The edge sequence stated above is not a way but the following is a way from a_1 to a_6: $(a_1, a_5) (a_5, a_3) (a_3, a_4) (a_4, a_6)$. Note that it is convenient to use the phrase " way from " even though no direction is indicated, it would be just as useful to refer to this way as a way from a_6 to a_1. In the case of digraphs the phrase "way from" has a more rigid meaning and depends on the use of directed edges making up the appropriate edge sequence, in such cases one can speak of directed paths or directed ways.

The concept of length of an edge sequence is most useful in cases where no edge is used more than once. Thus the length of a path or of a way represents the number of distinct edges used in the formation of the sequence.

A path of length one is necessarily a way and consists of a single edge. The two vertices so connected are said to be _adjacent vertices_. Similarly two edges are said to be adjacent edges if they are incident on the same vertex. An edge sequence consists of adjacent edges.

When the initial and terminal vertices of an edge sequence are the same the sequence is called a _cycle._ A _simple cycle_ uses each edge only once. A _circuit_ is a simple cycle using each vertex once. A circuit may also be called a simple cyclic way. A way is a path without cycles.

In Figure 4 the edges have been assigned numerical labels so that, e.g., edge (a_5, a_6) is denoted by 1, edge (a_2, a_3) is denoted by 12 and so forth. From Figure 4 the following edge sequences are respectively a cycle, a simple cycle, and a circuit: (11, 9, 10, 11, 12, 7, 4, 10); (11, 8, 2, 3, 5, 4, 10); (11, 8, 4, 10).

It is not necessarily true that an infinite sequence of edges must contain an infinite number of distinct edges. For example if e_1 and e_2 are adjacent edges of a graph which form a 3-cycle with a third edge e_3, one may form the infinite sequence $e_1 \, e_2 \, e_3 \, e_1 \, e_2 \, e_3 \, \cdots$ which is one-way infinite since it must start with some initial edge (of course it is equivalent to the two other possible sequences of this type). Such infinite sequences lead one to consider infinite edge sequences in finite graphs. Though such sequences might be expected to have periodic structure (as illustrated by the three cycle above) this need not be the case. Consider a graph with 10 vertices in which every vertex pair is connected by one edge. A nonperiodic infinite edge sequence may be constructed on this 10-graph by selecting each edge of the sequence as that edge connecting the vertices numbered by two consecutive digits of an irrational number (such as π) (due to R. Flynn). This edge sequence would start with the following edges (specified by vertex numbers): (3, 1), (1, 4), (4, 1), (1, 5), (5, 9), (9, 2), $\cdots$.

In a digraph all the concepts described above may be applied except particular attention must be paid to the direction of the edges. The paths must consist of edges such that the initial vertex of one edge is the terminal vertex of the preceding edge. One can not allow reversal of any edges in the sequence. In Figure 5, $(a_1, a_2)\,(a_5, a_2)\,(a_5, a_1)$ is not a circuit but $(a_1, a_2)\,(a_2, a_3)\,(a_3, a_5)\,(a_5, a_1)$ is a circuit. The clearest means of

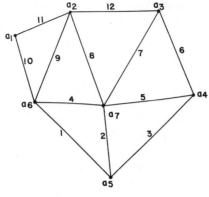

Figure 4

utilizing these concepts in digraphs is to use the same term preceded by "directed"; thus one speaks of directed cycles, directed circuits, and so on. Other terms are already widely used in the literature and cause no problem in specific areas but for a general presentation of these basic concepts it is desirable to employ as few terms as possible and give them clear description.

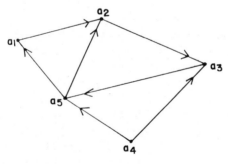

Figure 5

If there exists at least one path between every pair of vertices in a graph G then G is said to be _connected_. The concept of connectedness may be expressed in terms of more general edge sequences but the reader will observe that such definitions imply the usual path connection. In a digraph D the connectedness may be of two types, if a directed path exists between every pair of vertices then D is said to be _strongly connected_, if this is not the case but D is connected when considered as an undirected graph then D is said to be _weakly connected_. Figure 6 shows a connected graph G_1, an unconnected graph G_2, a strongly

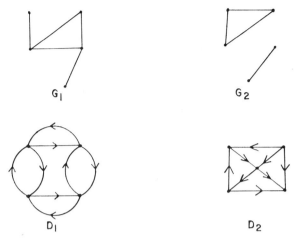

Figure 6

connected digraph D_1 and a weakly connected digraph D_2. A strongly connected digraph is necessarily connected. One may observe that a sufficient condition for a digraph to be strongly connected is that between every pair of vertices there is a directed edge in either direction. Such a graph is a complete directed graph.

The condition is certainly not necessary as illustrated by D_1 and by the strongly connected digraph shown in Figure 7.

Figure 7

The idea of connectedness, particularly in digraphs is sometimes expressed in terms of one vertex being *reachable* from another. In a digraph vertex B is reachable from vertex A if there exists a directed path whose initial vertex is A and whose terminal vertex is B. The pair of vertices are mutually reachable if B is reachable from A and A is reachable from B. In a strongly connected digraph every pair of vertices are mutually reachable. When every pair of vertices in a graph G are connected by at least m paths having no other vertice of G in common the graph G is called _m-connected._ A single circuit is a 2-connected graph and $m = 1$ corresponds to the normal concept of a connected graph. In a digraph the connections are in terms of directed paths and the equivalent concept is called strongly m-connected.

When a graph G is not connected it is necessarily composed of *connected components* each of which is a connected graph. The number of such connected components will be finite if G is finite; otherwise they may be infinite in number. A single vertex without edges incident upon it may be a component of a graph G; it is a single vertex component. Though most situations would not benefit from inclusion of such components it may prove necessary or useful to allow them in some cases. The connected components, or components for short, serve as a useful partition of a graph G into simpler elements. Denote the ith component of G by C_i and its vertex set by $V(C_i)$. If the vertex set of G is $V(G)$ then

$$V(G) = \underset{i}{U} V(C_i)$$

where $V(C_i) \cap V(C_j) = \Phi$ for $i \neq j$; thus the vertices are partitioned in the set theory sense. It is convenient to write $G = \sum_i C_i$, though the notation is formal and no algebra of graphs has been defined, the meaning here is clear enough. To every component C_i of a graph G is associated a *complement graph* which is all of G except for C_i. The complement graph of C_i is denoted by $\bar{C_i}$ and its vertex set is $\overline{V(C_i)}$. Though the complement graphs are not disjoint they are sometimes useful in expressing aspects of G.

MATRIX REPRESENTATIONS OF GRAPHS

The definitions and representations of a linear graph presented thus far are not analytic in nature but descriptive. Though a great deal can be developed on the basis of such representations it is also important for certain aspects of graph theory to have analytic representations available. By this we mean representations in terms of objects upon which we can carry out mathematical operations involving numbers. Two such representations will be given, both in terms of appropriately defined matrices which have real numbers as their elements.

For one type of numerical representation of a graph it is necessary to label the edges as well as the vertices. In many applications the edge is identified with physical quantities, hence achieves labels as part of the graph model formulation; in other cases we may simply assign labels for convenience. With both vertices and edges labeled one can define the *incidence matrix* representation of a graph. We may denote the incidence matrix of a graph by I_g where the graph is $G = (N, C)$. The elements of I_g are denoted by g_{ij} and are defined as follows:

$$g_{ij} = 1 \quad \text{if edge } j \text{ is incident on vertex } i,$$
$$g_{ij} = 0 \quad \text{otherwise.}$$

We say an edge is incident on (or to) a vertex if it has that vertex as one of its end points. Thus each row of I_g corresponds to a different vertex of G and each column of I_g corresponds to a distinct edge of G. Thus if N has k elements and C has m elements, I_g is a $k \times m$ matrix whose elements are zero or one. The incidence matrices representing each of three examples in Figure 1 are shown in (1).

$$
\begin{array}{c}
\begin{array}{cccccccc} 1 & 2 & 3 & 4 & 5 & 6 & 7 & 8 \end{array} \\
\begin{array}{c} a \\ b \\ c \\ d \\ e \end{array}
\begin{bmatrix}
1 & 0 & 0 & 0 & 0 & 0 & 1 & 1 \\
1 & 1 & 1 & 0 & 0 & 0 & 0 & 0 \\
0 & 0 & 1 & 1 & 1 & 0 & 0 & 0 \\
0 & 0 & 0 & 1 & 1 & 1 & 0 & 0 \\
0 & 0 & 0 & 0 & 1 & 1 & 1 & 1
\end{bmatrix}
\end{array}
\begin{array}{c}
\begin{array}{cccc} 1 & 2 & 3 & 4 \end{array} \\
\begin{bmatrix}
1 & 0 & 0 & 0 \\
0 & 1 & 1 & 0 \\
0 & 0 & 1 & 1 \\
0 & 1 & 0 & 1 \\
1 & 0 & 0 & 0
\end{bmatrix}
\end{array},
\begin{array}{c}
\begin{array}{cccccc} 1 & 2 & 3 & 4 & 5 & 6 \end{array} \\
\begin{bmatrix}
1 & -1 & 0 & 0 & 0 & 0 \\
0 & 1 & -1 & 0 & 0 & 0 \\
0 & 0 & 1 & -1 & 0 & 0 \\
0 & 0 & 0 & 1 & -1 & 1 \\
-1 & 0 & 0 & 0 & 1 & -1
\end{bmatrix}
\end{array}.
$$

In Figure 1 the edges have been assigned numerical labels. For digraphs the third example shows that the definition is modified as follows:

$g_{ij} = 1$ if edge j is directed away from vertex i and is not a loop,
$g_{ij} = -1$ if edge j is directed toward vertex i and is not a loop,
$g_{ij} = 0$ in all other cases (including edge j forming a loop at vertex i).

One can remark several elementary properties about these incidence matrices (often called *edge incidence matrices*). For digraphs every column sum is zero. For undirected graphs without loops every column sum is 2, every row sum equals the number of edges incident to the corresponding vertex for that row. In digraphs the row sum can be positive, negative or zero and indicates the excess of outward directed edges over inward directed edges when positive and the converse excess when negative. We see that by using various numerical properties of incidence matrices corresponding features of the graph may be deduced. It is clear that the presence of loops can complicate such consideration and in many applications we do not need to allow loops; it is always desirable to not allow them when our application allows them to be omitted. Thus many treatments of graph theory rule out loops at once. In particular applications and theoretical results we will state when we are allowing loops and when we are not.

The incidence matrix of a graph G is a special case of the class H of all matrices having 0, 1 or -1 for elements. Such matrices have many important, special properties. A matrix $B = (b_{ij})$ is said to be *unimodular* if the determinant of any square submatrix of B has value of 0, 1, or -1. It is clear that every unimodular matrix is of the general class H defined above. Those members of H which enjoy the unimodular property are characterized to some extent by the following theorem (p. 142 in [6], [7]).

Theorem (due to A. J. Hoffman; see Heller and Tompkins [7]). Let $B \in H$ such that every column contains at most two nonzero elements; if the rows of B can be divided into two disjoint sets $R1$ and $R2$ such that

(a) if the two nonzero elements of a column are of the same sign, one is in $R1$ and the other is in $R2$;

(b) if the two non-zero elements of a column are of opposite sign, both are in $R1$ or both are in $R2$, then B is unimodular.

Proof. It is required to prove that the matrix B has the unimodular property.

The hypothesized division of the rows of B into the classes $R1$ and $R2$ imposes a similar hypothesis on any square submatrix of B. Hence induction may be utilized on the square submatrices of B to establish the result.

The elements of B are -1, 0 or 1 so that every one by one submatrix of B has determinant -1, 0 or 1 establishing the unimodular property for the first order matrices.

Assume the theorem is true for submatrices of order $m - 1$ as an induction hypothesis.

Consider a submatrix of order m denoted by A. It falls into one of three cases (the last two are not necessarily mutually exclusive but this has no effect on the argument employed):

CASE 1. Every column of A has exactly two non-zero elements. Let $\bar{a}_i$ denote the ith row of A and $R1$ and $R2$ be represented by the row indices corresponding to the disjoint division of rows. Since only elements of opposite sign in a column can occur in the same row index set one observes that

$$\sum_{i \in R1} \bar{a}_i = \sum_{i \in R2} \bar{a}_i$$

This relation shows that the row vectors of A are not independent and hence the determinant of A is zero.

CASE 2. Some column of A has only zero elements. In this case it follows at once that the determinant of A is zero.

CASE 3. Some column of A has exactly one non-zero element. The determinant of A is expanded using the elements of this special column. Since there is only one non-zero element the expansion has only one term.

$$|A| = \pm 1 |A_1|$$

where A_1 is the cofactor of the single non-zero element. Since A_1 is of order $m - 1$ it has the unimodular property by the induction hypothesis. Thus $|A|$ is -1, 0 or 1 as required.

Since the above argument holds for any submatrix of order m the unimodular property is established for *B*.   ▲

Corollary. The incidence matrix of a digraph has the unimodular property.

Proof. Every column of such an incidence matrix contains exactly one $+1$ element and exactly one -1 element. Thus we can put every row into the set $R1$ and let $R2$ be the null set. This decomposition satisfies the theorem and the unimodular result follows.   ▲

The incidence matrix I_g of a linear graph G may or may not have the unimodular property. Those having the property are characterized by a second corollary to the theorem above.

Corollary. If I_g is the incidence matrix of a linear graph G then I_g has the unimodular property if and only if G has no circuits (simple cyclic ways) of odd length (i.e., with an odd number of vertices).

The incidence matrix (2) is unimodular.

$$\begin{pmatrix} 1 & 1 & 0 & 0 & 0 & 0 \\ 0 & 1 & 1 & 0 & 0 & 0 \\ 1 & 0 & 0 & 1 & 0 & 0 \\ 0 & 0 & 1 & 1 & 0 & 0 \\ 0 & 0 & 0 & 0 & 1 & 1 \\ 0 & 0 & 0 & 0 & 1 & 1 \end{pmatrix} \tag{2}$$

We may remark that multiple edges have no effect on this corollary, they may or may not occur in *G*. However self loops are excluded from *G* by the corollary since such a loop is a circuit of length one (hence odd). A simple proof of this corollary together with a number of additional results on unimodular matrices and their occurrence in graph theory may be found in Hoffman and Kruskal, [8].

Many properties of unimodular matrices and matrices of elements 0, 1 or -1 can be developed but further detail of these systems is outside our present scope of interest. For further discussion see Ryser or Williamson [9] and [10]. An interesting application to switching networks is given by Berge (page 150, [6]).

Another type of matrix representation for a graph is by means of the *adjacency matrix* A with elements a_{ij} where a_{ij} is equal to the number of edges going from vertex i to vertex j. Thus A is a $k \times k$ square matrix when G is a k-graph. For digraphs the wording defining a_{ij} is to be interpreted exactly as stated with direction indicated by the words "from" and "to." For undirected graphs the specification of edges is to be taken in the sense of lying between the two vertices. For the adjacency matrix formulation we need not label the edges. The three previous examples (Figure 1) have the adjacency matrix representations (3).

$$
\begin{array}{c}
\begin{array}{ccccc} a & b & c & d & e \end{array} \\
\begin{array}{c} a \\ b \\ c \\ d \\ e \end{array}
\begin{bmatrix}
0 & 1 & 0 & 0 & 2 \\
1 & 1 & 1 & 0 & 0 \\
0 & 1 & 0 & 2 & 0 \\
0 & 0 & 2 & 0 & 1 \\
2 & 0 & 0 & 1 & 0
\end{bmatrix},
\end{array}
\quad
\begin{array}{c}
\begin{array}{ccccc} a & b & c & d & e \end{array} \\
\begin{array}{c} a \\ b \\ c \\ d \\ e \end{array}
\begin{bmatrix}
0 & 0 & 0 & 0 & 1 \\
0 & 0 & 1 & 1 & 0 \\
0 & 1 & 0 & 1 & 0 \\
0 & 1 & 1 & 0 & 0 \\
1 & 0 & 0 & 0 & 0
\end{bmatrix},
\end{array}
\quad
\begin{array}{c}
\begin{array}{ccccc} a & b & c & d & e \end{array} \\
\begin{array}{c} a \\ b \\ c \\ d \\ e \end{array}
\begin{bmatrix}
0 & 0 & 0 & 0 & 1 \\
1 & 0 & 0 & 0 & 0 \\
0 & 1 & 0 & 0 & 0 \\
0 & 0 & 1 & 0 & 1 \\
0 & 0 & 0 & 1 & 0
\end{bmatrix}.
\end{array}
\qquad (3)
$$

We remark that the vertex (or edge) labels on incidence and adjacency matrices need not be written. They are included above to show how the matrices are formed and to clearly identify the rows and columns as representing specific vertices (or edges). Changes of labels are equivalent to permuting rows or column in the matrix representations and the graph itself is unchanged by such operations.

In some areas of application where at most one edge is allowed between any pair of vertices (as in some electrical network theory) adjacency matrices belong to the class H, in fact they form a special case since they have only 0 or 1 as elements. In the general theory and range of application we will allow multiple edges and therefore cannot suppose A to belong to the class H in general. If no loops are allowed the diagonal elements of A are zero; in fact the trace of A (sum of its diagonal elements) is equal to the number of loops in G.

For undirected graphs A is symmetric and can be divided into an upper triangular part $\overline{\Delta A}$ with elements $a_{ij}(i < j)$, a lower triangular part $\underline{\Delta A}$ with elements $a_{ij}(i > j)$ and the diagonal elements a_{ii}. The number of edges in G (elements of C) is given by the trace of A plus the sum of elements in $\overline{\Delta A}$. For digraphs A is not symmetric in general and the number of edges in G is the sum of all the elements in A.

A useful concept directly related to the matrix representation of linear graphs or digraphs is the *reachability* of one vertex from another. A vertex v_j is said to be reachable from a vertex v_i in a linear graph (digraph) if there exists an edge sequence (directed edge sequence) from v_i to v_j. The reachability structure of a graph G can be clearly expressed by means of its *reachability matrix* $R(G) = (r_{ij})$ in which $r_{ij} = 1$ if v_j is reachable from v_i and otherwise $r_{ij} = 0$. In a digraph the word "from" is to be taken in a strict sense. Thus $R(G)$ is symmetric if G is a linear graph but $R(G)$ need not be symmetric when G is a digraph. For completeness we stipulate that each vertex is reachable from itself so that $r_{ii} = 1$ for all i in any graph (this gives logical consistency when self loops are present and is satisfying to our intuitive concept of reachability in all other cases as well). Further developments of the reachability concept and of the concept of distance between vertices in a graph will occur in Chapter 2 and be applied, for example, in Chapter 9.

SPECIAL GRAPHS AND SUBGRAPHS

A number of special graphs and types of graphs recur throughout the theory and applications of linear graphs and it is convenient to provide the reader with descriptions of those graphs in this section.

A graph having no vertices and therefore necessarily no edges, that is, $N(G) = \Phi$ and $C(G) = \Phi$, is called the *null graph*, following the formal development of graphs by Tutte[11]. A graph for which $N(G) \neq \Phi$ but $C(G) = \Phi$ has no edges and is called a *vertex graph* (the term null graph is also widely used for such graphs but our preference is for the more precise usage introduced by Tutte). A simple graph on n vertices having the largest possible number of edges is a *complete graph* or complete n-graph denoted by K_n.

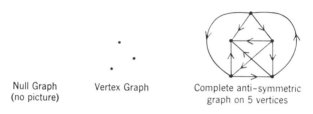

| Null Graph (no picture) | Vertex Graph | Complete anti-symmetric graph on 5 vertices |

Figure 8

Complete graphs are also called cliques or n-cliques. We shall discuss the concept of cliques more fully later in this chapter. In a complete graph every pair of vertices is connected by an edge. One can see that the number of edges is $n(n-1)/2$ since any vertex can be connected to $n-1$ other vertices and in the product $n(n-1)$ each edge is counted twice (once from each of its defining vertices). For directed graphs one has the concept of a *complete antisymmetric graph* in which every pair of vertices determines exactly one directed edge. Thus vertices A and B determine (A, B) or (B, A) but not both for any pair of vertices. This type of digraph may also be said to be *d-complete*. The concepts introduced above are illustrated in Figure 8 and in Figure 9 which gives the complete n-graphs for $n \leq 6$. Note that $n = 1$ is trivial having no edges; it is a vertex graph. The $n = 5$ case is particularly interesting. It is sometimes called the pentangle. Sir Gawain carried the pentangle emblem

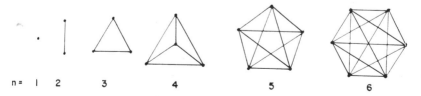

$n = 1 \quad 2 \quad 3 \quad 4 \quad 5 \quad 6$

Figure 9. Complete graphs.

upon the inner face of his shield to remind him of the power of his five virtues when their close interrelation was considered [12]; a philosophy which carries over to more modern applications of graphs as well (as we shall see, e.g., in Chapter 9).

Before discussing additional special types of graphs the concept of a subgraph will be introduced. Let V_g and E_g denote the set of vertices and the collection of edges of a graph G where multiple edges are allowed. A graph F is a *subgraph of G* if its vertex set V_f is a subset of V_g and its edge collection E_f is a subcollection of E_g. A particularly useful type of subgraph occurs when the vertex set V_f completely defines the graph F, i.e., all edges incident on the vertices of V_f are included in E_f. Subgraphs of this type are called *section graphs*. They may be thought of as sections taken directly out of the graph G without leaving any edges of G out. Figure 10 shows a graph G, a subgraph F of G which is not a section graph, and a section graph S based on the same vertex set as F.

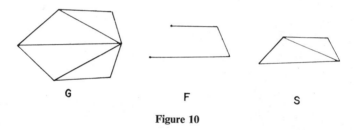

G   F   S

Figure 10

We will now introduce a very important special type of linear graph known as a tree. A connected graph without cycles is called a *tree*. Since there are no cycles there are no odd cycles and by our results on unimodular matrices the incidence matrix of a tree has the unimodular property. At a vertex v of a tree T there may be several edges incident, these may be denoted by e_1, $e_2, \cdots$. Any such edge, say e_i, defines a subgraph of T consisting of the section graph of T whose vertex set contains those vertices reachable from the terminal vertex of e_i which has initial vertex v, together with e_i itself. Such a subgraph is also a tree and therefore is a *subtree*. It is called the *branch* defined by e_i. Vertices of T which have only one edge incident on them and therefore define only one branch, namely T itself are called *terminal vertices* of T. All other vertices define at least two branches. The unique edge at a terminal vertex is called a *terminal edge*.

Theorem. Any two vertices of a tree are connected by a unique path.

Proof. Assume two vertices exist in a tree T such that they can be connected by two different paths A_1 and A_2. Since the paths are different there must be a sub path A_1^* of A_1 that has no edge in common with the other path A_2. Let a and b denote the initial and terminal vertices of A_1^*. The vertices a and b determine a subpath A_2^* of A_2 and the subgraph $A_1^* + A_2^*$ of T is a cycle. This contradiction proves the theorem. ▲

This proof illustrates a standard technique in the study of graph properties where one wishes to carefully establish results that often seem clearly true in the pictorial representation.

A *stem* is any path in a tree T. It is a subtree with simple structure, each of its interior vertices being incident on exactly two edges and containing exactly two terminal vertices if T is finite. When T is not finite a stem may be finite or one or two way infinite. Once a stem has been selected in a tree T every vertex in T is connected by a unique path to a nearest vertex in the stem. This vertex is determined by finding the path of least length from the vertex in question to the stem. Figure 11 shows a tree T with a stem selected whose edges are b_1, b_2, b_3, terminal vertices are marked $t_1, \ldots, t_7$ and a subtree consisting of the branch at vertex v defined by edge b_3 is also shown.

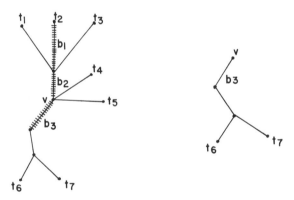

Figure 11

In any graph G the *distance between two connected vertices* u and v is the number of edges in the path of shortest length connecting u and v, it is denoted by $d(u, v)$. The path yielding the distance value need not be unique. A path whose distance is the maximum of all path distances in G is called a *diametral* path of G and its distance d^* is the *diameter* of G.

Diametral paths are especially useful in studying trees since they can be used as basic stems upon which the other parts of the tree can be developed. Thus one selects a diametral path D of a tree T as a stem. The length d^* of D is either even or odd, the two possibilities serving as an important classification of trees into two groups, those having a center and those having a bicenter.

The case of d^* even means there exists a single vertex on D which is equidistant from either terminal vertex of D, this vertex is called the *center* of the tree. In case d^* is odd there are two adjacent vertices on D such that each is the same distance from its closest terminal vertex as the other. This pair of vertices is called the *bicenter* of T and its two vertices together define the *central edge* of T. It is clear that every finite non-trivial (i.e., not a vertex graph) tree has either a center or bicenter but not both. For infinite trees the concept of center is not defined. The major properties of centers and bicenters are given by the following theorem.

Theorem. Let D be a diametral path of length d^* of a tree T. When d^* is even let c denote the center vertex, otherwise let the bicenter be specified by c_0 and c_1.

If d^* is even every diametral path contains c and is the sum of two paths from c of length $d^*/2$.

If d^* is odd every diametral path contains c_0, c_1 and the central edge (c_0, c_1) and is the sum of two paths of length $(d^* - 1)/2$ and the central edge.

Proof. In case d^* is even assume there is a diametral path R which does not contain c. Any such path must contain some vertex of D since D is a maximal length stem and every diametral path must contain at least one vertex of such a stem. Let w denote the vertex of D which is contained in R. Vertex w divides R into two parts which will in general be of unequal length, call these R_1 and R_2 where R_2 has the greater length. Since w is not c it divides D into two parts of unequal length say D_1 and D_2 with D_2 having the greater length. Of course the length of R is equal to the length of D namely d^*. If the only common vertex between R and D is w the result is simple to establish. The path $R_2 + D_2$ must have length greater than d^* which is a contradiction hence there can exist no such vertex w as the only vertex held in common by R and D. They must hold c in common and the above argument shows that c lies at the center of R as well as of D. A similar argument establishes the situation for the case in which R contains a segment of D but is not as direct. This case and the result for the bicenter when d^* is odd are left as exercises for the reader.   ▲

One should note that the center or bicenter of a tree T is unique and therefore is a fundamental characteristic of T. Two trees are shown in Figure 12 with their center or bicenter marked.

It may be useful to distinguish a particular vertex in a tree T; such a special vertex is called a *root*. Thus a rooted tree is a tree in which one vertex has been designated to be the root. Such a vertex need not be a terminal vertex though in many applications one might find it a natural aspect of the mathematical model to designate a terminal vertex as root.

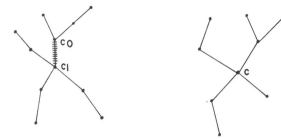

Figure 12

A graph G without cycles is called a *forest* since all of its connected components are trees. If the number of vertices in a forest G is n and there are m connected components there must be $n - m$ edges in G. This follows from the fact that a tree on k vertices has $k - 1$ edges. The reader may establish this simple combinatorial result by induction.

A directed tree with root at a non terminal vertex v is called an *arborescence with root* v if every vertex except v is the terminal vertex of a unique directed path from v. An arborescence is *bifurcating* if every interior vertex is incident on exactly three edges except the root which is incident on two. The terminal vertices are incident on exactly one edge. If we denote the number of vertices and edges by V and E respectively we have $E = V - 1$ since an arborescence is a tree. For a bifurcating arborescence with n terminal vertices there are $V - n - 1$ vertices incident on three edges, n vertices incident on one edge and one vertex (the root) incident on two edges hence $2E = 3(V - n - 1) + n + 2$ since the sum on the right counts each edge twice. Using the previous value $V - 1$ for E we find $V = 2n - 1$. Hence in a bifurcating arborescence with n terminal vertices there are $n - 3$ internal vertices and the root. Figure 13 shows an arborescence and a bifurcating arborescence.

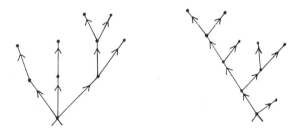

Figure 13.   Arborescences.

For any graph *G* a subgraph which is a tree may be called a *subtree* of *G*. When *G* is a tree this usage conforms to our standard use of the prefix "sub," otherwise it has a more general meaning. In case the subtree *T* of a graph *G* has the full vertex set of *G* as its vertex set, *T* is called a *spanning subtree* of *G*. Such spanning subtrees are particularly useful in expressing a graph *G* in terms of its subgraphs. We shall see that they are useful in formulating two types of matrix representations associated with the graph which they span. When *G* is not connected it has no spanning subtree but one may define a spanning subforest whose components are spanning subtrees of the components of *G*.

In concluding our discussion of trees it should be observed that we have in fact been describing what are known as Cayley trees. These form a special case of more general trees which are sometimes useful in classifying graphs. Some simple graph may be used to form a larger graph in the same way lines are joined to form a Cayley tree. When triangles are used in this way the result is called a cactus. When we speak of trees it will always be in the sense of Cayley. If the points of a tree are labelled in various ways the result may be called a chemical tree as has been done by Riordan (page 160, [13]).

Some other types of special graphs will now be described.

A *star graph* (In some areas of application, e.g., physics, it is common practice to use the term star or star graph for any connected graph or connected subgraph. Our definition is rather standard in graph theory itself.) consists of one vertex *v* which is incident with every edge and all other vertices incident on exactly one edge whose other vertex is *v*. Though self loops may be allowed at *v* they are usually not considered when one speaks of a star graph. A *star subgraph* of a graph *G* is a subgraph of *G* which is a star. When self loops at *v* are excluded a star graph is a tree with center *v* having diameter two.

Another special type of graph is a wheel. A wheel has one special vertex connected by an edge to every other vertex. The other vertices together form a circuit (when taken as a section graph of the wheel) so that each is connected to exactly two others and also to *v*. If a wheel has *n* vertices it must have $2n - 2$ edges. Figure 14 shows a star on 7 vertices and a wheel on 6 vertices.

Figure 14.   Star and wheel.

If the vertices of a graph G can be divided into two disjoint sets where every edge has one of its vertices in one set and its other vertex in the other set then G is called a *bipartite* graph. Two vertices in the same set of a bipartite graph are never connected by an edge. A bipartite graph can contain no circuits of odd length (hence no self-loops) and must therefore have only circuits of even length or no circuits at all. Every forest is a bipartite graph with no circuits. Therefore by our previous results on matrices with the unimodular property we may conclude that the incidence matrix of a bipartite graph has the unimodular property.

If the vertices of G can be divided into m disjoint sets in such a way that vertices in the same set are not adjacent then G is said to be $m - partite$.

When every possible edge is present in a (simple) m-*partite* graph it is said to be a *complete m-partite* graph and is given the standard notation $k(r_1, r_2, \ldots, r_m)$ or $k_{r_1, r_2, \ldots, r_m}$ in which r_i is the number of vertices in the ith part (disjoint vertex set).

The complete 3-partite graph $k_{2,2,2}$ is shown in Figure 15.

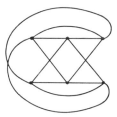

Figure 15

B. M. Stewart [14] has introduced an interesting classification of graphs under the general heading of magic graphs. We shall give a very brief introduction to that topic so as to arrive at special results that apply to complete bipartite graphs. Such results provide a connection between graph theory and number theory.

The graphs considered by Stewart are finite with n vertices v_i and E edges e_j. They have no self loops, no multiple edges, and no isolated vertices. Let $f(\cdot)$ be a function defined on the edges of such a graph, G, with values in the real field R. The set $A(G)$ of all such functions f is a vector space over R under the operation of addition, that is, $(x_1 f_1 + x_2 f_2)(e) = x_1 f_1(e) + x_2 f_2(e)$ for all $x_1, x_2 \in R$, all $f_1, f_2 \in A(G)$, and all e of G. Then $A(G)$ is said to be the edge space of G. The functions ε_i defined by

$$\varepsilon_i(e_j) = \delta_{ij} \qquad \text{for} \quad i, j = 1, 2, \ldots, E$$

form a basis for $A(G)$ and the dimension of $A(G)$ is E.

The edges of G are partitioned by any f in $A(G)$ into classes. Edges e_i and e_j are in the same class if and only if $f(e_i) = f(e_j)$. Thus a double subscript notation e_{uv} can be used to order the edges of G, where u denotes the class and v denotes the ordering of edges within the class (both orderings being arbitrary). It is unnecessary to develop this aspect of Stewart's work for our brief survey of his results. However the ordering concept is interesting in itself and should be recalled by the reader when related topics occur later in the book (e.g., Information content of graphs, Chapter 8, and ordering of graphs for physical applications, Chapter 10).

For each vertex v_i in G one may define the vertex sum $\sigma_f(v_i) = \sum f(e_j)$ where j assumes all values in the index set E_i of edges incident to v_i. Thus each element of $A(G)$ specifies n vertex sums for G (one at each vertex). Clearly $\sigma_{x_1 f_1 + x_2 f_2}(v_i) = x_1 \sigma_{f_1}(v_i) + x_2 \sigma_{f_2}(v_i)$.

The semimagic condition is defined as M1.

$$\sigma_f(v_i) = \sigma_f \qquad \text{for} \quad i = 1, \ldots, n$$

(the vertex sums are all equal for the function f).

Let $S(G)$ denote the set of elements of $A(G)$ that satisfy the semimagic condition ($M1$). Then $S(G)$ is a subspace of $A(G)$ called the semimagic space of G. If G is connected, then $E - n + 1 \le \dim S(G) \le E - n + 2$. For a connected graph the lower bound $E - n + 1$ is the cyclomatic number (number of independent cycles or cycle rank) $c(G)$.

Some further results are as follows.

If the graph is a tree T with n vertices and n is odd, $n \ge 3$, then dim $S(T) = 0$. More generally if G is connected with n odd, $n \ge 3$, then dim $S(G) = c(G)$.

For a complete graph K_n, $n \ge 3$, $\dim S(K_n) = \binom{n-1}{2}$.

Let $Z(G)$ be the set of all f in $S(G)$ such that: ($M2$) $\sigma(f) = 0$. Then $Z(G)$ is a subpace of $S(G)$ called the zero-magic space of G. One has the result: $\dim Z(G) \le \dim S(G) \le 1 + \dim Z(G)$.

On the basis of the concepts introduced above Stewart classifies a graph G as being:

    G1. Trivially Magic   if and only if dim $S(G) = 0$
    G2. Zero magic   if and only if dim $S(G) = \dim Z(G) > 0$
    G3. Semimagic   if and only if dim $S(G) > \dim Z(G)$.

Examples of each of these types of graphs are given in [14].

A further classification of graphs is achieved by introducing the following conditions:

M3. For all pairs of edges in G, $e_i \ne e_j$ implies $f(e_i) \ne f(e_j)$ This is called the distinctness condition.

M4. For every edge e in G, $f(e) \geq 0$, called the positiveness condition.
In terms of these a graph G is as follows.
G4. Pseudomagic if and only if there exists an f in $S(G)$ which satisfies $(M3)$ (has distinctness).
G5. Magic if and only if there exists an f in $S(G)$ satisfying $(M3)$ and $(M4)$ (is positive and distinct).

A proper subgraph U of a graph G is called a skeleton for G if it contains every vertex of G and has no single vertex components. A major result from [14] is the following.

Theorem. If G is pseudomagic, a sufficient condition that G be magic is that for each edge e in G there exists a skeleton containing e whose components are semimagic and have property $(M4)$ under functions having rational values.
Some results of specific interest are the following:

1. A complete graph K_n is magic for $n = 2$ and $n \geq 5$.
2. A complete bipartite graph $K_{n,\,m}$ with $n \geq m \geq 1$ is semimagic if and only if $n = m$.
3. A complete bipartite graph $K_{n,\,n}$ is magic for $n \geq 3$.

It is the third result that is the major reason for introducing the above concepts from [14] here. Let $v_1, \ldots, v_n$ and $w_1, \ldots, w_n$ denote the two sets of vertices comprising $K_{n,\,n}$. Consider a square table of numbers m_{ij} such that:

$$\sigma_f(v_i) = \sum_{j=1}^{n} m_{ij},$$

$$\sigma_f(w_j) = \sum_{i=1}^{n} m_{ij}.$$

Then if $K_{n,\,n}$ is magic under f the table (m_{ij}) is a weakly-magic square (of number theory). Result (3) above establishes the existence of weakly magic squares of all orders $n \geq 3$. To obtain a magic square one requires that the numbers be consecutive integers. This concept leads to: G6. A graph G is super–magic if and only if there exists an f in $S(G)$ such that
M5. The set $\{f(e_i)\}$ consists of consecutive positive integers.
The known properties of magic squares assures us that $K_{n,\,n}$ is supermagic for $n \geq 3$.

The results above show how the topic of magic graphs relates to complete bipartite graphs. Additional material on magic graphs and related topics may be found in [14].

For a simple graph G each edge can be considered to be a vertex of an associated graph G' called the *interchange graph* of G (as defined by Ore [15]). In G' two vertices are connected by an edge if the edges of G defining these vertices are incident on a vertex of G. The interchange graph of a star graph on n vertices is the complete $n - 1$ graph. A graph and its interchange graph are shown in Figure 16. The reader will observe that the interchange graph of an interchange graph is not the original graph in general. The interchange graph of G' in Figure 16 will have nine vertices so it cannot be G. It is interesting to observe that the interchange graph of a simple cycle of length k is the cycle itself. The interchange graph of a graph G is also called the *line graph* of (associated with) G. This name derives from that body of notation which refers to edges as lines.

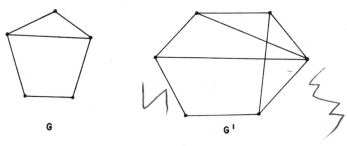

G   G'

Figure 16

A vertex v of a graph G is said to form (or belong to) a clique Q in G if each pair of vertices in Q that are edge connected to v are themselves edge connected (i.e., neighboring vertices). Note that one need not include all neighboring vertices of v. In [16] Fulkerson and Gross utilize this concept of a clique to introduce a matrix representation of G differing from the representations that we have previously defined. They employ the representation to study topics of rather wide interest and we shall present some of these topics here.

The collection of all cliques (sets of vertices) may be partially ordered by set inclusion. Maximal sets in such a partial ordering are called dominant cliques of G. One then defines the dominant clique vs. vertex matrix (or clique matrix) of G as $C(G) = (c_{ij})$ where $c_{ij} = 1$ if vertex j belongs to dominant clique i and $c_{ij} = 0$ otherwise. The matrix $C(G)$ specifies G completely. We now introduce several topics from [16], interesting in themselves, which are brought into particularly close association by means of the clique matrix.

Consider a $(0, 1)$-matrix M with columns (as vectors) $m_1, m_2, \ldots, m_n$. Form a linear graph $G(M)$ whose vertices $\{v_i\}$ correspond to the n-columns m_i of M. The edge (v_i, v_j) is in $G(M)$ if and only if $0 < m_i^t m_j < \min(m_i^t m_i, m_j^t m_j)$. Then m_i and m_j are said to overlap and $G(M)$ is called the overlap graph of M.

The major topic considered in [16] is the consecutive 1's property for $(0, 1)$

matrices. A matrix M whose elements are either zero or one, is said to have the consecutive 1's property (for columns) if there exists a permutation of the rows of M which will result in the 1's in each column occurring consecutively. Fulkerson and Gross discuss the theory of such matrices and give several theorems characterizing them. To state some of their results we may suppose a $(0, 1)$-matrix M is expressed in the form $M = (M_1, M_2, \ldots, M_k)$, where each submatrix M_i, $i = 1, 2, \ldots, k$ corresponds to a component X_i of the overlap graph $G(M)$ of M (columns being rearranged when necessary). This is called the overlap decomposition of M and is unique (except for indexing of the submatrices). The M_i are referred to as components of M. If M has exactly one component it is called a connected matrix. Two results characterizing the consecutive 1's property are as follows.

A $(0, 1)$-matrix has the consecutive 1's property if and only if each of its components has that property.

If each component of a $(0, 1)$-matrix M has at most two columns, then M has the consecutive 1's property.

If M has the consecutive 1's property, then M is unimodular. This relates the present topic to the study of unimodular matrices previously discussed in this chapter. It also provides a connection with linear inequality theory and linear programming (particularly integer programming and network flows).

Fulkerson and Gross also give a technique for testing a connected matrix for the consecutive 1's property. We do not discuss that technique but turn our attention to some particular types of linear graphs utilized in [16].

Consider finite, undirected graphs without self loops or multiple edges. A graph G is called a rigid cycle (or circuit) graph if every cycle (circuit) of G with more than three vertices has a chord. A chord is an edge not in the cycle but incident to vertices of the cycle. It is possible to test a graph for the rigid cycle property. Search for a clique vertex; if one is found remove it and its incident edges and repeat the search in the reduced graph. The graph is a rigid cycle graph if and only if the process ends in the deletion of all vertices.

The rigid cycle graphs are particularly interesting from the point of view of their clique matrix representation. The dominant cliques can be determined as the above test procedure is carried out. When a clique vertex is removed, list it and its neighboring (edge connected) vertices as a clique. When the process shows an n-graph to have the rigid cycle property it will terminate in the removal of n vertices. Thus n cliques will have been recorded. From that collection one obtains the maximal sets as the dominating cliques of the graph.

Consider a collection of sets $S_1, \ldots, S_n$. The intersection graph of the collection is formed by associating a vertex with each set of the collection $(v_i \sim S_i)$. The edge (v_i, v_j) is in the graph if and only if $S_i \cap S_j$ is not empty. Any finite graph can be considered as the intersection graph of a collection of sets (a contrasting result is given by Marczewski's Theorem in Chapter 6). If each set of the associated collection can be considered to be an interval on the real line, the graph is called an interval graph. A theoretical question, with possible applications (e.g., to genetics as mentioned in Chapter 9), is the determination of whether or not a given graph is an interval graph. In this connection a major result of [16] is the following.

Theorem. A graph G is an interval graph if and only if the clique matrix $C(G)$ has the consecutive 1's property.

Linear programming theory is employed in [16] to establish the following result.

Let G be an interval graph. The minimum number of vertices of G required to represent all dominant cliques is equal to the maximum number of dominant cliques that are mutually disjoint.

A set of vertices represents the cliques if each clique has some vertex in the set. This topic is closely related to systems of distinct representatives as discussed in Chapter 6. The above result will be seen to be very much in the spirit of that topic.

Another special type of intersection graph occurs when the sets S_i constitute a collection of arcs on a circle. Such an intersection graph is called a circular-arc graph. Such graphs are clearly closely related to interval graphs. They are discussed by Tucker [17] who gives the following definition and characterization theorem.

A (0, 1)-matrix is said to have the circular 1's property for columns if the rows can be permuted so that the 1's in each column are circular. That is if the top of the matrix is joined to the bottom to form a cylinder the 1's in each column will be consecutive on the cylinder.

Let M be a (0, 1)-matrix and form a matrix M_1 by taking the complement (i.e., interchanging 0's and 1's) of each column of M with a 1 in the first row. Then M has the circular 1's property for columns if and only if M_1 has the consecutive 1's property for columns.

Let $A(G)$ denote the adjacency matrix of a graph G(finite, undirected) without self loops or multiple edges. The augmented adjacency matrix $A^*(G)$ is formed from $A(G)$ by placing 1 in each main diagonal position. Then Tucker shows that: G is a circular-arc graph if $A^*(G)$ has the circular 1's property for columns.

One may introduce a useful numerical measure of the degree of incidence present at each vertex of a graph G by defining the _degree at a vertex v_ as the number of edges incident at v. This is written $d(v)$ and is sometimes called the local degree to emphasize the fact that it characterizes the incidence situation at a particular vertex. When $d(v)$ is finite for all vertices v of a graph G we say G is _locally finite_. If $d(v) = 0$ the vertex v is said to be _isolated_. The concept of degree is widely used in graph theory, one of the most elementary and interesting results being the following theorem.

Theorem. In a finite graph G the number of vertices with odd degree must be even.

The reader will note that zero is considered to be an even number.

Proof. Let E denote the number of edges in G. If we add all degrees for the vertices of G we must get twice the number of edges since each edge is added in two times. Thus $2E = \sum_v d(v)$ for all v in G. This shows that the sum of the degrees is even hence by familiar properties of the integers the theorem is established.   ▲

Though this theorem is elementary it shows how a rather rigid structure exists in any graph and may come as a surprise in view of the very general way in which graphs can be formed.

When G is a digraph there are two numbers associated with each vertex. The *indegree* of v is the number of edges directed toward and incident upon v. The *outdegree* is the number of edges directed away from and incident upon v. These may also be called inward demi-degree or outward demi-degree as is done by Berge (page 86, [6]) and others. A vertex with indegree zero is called a transmitter and a vertex with outdegree zero is called a receiver (page 409, [4]).

A graph G is said to be *regular* when the local degree is a constant over all vertices of G. Thus $d(v) = d$ for all v which are vertices of G. By the above theorem we see that every regular graph on an odd number of vertices must have an even value for its degree d. The complete graph on n vertices is regular with degree $n - 1$. One may also say that G is regular of degree d or that it is d-regular, thus the complete n-graph is $n - 1$ regular. An interesting set of regular graphs is the set of regular polyhedral graphs which are formed by taking the vertices of the five Platonic solids and using them to construct graphs which have the same connection between vertices as occur in the corresponding solid. The regular polyhedral graphs are shown in Figure 17. The corresponding solids are such that in a particular solid every face is the same regular figure, moreover all edges and internal angles are the same in a given solid.

A digraph G is called *symmetric* if every pair of vertices that are connected by an edge are connected by two directed edges, one in each direction. In a symmetric graph the local indegree is equal to the local outdegree. The digraph is *antisymmetric* if every pair of vertices that is connected by an edge is connected by exactly one directed edge. Such graphs may also be called *asymmetric*. When the indegree is equal to the outdegree for every vertex of a digraph G we say G is *pseudosymmetric*. Thus every symmetric graph is

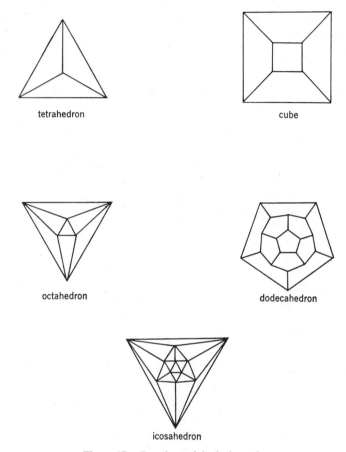

tetrahedron

cube

octahedron

dodecahedron

icosahedron

Figure 17.   Regular polyhedral graphs.

pseudo-symmetric but the converse is certainly not true as is illustrated in Figure 18 where G is symmetric and F is pseudosymmetric.

Complete antisymmetric digraphs are sometimes called *tournaments* because the edges of such graphs can represent pairings of the contestants in tournaments. Considerable study of the properties of tournament graphs is possible, the interested reader can see Harary and Moser [18] or Moon [19].

When the vertex set $N(G)$ consists of lattice points from a euclidean k space, G is called a *lattice graph*. Such lattice points are k-tuples of integers and for most utilizations of the lattice graph concept the integer components may be taken to be nonnegative. The most important class of lattice graphs are based on lattice points in the first quadrant of two dimensional euclidean space. The definition of lattice graphs

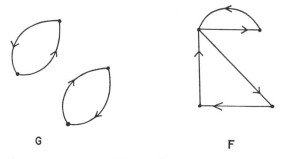

G   F

Figure 18

imposes a special pictorial representation on them in which the vertices are all lattice points and the edges are straight line intervals connecting lattice points and lying parallel or perpendicular to each other. In such graphs only nearest neighbor vertices (in the euclidean space sense) can be adjacent (i.e., edge connected) and such connection is always by means of the interval of smallest euclidean distance, that is, no "diagonal" edges are allowed. Figure 19 shows two lattice graphs, isolated vertices are included for pictorial clarity. It should be noted that lattice graphs may be finite or infinite, they may be connected or not, and they are necessarily simple graphs. They are very important in certain applications to physics as discussed in Chapter 10 and occur at various points in the study and application of linear graphs.

Figure 19.   Lattice graphs.

EULER GRAPHS

A graph which contains a cyclic path utilizing every edge is called an *euler graph*. The reader will recall that a path contains a given edge at most once so that any cyclic path of the type required in an euler graph contains every edge of the graph exactly once. Any such cyclic path is called a *cyclic euler path*. Thus one may say that G is an euler graph or equivalently that G contains a cyclic euler path. One of the most interesting things about this important special class of graph is that they may be easily recognized as shown by the following theorem.

Theorem. A finite connected graph G is an euler graph if and only if all local degrees are even.

Proof. Any cyclic euler path must leave a vertex by an edge other than the edge by which it arrived at that vertex since any edge is used exactly once. Thus if G is an euler graph every local degree is even.

Consider a finite connected graph G which has all local degrees even. Start a path at some vertex v and move along an unused edge to a next vertex until a cyclic path is obtained which returns to v. If all edges have been used the desired cyclic euler path has been established. If all edges have not been used the construction must have (by connectivity) encountered a vertex where another edge could have been taken. Call such a vertex w. We may now return to w and start from w using an edge not previously taken and move along in the same way returning to w (since the degree at w is even). If all edges are now utilized one can construct a cyclic euler path from v by proceeding as before to w then following the cyclic path just established from w back to w then move from w to complete the original cyclic path to v. The reader will observe that this process can be continued and that in a finite number of steps a cyclic euler path from v will be obtained.   ▲

Note that in counting local degrees self loops add a count of 2 to the local degree.

Though the above theorem was stated in such a way that the connectivity of G was assumed rather than deduced the reader will observe that connectivity is easily shown to be necessary for G to be an euler graph.

Though a graph G may fail to have a cyclic euler path it may have an euler path which is not cyclic. This more general type of euler path contains every edge exactly once but does not return to its initial vertex (without employing an edge more than once.) Of course an euler path may utilize any vertex many times. Figure 20 shows an euler graph at A, a graph with an euler path at B and a graph with no euler path at C.

The famous Königsberg bridge problem is solved by using the above criterion theorem for euler graphs. In Figure 21 the vertices represent four

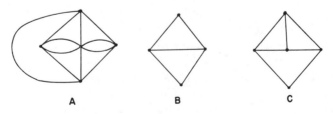

A   B   C

Figure 20

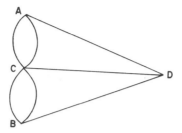

Figure 21

parts of the city, A and B being two sides of the Pregel River and C and D being islands in the river. The edges represent bridges. It is desired to find a cyclic path crossing each bridge exactly once. The demonstration by Euler that there was no such path is often considered to be the beginning of topological concepts in mathematics. It is certainly one of the first formulations of a graph theory type analysis and is the reason for giving the name of euler graphs to the class of graphs having the desired cyclic paths. Since the graph representation of the Königsberg bridge problem shown in Figure 21 does not have all even local degrees the many people who endeavored to discover a cyclic walk of the desired kind though Königsberg searched in vain.

When a finite connected graph G does not possess an euler path it can be covered by some collection of edge disjoint paths. This covering is meant to include every edge of G exactly once in some path of the collection. There is always such a covering since each edge can be taken to be a path and the collection of all edges becomes the desired collection of covering paths. The interesting aspect of such coverings is to find the minimal number of paths which cover G. This number is one if G possesses an euler path. The general situation is expressed by the following theorem.

Theorem. The minimal number of edge disjoint paths which will cover a finite connected graph G having n odd degree vertices is $n/2$ for $n > 0$.

Proof. By a previous theorem n must be even. If n is zero, G is an euler graph and the theorem does not apply, the number of covering paths required of course is one. Let H denote the minimal number of covering paths.

Each odd vertex must be a terminal vertex for at least one of the H covering paths so that $H \geq n/2$.

Let each pair of odd vertices be connected by an additional edge by increasing the edges of G. This is accomplished by adding $n/2$ edges to G. The modified graph has only vertices of even degree, it is an euler graph and therefore possesses a cyclic euler path. When the extra $n/2$ edges are removed the cyclic euler path splits into at most $n/2$ segments which certainly cover the edges of G. Thus $H \leq n/2$ so that $H = n/2$. ▲

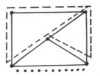

Figure 22

Figure 22 shows a 5 graph with 4 odd vertices. It can be covered by 2 edge disjoint paths as indicated by the one solution shown.

Graphs *B* and *C* of Figure 20 require one and two disjoint paths respectively to cover their edges. Note that the above theorem gives a condition for the existence of an euler path which we may state as follows. A finite connected graph *G* which is not an euler graph possesses an euler path if and only if it has exactly two vertices of odd local degree.

For infinite connected graphs the existence of euler paths (cyclic or non-cyclic) is harder to characterize. Some results on infinite euler graphs are to be found in the literature; in particular, necessary and sufficient conditions have been obtained. The interested reader should consult Erdös, Grunwald, and Vazsonyi [20] and a discussion will also be found in Ore (Section 3.2, [15]).

Cyclic euler paths on digraphs have a theory similar to that for undirected paths. In digraphs the edges must be traversed in the assigned direction, otherwise the concept of cyclic euler paths is the same. As in the undirected case there is a simple characterization theorem for cyclic euler paths in a digraph *G*.

Theorem. A finite connected digraph *G* contains a directed cyclic euler path if and only if at every vertex the indegree is equal to the outdegree.

The proof of this theorem is identical to the proof of the characterizing thorem for undirected cyclic euler paths.

Corollary. It is always possible to construct a cyclic directed path passing through each edge exactly once in each direction on any finite connected graph.

Proof. Each edge of a finite connected graph *G* can be split into two edges one being assigned one of the two possible directions and the other assigned the remaining direction. The digraph *G** formed in this way satisfies the above characterization theorem. Any directed cyclic euler path in *G** will be a path of the desired type. ▲

A useful method for producing a path of the type whose existence is insured by the corollary is known as *Tarry's construction* which we express by means of a flow diagram in Figure 23. The reader may assure himself that the

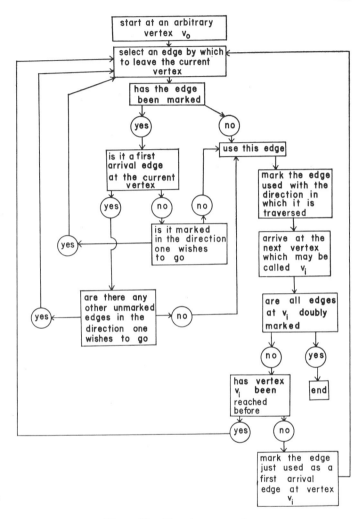

Figure 23. Tarry's construction.

process described does in fact yield a path of the desired type. An example is shown in Figure 24, in which the edges are numbered in order of their addition to the euler path following the Tarry construction. A first arrival edge is labelled with an f and v_0 is the starting vertex.

In the study of digraphs an important role is played by directed spanning trees and in particular those which are arborescences. A directed tree requires that the tree taken as a digraph be strongly connected. The reader may recall that an arborescence with root v is such a directed tree in which there is

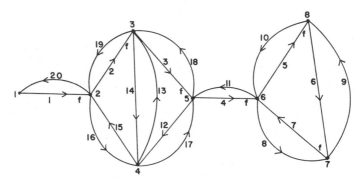

Figure 24.   Directed cyclic euler path.

exactly one inward arc at every vertex except the root v which has only outward arcs.

For pseudo-symmetric digraphs (i.e., digraphs such that at each vertex the indegree equals the outdegree), there exists two important combinatorial results. The first of these is given by the following theorem.

Theorem. (Aardenne-Ehrenfest, de Bruijn [21]). In a finite connected pseudo-symmetric digraph G let A_d denote the number of spanning arborescences with root v_1. Let n be the number of vertices in G and r_k, $k = 1, \ldots, n$ denote the outdegree at vertex v_k. The number of distinct directed cyclic euler paths in G is equal to

$$A_d \prod_{k=1}^{n} (r_k - 1)!.$$

Note two directed cyclic euler paths are distinct if one can not be obtained from the other by a circular permutation of edges (i.e., of the labels assigned to the edges of G).

Our proof of this theorem is similar to that given by Berge (page 169, [6]). It is less detailed than the proof of Aardenne-Ehrenfest, de Bruijn (1951); however, that paper contains a number of interesting results falling beyond the scope of this text.

Proof. Consider a specific arborescence A with root v_1. We shall enumerate the number of distinct directed cyclic euler paths in G determined by A. At each vertex v_k, $k \neq 1$ the edges directed into v_k may be numbered from 1 to r_k (since the indegree is equal to the outdegree). There will be an edge of A directed into each vertex except v_1. Let the numbering of inward edges at a vertex be such that the edge at v_k belonging to A is numbered r_k. Thus at each vertex we have a set of edges $E_k = \{e_{k1}, \ldots, e_{kr_k}\}$ where $e_{kr_k} \in A$ and we may assign the numbers $1, \ldots, r_k - 1$ in any way among the other edges. There are $(r_k - 1)!$ ways to assign the numbers to edges at v_k. Thus if we fix the first edge leading into v_1 (since there are no edges of A leading into v_1) there are

$$\prod_{k=1}^{n} (r_k - 1)!$$

ways to number the edges which are free to have numbers assigned (in terms of those edges which are fixed by A). A given numbering of edges yields a unique directed cyclic euler path since at each vertex one may proceed to the next by moving against the indirected edge of lowest number which has not previously been used. The first step in any such path is to move from e_{11}. In such a path no edge can be used more than once. Moreover every edge is used, as shown by the following argument. Suppose some edge e_{ki} at a vertex v_k has not been used. By the numbering system, e_{kk} has also not been used since this edge is always the last to be used at a vertex (by the path generation procedure). This means that the initial vertex of e_{kk} has an unused edge and therefore has an unused edge of A. One can proceed in this way to arrive by edges of A back to v_1. This is a contradiction since the path procedure terminated only when there were no unused edges at v_1. Thus each numbering corresponds to a directed cyclic euler path of G.

The numberings corresponding to A all give different paths when the above procedure is used. Any renumbering must give a non-cyclic rearrangement of the order in which the edges are utilized.

Moreover paths corresponding to different arborescences yield different paths. Since the last edge by which any path leaves a vertex is fixed by A no path determined by A can be a cyclic permutation of a path determined by some other arborescence of G. ▲

Corollary. The number of spanning arborescences with root v in a digraph G is independent of the choice of v.

Proof. The number of directed cyclic euler paths in G is fixed as is the numeric factor $\prod_{k=1}^{n} (r_k - 1)!$. Thus by the above theorem A_d is a fixed number independent of the root v. ▲

The second result deals with the enumeration of the number A_d of spanning arborescences with a fixed root v in a digraph G without self loops. Only the enumerative procedure will be given. For a deeper discussion one may refer to Berge (Chapt 16, [6]). Let A denote the adjacency matrix of an n-vertex digraph G without self loops. Denote the elements of A by a_{ij}. Let $H = (h_{ij})$ be a matrix defined as follows:

$$h_{ij} = 0 \qquad \text{if } i \neq j,$$

$$= \sum_{\substack{k=1 \\ k \neq i}}^{n} a_{ik} \qquad \text{if } i = j.$$

Define B as the matrix $H - A$ so that (4) is true.

$$B = \begin{bmatrix} \sum_{k=2}^{n} a_{1k} & -a_{12} & \cdots & -a_{1n} \\ -a_{21} & \sum_{\substack{k=1 \\ k \neq 2}}^{n} a_{2k} & \cdots & -a_{2n} \\ \cdot & \cdot & & \cdot \\ \cdot & \cdot & & \cdot \\ \cdot & \cdot & & \cdot \\ -a_{n1} & -a_{n2} & \cdots & \sum_{k=1}^{n-1} a_{nk} \end{bmatrix} \qquad (4)$$

It may be remarked that B is a singular matrix since its columns are not linearly independent (n-vectors). The enumeration of A_d is as follows: let M_i denote the determinant of the minor of B found by deleting the ith row and the ith column from B.

Theorem. Let G be a digraph without self loops and having adjacency matrix A. The number of spanning arborescences with root at vertex v_i is equal to M_i.

This theorem together with the previous theorem can be used to enumerate the cyclic euler paths in a connected pseudo-symmetric digraph G. Moreover reference to those two theorems shows that in a pseudo-symmetric digraph the number of spanning rooted arborescences is independent of the vertex selected as root. This follows from the observation that the number of cyclic euler paths in a graph can not depend on the particular manner in which they are enumerated.

As an example of the above concepts consider the pseudo-symmetric digraph G with 5-vertices shown in Figure 25 for which the adjacency matrix is shown in (5).

$$A = \begin{bmatrix} 0 & 1 & 1 & 0 & 0 \\ 0 & 0 & 0 & 1 & 1 \\ 0 & 1 & 0 & 0 & 1 \\ 1 & 0 & 1 & 0 & 0 \\ 1 & 0 & 0 & 1 & 0 \end{bmatrix}. \tag{5}$$

In this case if we let the determinant of M_1 be denoted by $|M_1|$ we get (6)

$$|M_1| = \begin{vmatrix} 2 & 0 & -1 & -1 \\ -1 & 2 & 0 & -1 \\ 0 & -1 & 2 & 0 \\ 0 & 0 & -1 & 2 \end{vmatrix} = 11. \tag{6}$$

Hence $A_d = 11$. The outdegree at each vertex is 2 so that

$$\prod_{j=1}^{5} (r_j - 1)! = 1$$

and the number of cyclic euler paths is equal to 11. These are listed in Table 1 where the ten directed edges of G are labeled as shown in Figure 25. Of course cyclic permutations are not counted as different paths.

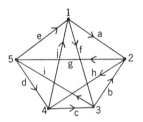

Figure 25

Table 1

a	h	c	i	e	f	b	g	d	j		a	g	e	f	b	h	c	i	d	j
a	h	c	i	d	j	f	b	g	e		a	g	e	f	i	d	c	b	h	j
a	h	c	b	g	d	j	f	i	e		a	g	d	j	f	b	h	c	i	e
a	h	c	b	g	e	f	i	d	j		a	g	d	c	b	h	j	f	i	e
a	h	j	f	b	g	d	c	i	e		a	g	d	c	i	e	f	b	h	j
a	h	j	f	i	d	c	b	g	e											

The enumeration discussed above can be utilized to solve the problem of enumerating the number of possible arrangements of dominos. A domino consists of a rectangle one unit wide by two units long forming two distinct squares. Each square is assigned a number so for example a set of dominos specified by the integers 0, 1, 2, 3, 4 consists of fifteen dominos of the form

(i, j) where i may equal j and where i and j assume values $0, \ldots, 4$. We shall consider the possible arrangements of these 15 dominos in a straight line where adjacent dominos must have the same numbers adjacent. A mathematical representation of this problem is provided by the pseudo-symmetric 5-graphs such as the one shown in Figure 25. An euler cycle in such a graph represents the part of a domino arrangement formed by the dominos having different numbers, for example, the edge going from vertex 2 to vertex 4 represents a domino having the number pair $(2, 4)$. The five double dominos of the form (i, i) $i = 0.1, 2, 3, 4$ must each go between dominos having the common number. In any arrangement of the 10 non-double dominos there are just two places where a given double may be placed hence corresponding to every arrangement of 10 there are 2^5 arrangements of the full set of 15 dominos. As we have seen above for each distinct 5-vertex pseudo symmetric digraph with indegree 2 and outdegree 2 there are 11 distinct arrangements of the 10 non-double dominos or (2^5) (11) arrangements of the 15 dominos. The reader may show that there are 24 distinct graphs of the required type so that the total number of domino arrangements is (2^5) (264). This value can be obtained by entirely different methods but the graph methods illustrated above may be suggestive of other applications.

CYCLE STRUCTURE

Let G be a linear graph in which self loops and multiple edges are allowed. Let n denote the number of vertices, m the number of edges and p the number of connected components. One then defines the *cyclomatic number* (sometimes called nullity or degrees of freedom) of G as

$$c(G) = m - n + p$$

When every connected component of G is a simple cycle the graph has p simple cycles. In this case $c(G)$ enumerates the simple cycles. Such cycles use an edge at most once. When multiple use of edges is allowed in forming cycles every graph that is not a forest has an infinite number of cycles. One might be tempted to associate the number of distinct simple cycles with $c(G)$ and this can indeed be done but not in a trivial direct way as Figure 26 shows. For this graph $c(G)$ is 2 but there are 3 distinct simple cycles.

By contemplating even very simple graphs the reader will observe that the study of cycles is complicated. One approach to their study is by means of *independent cycles* which for finite graphs (and special infinite graphs) allows the representation of any cycle in terms of a finite collection of special cycles. The concept imposes an equivalence class structure on the cycles in which any member of a class can be taken as the representative. All other cycles not

Figure 26

belonging to the special classes are then represented as composite cycles. The situation is exactly the same as the representation of a vector space in which every vector can be represented as a composite of vectors from the independent classes of basic vectors. In fact this analogy is utilized in defining independent cycles. In the procedure to be given a vector in m-space (where m is the number of edges in G) is associated with every cycle. The converse is clearly not true and hence the requirement for an equivalence class structure. All cycles associated with the same vector are equivalent. It is this situation that limits the use of the independent cycle technique in the study of cycles but the theory is nevertheless sufficiently rich to be of considerable use. The assignment procedure requires the assignment of a direction to each edge. This is carried out in a completely arbitrary way as particular assignments of direction do not affect the results at all. An m-vector $(a_1, \ldots, a_m)$ is associated with a given cycle as follows: if the edge k is traversed f_k times in the direction of its orientation (forward) and r_k times in the opposite direction (reverse) as the cycle is expressed as an ordered sequence of edges then $a_k = f_k - r_k$. Thus many cycles can give exactly the same associated vector but each yields a unique vector. Cycles are said to be independent if their associated vectors are linearly independent in the vector space sense. Any set containing a maximum number of independent cycles of G is said to form a *cyclic basis* for G, their associated vectors constitute a basis for some subspace of the m-space of all cycle associated m-vectors. It is important to understand that we have not established an equivalence between the cycles and their associated vectors. There are many vectors which correspond to no cycle (e.g., only vectors with integral coordinates can correspond to cycles), and a given vector corresponds to an infinite number of cycles in general. We do have a method of setting up a cycle basis of independent cycles and expressing other cycles in terms of these. The most convenient way to do this is in the associated vector notation in terms of the cycles themselves so that the nonunique character of the association does not present any difficulty. Thus we speak of a cycle and write its associated vector but it is the cycle we are working with and not the vector. For example the graph in Figure 26 has two independent cycles which we may take as (1 2 5) and (3 4 5) written as edge sequences. The associated

5-vectors are $(1\ 1\ 0\ 0\ -1)$ and $(0\ 0\ 1\ 1\ 1)$. Cycle $(1\ 2\ 3\ 4)$ has associated vector $(1\ 1\ 1\ 1\ 0)$ and it is therefore the (cyclic basis) sum of the other two cycles.

From the definition of cyclomatic number we see that adding a new edge to a graph G will increase the cyclomatic number by one if and only if the terminal vertices of the new edge belong to the same connected component of G. This includes the case where both vertices are the same (the new edge is a self-loop). In case the vertices belong to different connected components $c(G)$ will be unchanged. The reader will be convinced of these facts by observing that the addition of an edge does not alter n while it increases m by one. Thus if the components remain the same in number $c(G)$ increases by one and otherwise it is unchanged. Since addition of an edge can do away with at most one component (i.e., connect two components), we see that $c(G) \geq 0$ as well. These are the basic facts about the cyclomatic number and its meaning is expressed by the following result.

Theorem. The cyclomatic number of a graph G is equal to the maximum number of independent cycles in G.

Proof. One may start with n vertices of G and no edges. The edges are added one at a time until the total number m are present. In this process the only time the cyclomatic number of the configuration changes is when an edge is added within a connected component as we have remarked above. The addition of such an edge produces a cycle which is independent of all independent cycles which were present prior to the introduction of the new edge. Denote the new edge by index k then every independent cycle previously present has $a_k = 0$ since edge k can be present in none of them. Of all the new cycles which are produced by the introduction of edge k at least one has $a_k \neq 0$. Since we added edge k to a connected component its terminal vertices have a connecting path which together with the new edge is a cycle in which edge k is traversed once. Thus no matter what orientation we assign to edge k there is at least one cycle with $a_k = 1$. This cycle is independent of all previous independent cycles and when combined with them forms a cycle basis for the cycles of the configuration with edge k just added. When the new edge joins different connected components no new independent cycles occur since for every cycle including edge k we have $a_k = 0$. Here again we see the rather special nature of the concept of independent cycles and the cycle associated m-vectors.   ▲

As an immediate consequence of the above theorem we have the following corollary.

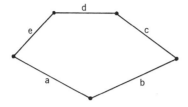

Figure 27

Corollary. A graph G contains no cycles if and only if $c(G) = 0$ and contains exactly one simple cycle if and only if $c(G) = 1$.

We remark that the corollary requires the term simple cycle in its second statement for if non-simple cycles are allowed a graph with $c(G) = 1$ has an infinite number of cycles. For example we construct the following cycles in edge sequence representation for the graph of Figure 27: $(abcde)$, $(abcdeabcde)$, $(abcdeabcdeabcde)$, and so on.

The cycle structure of a graph G may be expressed in terms of the *cycle matrix* (or circuit matrix where the term circuit means simple cycle). The cycle matrix $c(G)$ has a row corresponding to each simple cycle of G and a column corresponding to each edge of G. If $C = (c_{ij})$ then $c_{ij} = 1$ if simple cycle i contains edge j and otherwise $c_{ij} = 0$.

It is interesting to note that if the incidence matrix I_g of G if formed using the same numbering of edges used to form the cycle matrix C then $I_g C^T = 0$ when the arithmetical operations are performed modulo 2. This can serve as a useful check on the matrix $C(G)$, which is involved to evaluate in any but extremely simple cases (due to the necessity of forming each simple cycle in a graph). In practice $C(G)$ is often too hard to form for it to be useful but it can be valuable in theoretical or conceptual studies. The above concepts are illustrated by reference to Figure 28 in which the graph shown has 6 vertices, 9 edges, and 10 simple cycles. The cyclomatic number is 4 and the cycles: $(1, 2, 9)$, $(3, 9, 4)$, $(5, 6, 7)$, and $(7, 8)$ constitute a set of independent simple

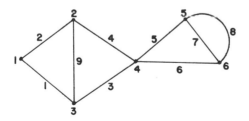

Figure 28

cycles. The cycle matrix specifies the edges that constitute each simple cycle where e_j denotes edge j, c_i denotes simple cycle i, and v_i denotes vertex i.

$$
C = \begin{array}{c}
\\ c_1 \\ c_2 \\ c_3 \\ c_4 \\ c_5 \\ c_6 \\ c_7 \\ c_8 \\ c_9 \\ c_{10}
\end{array}
\begin{array}{c}
\begin{array}{ccccccccc} e_1 & e_2 & e_3 & e_4 & e_5 & e_6 & e_7 & e_8 & e_9 \end{array} \\
\left[
\begin{array}{ccccccccc}
1 & 1 & 0 & 0 & 0 & 0 & 0 & 0 & 1 \\
0 & 0 & 1 & 1 & 0 & 0 & 0 & 0 & 1 \\
0 & 0 & 0 & 0 & 1 & 1 & 1 & 0 & 0 \\
0 & 0 & 0 & 0 & 0 & 0 & 1 & 1 & 0 \\
1 & 1 & 1 & 1 & 0 & 0 & 0 & 0 & 0 \\
0 & 0 & 0 & 0 & 1 & 1 & 0 & 1 & 0 \\
1 & 1 & 1 & 1 & 1 & 1 & 0 & 1 & 0 \\
0 & 0 & 1 & 1 & 1 & 1 & 0 & 1 & 1 \\
1 & 1 & 1 & 1 & 1 & 1 & 1 & 0 & 0 \\
0 & 0 & 1 & 1 & 1 & 1 & 1 & 0 & 1
\end{array}
\right]
\end{array}
\tag{7}
$$

The incidence matrix is as shown in (8).

$$
I_g = \begin{array}{c}
\\ v_1 \\ v_2 \\ v_3 \\ v_4 \\ v_5 \\ v_6
\end{array}
\begin{array}{c}
\begin{array}{ccccccccc} e_1 & e_2 & e_3 & e_4 & e_5 & e_6 & e_7 & e_8 & e_9 \end{array} \\
\left[
\begin{array}{ccccccccc}
1 & 1 & 0 & 0 & 0 & 0 & 0 & 0 & 0 \\
0 & 1 & 0 & 1 & 0 & 0 & 0 & 0 & 1 \\
1 & 0 & 1 & 0 & 0 & 0 & 0 & 0 & 1 \\
0 & 0 & 1 & 1 & 1 & 1 & 0 & 0 & 0 \\
0 & 0 & 0 & 0 & 1 & 0 & 1 & 1 & 0 \\
0 & 0 & 0 & 0 & 0 & 1 & 1 & 1 & 0
\end{array}
\right]
\end{array}.
\tag{8}
$$

The reader may verify that $I_g c^T = 0$ (modulo 2).

A cycle basis matrix has as many rows as there are cycles in a basic set of cycles. The columns of such a matrix correspond to the edges of the graph and the rows correspond to the cycle bassis. A matrix element is one if the corresponding edge lies on the cycle and otherwise is zero. For a connected graph G every spanning subtree T of G determines a unique cycle basis for G, hence also determines a cycle basis matrix for G. Each edge of G that is not in T determines a unique cycle that is a member of the cycle basis for G determined by T. That cycle consists of the edge e external to T together with the unique path in T connecting the terminal vertices of e. Thus the number of cycles in a cycle basis is equal to the number of external edges relative to a spanning subtree. This is simply another way to express the cyclomatic number $c(G)$ since the number of edges in G that are not in a spanning subtree T is equal to $m - (n - 1)$, the cyclomatic number for a connected graph. This relation between a spanning subtree and its associated cycle basis is useful in many applications. In using the relation one should observe that any column of the matrix that corresponds to an external edge has exactly one 1 with all other elements zero. The converse however is not true.

EQUIVALENCE AND MAPPINGS

Equivalence of graphs is defined in such a way as to equate graphs that have identical behavior in terms of major graph properties. The graph properties used for the standard definition of equivalence depend on how the vertices are connected and are not at all dependent on the physical shape in a pictorial representation or upon the specific labeling in analytic representations. Of course other definitions of equivalence are sometimes required. Thus two pictorial graphs are said to be equivalent if their vertices can be brought into incidence in such a way that for every edge of one graph there is a corresponding edge of the other (two vertices connected by an edge in one graph are connected by an edge in the other). For digraphs the direction of corresponding edges must also agree.

When graphs are represented by incidence or adjacency matrices with the associated assignment of labels to vertices (or edges) the equivalence of graphs is defined in terms of permutations of the rows or columns of the graph which, in turn, correspond to changes in the assignment of labels. Clearly, such changes do not change the graph. Thus if G is a k-graph represented by an adjacency matrix A there are $k!$ equivalent graphs corresponding to the number of ways the k vertex labels can be applied. We can obtain the adjacency matrices for each of these by permuting the rows and column of A. Each such graph has the same pictorial representation and is the "same" graph as far as most graph theoretical properties are concerned, hence all are equivalent in this sense (e.g., they have the same cycle structure).

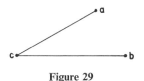

Figure 29

For example consider the 3-graph in Figure 29. That graph has adjacency matrix A:

$$
\begin{array}{c}
\\
a \\
b \\
c
\end{array}
\begin{array}{ccc}
a & b & c \\
\left[\begin{array}{ccc}
0 & 0 & 1 \\
0 & 0 & 1 \\
1 & 1 & 0
\end{array}\right],
\end{array}
\quad \text{equivalent graphs}
$$

with adjacency matrices and pictorial representations are shown in Figure 30, there are five in addition to the given one making a total of $3! = 6$.

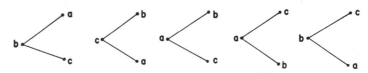

Figure 30

$$
\begin{array}{ccc} & a & b & c \end{array}
$$

$$
\begin{array}{c} a \\ b \\ c \end{array}
\begin{bmatrix} 0 & 1 & 0 \\ 1 & 0 & 1 \\ 0 & 1 & 0 \end{bmatrix},
\quad
\begin{array}{c} a \\ b \\ c \end{array}
\begin{bmatrix} 0 & 0 & 1 \\ 0 & 0 & 1 \\ 1 & 1 & 0 \end{bmatrix},
\quad
\begin{array}{c} a \\ b \\ c \end{array}
\begin{bmatrix} 0 & 1 & 1 \\ 1 & 0 & 0 \\ 1 & 0 & 0 \end{bmatrix},
\quad
\begin{array}{c} a \\ b \\ c \end{array}
\begin{bmatrix} 0 & 1 & 1 \\ 1 & 0 & 0 \\ 1 & 0 & 0 \end{bmatrix},
\quad
\begin{array}{c} a \\ b \\ c \end{array}
\begin{bmatrix} 0 & 1 & 0 \\ 1 & 0 & 1 \\ 0 & 1 & 0 \end{bmatrix}.
$$

Note that any of the matrices above can be obtained from the original A by permutations of rows and columns corresponding to interchanges of the labels. Of course the names of rows and columns must remain unchanged in this process since the transition is affected by the permutation within the matrix. For example, to obtain the third case, matrix

$$
\begin{bmatrix} 0 & 1 & 1 \\ 1 & 0 & 0 \\ 1 & 0 & 0 \end{bmatrix}
$$

from A, we observe that in this case b and c have been interchanged then b and a. Thus we proceed as follows; in each case we first permute rows then corresponding columns to yield a complete transformation, two such are required here:

$$
\begin{bmatrix} 0 & 0 & 1 \\ 0 & 0 & 1 \\ 1 & 1 & 0 \end{bmatrix}
\begin{bmatrix} 0 & 0 & 1 \\ 1 & 1 & 0 \\ 0 & 0 & 1 \end{bmatrix}
\begin{bmatrix} 0 & 1 & 0 \\ 1 & 0 & 1 \\ 0 & 1 & 0 \end{bmatrix}
\begin{bmatrix} 1 & 0 & 1 \\ 0 & 1 & 0 \\ 0 & 1 & 0 \end{bmatrix}
\begin{bmatrix} 0 & 1 & 1 \\ 1 & 0 & 0 \\ 1 & 0 & 0 \end{bmatrix}.
$$

In the pictorial (topological) equivalence sense all these are equivalent to the graphs in Figure 31, and so forth.

However, here we have some real problems of a structural combinatorial nature. Though for certain purposes we may be very content with the equivalence of all these pictorial graphs; in other cases we require a distinction

Figure 31

Figure 32

between them. Thus in some application or theoretical study we may wish to consider the graphs in Figure 32 as different.

In the enumeration of graphs the definition of equivalent cases becomes of extreme importance since we do not wish to count cases that are equivalent. There can be little doubt that the concept of equivalent graphs is one of the most difficult and important in both the theory and certain applications of graph theory. Different situations dictate different definitions of equivalence and we will discuss special situations and give appropriate definitions as required, so far as this is possible. The reader is warned that in some applications and theoretical investigations (e.g., double correspondence graphs) concepts of equivalence are, even now, far from being completely formulated. In many cases the simple concepts of equivalence given above will suffice. At this point we may ask the reader if we would care to state which of the graphs in Figure 33 are equivalent (our hope is that he sees the need to answer "it depends on what you want them to mean or what you are using them for" etc.):

Figure 33

Though we have indicated the problem of identifying equivalent graphs and have shown some possible ways of describing equivalence for the possible representations of a graph it remains to state a formal definition of equivalence widely used in graph theory.

Two graphs G and H are said to be equivalent or isomorphic if there exists a one to one correspondence between the vertex set of G and the vertex set of H called an *isomorphism* such that the existence of an edge in G implies the corresponding edge in H and conversely. These concepts apply most directly to graphs without self loops or multiple edges. When multiple edges occur the number of edges must of course be preserved in the equivalence. In the same way direction of edges must be preserved in equivalence of digraphs. The illustrations of equivalence given informally above are examples of this definition of isomorphic graphs. When one speaks of equivalent graphs it is this definition that is implied unless some special alternative is specifically stated. As an example of how the definition is employed we prove the graphs in Figure 34 are isomorphic by exhibiting a specific edge-preserving one to one correspondence (isomorphism) between them.

The graphs G and H are equivalent under correspondence (9).

$$
\begin{array}{c}
G\ H \\
1 \to 5 \\
2 \to 2 \\
3 \to 1 \\
4 \to 3 \\
5 \to 4
\end{array}
\tag{9}
$$

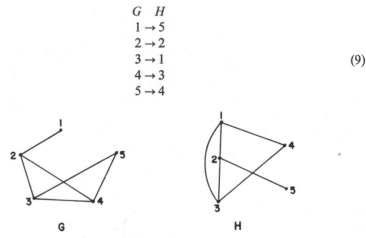

Figure 34

More general mappings of one graph into another will be discussed by defining the general mapping called a homomorphism. Let G be a linear graph without self loops or multiple edges and denote the set of vertices of G by N_g and its set of edges by E_g. Let N_r be another set of elements (vertices) and define a mapping H of N_g into N_r such that each element $v_g \in N_g$ correspond to some element $h(v_g) \in N_r$. A collection of edges E_r is associated with N_r as follows: if (a, b) is an edge in E_g with a and b vertices in N_g then $(h(a), h(b))$ is an element of the edge collection E_r. The collection N_r and E_r form a graph R called the *homomorphic* image of G under the *homomorphism H*. The graphs G and R are said to be homomorphic when a homomorphism can be defined between them. When E_g and E_r have the same number of elements and H preserves the edge incidence relation the graphs are isomorphic as described above. In general homomorphisms do not preserve the complete structure. For example every linear graph is homomorphic to a single isolated vertex. Every complete graph is homomorphic to the complete 2-graph. A star graph on n-vertices with v as its center is homomorphic to a star graph on $(n - 1)$-vertices with the homomorphic image $h(v)$ of v as its center.

A homomorphism H is called connected when the following is true: the set of vertices of N_g having the same image under the homomorphism H define a connected section graph in G. The star homomorphism referred to above is a connected homomorphism. A *connected homomorphism* is called a *contraction*. Such mappings are useful in a variety of places in detailed studies of graph theory (e.g., to investigations of graph coloring and of planar graphs), and for further discussion the reader is directed to [22]. Any contraction can be formed by a sequence of edge contractions in which the end vertices of an edge are associated with a single vertex.

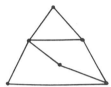

Figure 35.   Conformal graphs.

A simple but important contraction is one in which vertices of degree two are removed from a linear graph. This is done by associating such vertices with a nearest neighbor vertex until only vertices of degree other than two remain. Such a contraction is called a conformal homomorphism and when a graph R can be obtained from another G by means of that procedure the graphs R and G are said to be *conformal*. For example the graphs shown in Figure 35 are conformal. Note that the vertices remaining have the same degree in each graph. Sometimes conformal graphs are said to be isomorphic to within vertices of degree 2 (Chapter 4, [23]).

THE SPECTRA OF GRAPHS

In many ways the adjacency matrix $A(G)$ of a linear graph G characterizes G. One is therefore led to consider to what extent the spectrum of $A(G)$ characterizes G. That spectrum is the collection of characteristic values of the matrix $A(G)$. Two types of questions arise. On the one hand there is the question of what properties of G can be deduced from the spectrum of $A(G)$? On the other hand one asks to what extent does the spectrum of $A(G)$ specify G? The former is related to studies in the classical vibration of a membrane by Kac [24] and more recently to discrete vibrations in a lattice. The second question relates to problems of equivalence in graphs and was discussed by Harary in [25] and also by Fisher [26].

There seems to be very little information that one can obtain about a graph G from the spectrum of its adjacency matrix A. However, the basic ideas for studies of these questions are simple enough and the reader is invited to try and develop some results in this area of research. As a first step he may fix the idea of the spectrum of a graph in his mind by observing that the spectrum of the complete n-graph consists of $(n-1)$ repetitions of

(-1) and the number $(n-1)$ or in the notation of partitions it is the partition $((-1)^{n-1}, (n-1)^1)$, of 0 (where we have generalized the concept of partition to allow use of negative integers).

Two graphs G and R are isomorphic if and only if their adjacency matrices $A(G)$ and $A(R)$ are such that there exists a permutation matrix P such that $A(G) = PA(R)P^{-1}$. A permutation matrix being one that interchanges the rows of a matrix in the same pattern as its related permutation interchanges the row indices. Such a matrix is necessarily non-singular so that its inverse exists. Thus $A(G)$ and $A(R)$ are equivalent in the matrix theory sense and therefore have the same spectrum. We see that isomorphic graphs have the same spectrum. However matrices with the same spectrum need not be equivalent. Even so, considering the fact that adjacency matrices are special in form one asks whether two graphs must be isomorphic if they have the same spectrum. The answer is no and there has been some interesting research in showing that answer. The first result along these lines was the construction of various nonisomorphic pairs of linear graphs on 16 vertices where such pairs had adjacency matrices with the same spectra (due to R. C. Bose, R. H. Bruck, and A. J. Hoffman as reported in [25]). Then the question was to determine the smallest value of n such that there exists a pair of non-isomorphic n-graphs whose adjacency matrices have the same spectrum. For a time it was felt that $n=16$ was the smallest such value and detailed search showed that certainly $n>5$. However, consider the trees shown in Figure 36. Clearly T_1 and T_2 are not isomorphic. Their adjacency matrices are as shown in (10).

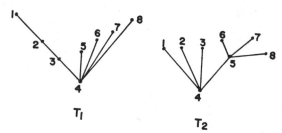

T_1

T_2

Figure 36

$$A_1 = \begin{bmatrix} 0 & 1 & 0 & 0 & 0 & 0 & 0 & 0 \\ 1 & 0 & 1 & 0 & 0 & 0 & 0 & 0 \\ 0 & 1 & 0 & 1 & 0 & 0 & 0 & 0 \\ 0 & 0 & 1 & 0 & 1 & 1 & 1 & 1 \\ 0 & 0 & 0 & 1 & 0 & 0 & 0 & 0 \\ 0 & 0 & 0 & 1 & 0 & 0 & 0 & 0 \\ 0 & 0 & 0 & 1 & 0 & 0 & 0 & 0 \\ 0 & 0 & 0 & 1 & 0 & 0 & 0 & 0 \end{bmatrix}, \quad A_2 = \begin{bmatrix} 0 & 0 & 0 & 1 & 0 & 0 & 0 & 0 \\ 0 & 0 & 0 & 1 & 0 & 0 & 0 & 0 \\ 0 & 0 & 0 & 1 & 0 & 0 & 0 & 0 \\ 1 & 1 & 1 & 0 & 1 & 0 & 0 & 0 \\ 0 & 0 & 0 & 1 & 0 & 1 & 1 & 1 \\ 0 & 0 & 0 & 0 & 1 & 0 & 0 & 0 \\ 0 & 0 & 0 & 0 & 1 & 0 & 0 & 0 \\ 0 & 0 & 0 & 0 & 1 & 0 & 0 & 0 \end{bmatrix}. \quad (10)$$

Both adjacency matrices A_1 and A_2 have the characteristic polynomial $x^4(x^4 - 7x^2 + 9)$ so that T_1 and T_2 have the same spectrum. Some further discussion of these questions may be found in [26] which also gives some more involved examples. A final result on the question of what may be called isospectral graphs is due to Baker [27].

By using detailed enumeration and study of graphs, Baker (and others) found various isospectral graphs. A major reason for such studies was their application to physics and we will discuss some related material in Chapter 10. In this section, we are interested in the general graph theory results rather than in applications. As an alternative to working directly with the spectra of characteristic values λ_j of an n-graph G one may study the first n spectral moments:

$$M_s = \sum_{j=1}^{n} \lambda_j^s,$$

these moments determine the coefficients of the characteristic polynomial and hence they determine the values λ_j as well. The relation between the spectral moments and the coefficients of the characteristic equation is given by Newton's identities (e.g., see Householder [28], Section 4.2). We are not concerned with the details of that relation but only in the fact that the moments are equivalent to the spectra. The matrix A^s is similar to a matrix with diagonal elements λ_j^s for each value of s (as a result of the reduction of a matrix to its canonical or Jordan form) and the trace of both matrices are equal (the trace or sum of diagonal elements is invariant under the similarity transformation between matrices). Thus $M_s = \text{trace}(A^s) = \sum_{j=1}^{n} a_{jj}^s$ where a_{jj}^s is the (j,j) element of A^s. We may observe that a_{jj}^s is equal to the total number of cycles of length s, containing vertex j in graph G with adjacency matrix A. All kinds of cycles are of course included in the total; vertices and edges may be repeated and so forth. This result shows an interesting relation between the number of cycles in a graph G and its spectrum. This relates to one of our questions about the information contained in the spectrum. The total number of cycles of length s is equal to the spectral moment M_s. It does not seem likely that the individual enumerations can be obtained from the spectrum alone. In fact, the existence of non-isomorphic isospectral graphs shows that the specific numbers of cycles between vertices are not specified in general by the spectrum.

A major result of Baker's enumerative study was the generation of a non-isomorphic pair of isospectral graphs on six vertices. Prior to that result the smallest number of vertices giving such a pair was seven as shown in Figure 36. Baker's six vertex pair is shown at the left in Figure 37. Another interesting result is the triple of nonisomorphic isospectral 9 vertex graphs also shown in Figure 37.

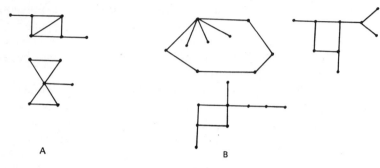

Figure 37

For the pair shown at A the spectral moments are $M_s = 0, 14, 12, 70, 120, 446$ for $s = 1, \ldots, 6$ respectively. The characteristic equation is $(1 + \lambda)^2 (1 - \lambda)$ $(\lambda^3 - \lambda^2 - 5\lambda + 1) = 0$. The spectral moments for the triple shown at B are $M_s = 0, 18, 0, 78, 0, 402, 0, 2214, 0$ for $s = 1, \ldots, 9$ respectively.

Both of the above illustrations of isospectral graphs show graphs with pendant edges. Indeed many of the examples encountered were reported to be of pendant type with either a pendant edge or an articulation point connecting two simpler graphs at a single vertex. However, not all examples of isospectral graphs are of that type. Figure 38 shows two pairs A and B of

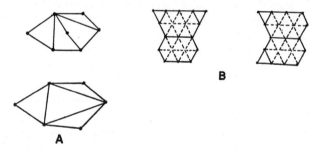

Figure 38

isospectral graphs that have no articulation points (are biconnected). Moreover B has the interesting property that it may be embedded in a triangular lattice as indicated by the dashed line lattice structure in the figure. Graphs with the embedding property are particularly interesting in applications (e.g., in physics).

As we have seen the spectrum of the adjacency matrix $A(G)$ does not fully characterize G. One is lead to consider the possibility of characterizing G by means of some set of functions of the matrix $A(G)$. That possibility has been studied by Turner [29] who found that the types of functions he considered (immanants) did not suffice to characterize G up to isomorphism. Thus the situation when using such matrix functions is similar to the spectral case.

GRAPH THEORY IN PROSPECTIVE

We have now defined and offered several representations of linear graphs and discussed the concept of equivalence between two such graphs. The basic objects of our studies have been introduced. Before proceeding to an expansion of graph theory concepts we shall discuss the position occupied by linear graphs in a wider context.

Within the subject matter of combinatorial topology (for an introduction to topology see Blackett [30]) a fundamental role is played by n-simplices or what may alternatively be called n-cells. These are objects of geometric character defined in Euclidian n-space. They may be treated as abstract objects entirely independent of any spacial configuration. Thus we utilize geometric n-space terminology to define an object but proceed to lift the resulting object out of its imbedded position and treat it as an abstract entity. A zero-cell or 0-simplex is a point. A one-cell or 1-simplex is a finite line segment, that is, a one-dimensional object *bounded* by (two) cells of one less dimension, 0-cells in this instance. This concept is extended to two dimensions by defining a 2-simplex as the convex two dimensional figure bounded by a minimum number of 1-simplices. (*A* region R in n-space is said to be convex if the occurrence of two points in the region implies that the straight line segment between the points lies within the region. Thus if a and b are any points in R then $x = \lambda a + (1 - \lambda)b$ is a point in R for all λ such that $0 \le \lambda \le 1$.) Since 1-simplices are line segments we see that a 2-simplex is a triangle. In general an n-simplex is the closed, bounded convex figure in n-space bounded by a minimum number of $(n - 1)$-simplices.

These n-simplices can intersect in many ways when considered as geometric objects. We may say that two n-simplices are incident when they share a single bounding $(n - 1)$-simplex which is common to both. An incidence hierarchy can be formulated in which the incidence just defined may be called n-incidence, the two n-simplices being called k-incident if they share a single $(k - 1)$ bounding simplex which is common to both (alternatively we may allow the n-simplices to share one or more $(k - 1)$, $k < n$, bounding simplices, however this leads to ambiguous situations as to the order of incidence).

With these rather heuristic descriptions of n-cells and their incidence hierarchy at hand one can formulate a very general concept of a graph. In this general context a graph is a representation of an incidence relation among a set of n-cells. We may for example consider a set of 2-cells (triangles) and a 1-incidence between various pairs of 2-cells (such pairs have a 0-cell or vertex point in common). For this example a pictorial representation is possible and a typical 2-cell, 1-incidence graph is shown as A in Figure 39.

A   **B**

Figure 39

The expression k-incidence means that the actual incidence is between k simplices and occurs by superimposing some of their bounding $k - 1$ cells. If the incidence is changed to 2-incidence for a set of 2-cells the pictorial representation is illustrated as B in Figure 39. In this case a pair of incident 2-cells share a common 1-cell (line segment). The reader should be cautioned that these graphs are *not linear graphs* even though the pictorial representations shown could be interpreted as linear graphs. The distinction is based on the types of objects for which incidence is defined and cannot always be deduced from pictorial representations alone. This is in fact a shortcoming of the pictorial representation when general types of graphs are being discussed.

Linear graphs are seen to have a special place in the general graph framework as they are the simplest possible such graphs. In this context we may describe a linear graph as a representation of a 1-incidence between 1-cells. Such a description gives us an additional way to think about a graph but one should realize that it is slightly different than our previous definitions and representations. In the previous discussions a single vertex could occur by itself but this cannot occur in the 1-incidence concept. To make all the presentations of linear graphs the same we must also include 0-cells in addition to 1-incidence of 1-cells and form the linear graph from a set of 1-cells and 0-cells. Such a procedure is in fact done in combinatorial topology and also in some applications. This description is very helpful when edges are removed from graphs since the edge is a 1-cell and therefore carries its bounding 0-cells with it. There is often an esthetic question as to what happens to the vertices when edges are removed and in other similar operations. The general structure outlined above may provide a useful interpretation of such operations.

We shall deal with linear graphs only and refer to them as graphs. Should some more general situation be needed the context will make clear what is intended. Thus our discussion of n-cells, incidence and general graphs has been presented only to place the subject of linear graphs in some perspective.

The subject of linear graphs has developed over many years along widely separated lines of investigation. Most of the early work on graphs was motivated by the desire to understand or formulate problems of mathematical interest. In this connection the pioneer studies of Euler which lead to aspects of modern topology and a great deal of the mathematical construction techniques by which map coloring problems were attacked stand out as major examples. Other lines of research on mathematical objects such as the studies of rigid linkage by A. B. Kempe (author of an important incorrect proof of the four color conjecture) in the third quarter of the 19th century lead to still other aspects of linear graphs ([31]). Rather late in the historical development of the subject some unification was introduced into the subject giving a connected theory, e.g., in the classical text of Konig [32]. While this solidification of a theory of linear graphs took place the subject of topology was undergoing extensive development. As we have indicated, topology does in a broad sense include linear graph theory and some aspects of graph theory are best treated in the general context of topology. An excellent introduction may be found in Aleksandrov [33]. However, by specializing linear graph theory a great deal can be gained in achieving specific detailed results which could not be achieved from too broad a theory. Thus while modern practice allows and utilizes linear graphs (and more general graphs as discussed above) in topology it also provides the theory of linear graphs as a special subject based on special definitions such as we have supplied in this chapter. It is this view of graph theory that is the subject of the text.

In addition to the development of graph theory ideas by mathematicians motivated by problems of primarily mathematical content the subject was vigorously investigated by physical scientists for many reasons. The occurrence of graph like systems and objects is a common place in such subjects as flow diagrams, electrical networks, road maps and so forth. The considerable interest shown by engineers of many types is therefore entirely understandable. More sophisticated studies of physical systems from the viewpoint of the structure of matter and statistical mechanical explanations of physical phenomena lead chemists and physicists to make powerful contributions to the subject matter of graph theory. Later in the text we shall discuss some of these aspects of the theory and their appropriate applications.

EXERCISES

For Exercises 1 through 11 let A denote the adjacency matrix of a graph G:

$$A = \begin{bmatrix}
0 & 2 & 0 & 1 & 0 & 0 & 0 & 0 & 0 & 0 & 1 & 0 & 0 \\
2 & 0 & 1 & 0 & 0 & 0 & 0 & 0 & 0 & 0 & 0 & 0 & 1 \\
0 & 1 & 0 & 2 & 0 & 0 & 1 & 0 & 0 & 0 & 0 & 0 & 0 \\
1 & 0 & 2 & 0 & 1 & 0 & 0 & 0 & 0 & 0 & 0 & 0 & 0 \\
0 & 0 & 0 & 1 & 0 & 1 & 0 & 0 & 0 & 1 & 0 & 0 & 0 \\
0 & 0 & 0 & 0 & 1 & 0 & 1 & 0 & 3 & 0 & 0 & 0 & 0 \\
0 & 0 & 1 & 0 & 0 & 1 & 0 & 1 & 0 & 0 & 0 & 0 & 0 \\
0 & 0 & 0 & 0 & 0 & 0 & 1 & 0 & 1 & 0 & 0 & 0 & 1 \\
0 & 0 & 0 & 0 & 0 & 3 & 0 & 1 & 0 & 1 & 0 & 3 & 0 \\
0 & 0 & 0 & 0 & 1 & 0 & 0 & 0 & 1 & 0 & 1 & 0 & 0 \\
1 & 0 & 0 & 0 & 0 & 0 & 0 & 0 & 0 & 1 & 0 & 1 & 0 \\
0 & 0 & 0 & 0 & 0 & 0 & 0 & 0 & 3 & 0 & 1 & 0 & 1 \\
0 & 1 & 0 & 0 & 0 & 0 & 0 & 1 & 0 & 0 & 0 & 1 & 0
\end{bmatrix}$$

1. From A find the degree of each vertex, the number of vertices of odd degree, the number of edges, and the number of components of G.

2. Draw a pictorial representation of G and form the vertex-edge incidence matrix. Does G have any circuit of odd length?

3. Assign a direction to each edge of G to form a digraph G^*. Is G^* strongly connected? Can you assign directions so as to make G^* strongly connected?

4. Form the section graph S of G determined by vertices 1, 2, 3, 4, 6, 8, 9, 10, 12 Is S connected? What kind of graph is the complement of S?

5. Form a spanning subtree of G. Determine its diameter. Does the tree have a center or a bicenter?

6. Find a star subgraph and a wheel subgraph of G.

7. What is the minimal number of edge disjoint paths that will cover G? Is G an euler graph?

8. What is the minimal number of edges that must be added to G to make it regular? What is the degree of the resulting graph?

9. Can one impose directions on the edges of G so that G^* will be pseudosymmetric?

10. Determine the spectrum of the section graph of G determined by vertices 5 through 13.

11. Apply Tarry's construction to produce a cyclic directed path passing through each edge of G exactly once in each direction.

12. Let $n = 2k - 1$ for $k = 1, 2, 3, \ldots$. Determine the spectrum of the star graph and of the wheel on n vertices.

For Exercises 13 through 20 let B denote the adjacency matrix of a digraph H.

$$B = \begin{pmatrix} 0 & 1 & 0 & 0 & 0 & 0 \\ 0 & 0 & 1 & 0 & 0 & 1 \\ 0 & 0 & 0 & 1 & 0 & 0 \\ 0 & 1 & 0 & 0 & 1 & 0 \\ 0 & 0 & 0 & 0 & 0 & 1 \\ 1 & 0 & 0 & 1 & 0 & 0 \end{pmatrix}$$

13. Determine the number of spanning arborescences of H with root at vertex v_1. Do the same for the root at v_4. Illustrate the fact that the number is independent of root location.

14. Determine the number of distinct euler paths in H.

15. List the distinct euler paths as edge sequences.

16. Let H^* denote the undirected graph derived by removing all directions from the edges of H. Write the adjacency matrix, vertex-edge incidence matrix, and cycle matrix for H^*.

17. Determine a cycle basis for H^*. Draw the cycles that belong to the basis.

18. Find the smallest graph (least number of vertices) to which H^* is homomorphic. Give a homomorphism that establishes your illustration. Is the homomorphism connected? Are the two graphs conformal?

19. Find the spectrum of H^*.

20. Interpret H^* in terms of edge incidence of 2-simplices rather than vertex incidence of 1-simplices. Show that such an interpretation produces a graph that is isomorphic to a star graph on four vertices.

Chapter 2

Connectivity and
Independence in Graphs

CONNECTIVITY

A connected graph has been defined as a graph in which every pair of vertices lie on at least one path. For a digraph to be strongly connected the vertices must lie on a directed path. When one looks more closely at the concept of connectedness, some graphs seem to be more connected than others. Such an intuitive impression can be formalized into mathematical description in many ways, some of which are discussed in this book. For a much deeper study of these concepts the reader is directed to Tutte's excellent mathematical treatment [11].

The extent to which a connected graph is connected can be studied by considering how the graph may be disconnected by the removal of vertices, of edges, or of both. A number of concepts have been defined and investigated relative to disconnection by removal of vertices (hence the edges incident to them) and relative to the removal of edges (the incident vertices being considered to remain in the graph). Several of these concepts are discussed, as is the removal of combinations of vertices and edges (e.g., section graphs);

however, there will be no further discussion of these ideas here since such concepts are new and their formalism remains to be developed. The emphasis will be on the removal of vertices but there will be some discussion of the very similar concepts that correspond to removal of edges.

Two special concepts are closely related to the basic idea of connectedness; one is a special edge and the other is a special vertex. An *isthmus* (or bridge) is an edge of a connected graph G whose removal divides G into two nonvertex graph connected components. A vertex A is an *articulation point* of a connected graph G if its removal disconnects G. If the resulting graph is such that each component has a single edge connecting it to the vertex A in G then the vertex A is called a *simple articulation point*. In Figure 40 graph G has three components, component C_2 has an isthmus at edge e and component C_3 has an articulation point at vertex A. One may observe that edge f in component C_1 is not an isthmus since its removal will produce a vertex graph component.

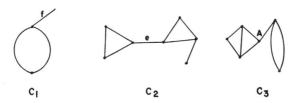

Figure 40. A three component graph G.

A graph G may be disconnected by removing various sets of vertices. Any such set of vertices will be called a *disconnecting vertex set* (sometimes called an articulation set). For a given graph G let $S(V, G)$ be the set of all disconnecting vertex sets of G, that is, $S(V, G) = \{V|$ removal of the vertices in V disconnects $G\}$. Denote the number of elements in a set V by $|V|$. Then the graph G is said to have *vertex connectivity* or, more commonly, connectivity (and sometimes connection number) $\kappa(G)$ where

$$\kappa(G) = \min_{V \in S(V, G)} |V|.$$

If $\kappa(G) = |V^*|$ then V^* is a minimal disconnecting (vertex) set of G. Such a set may be, but need not be unique for a graph G and any set of vertices of which V^* is a subset is a disconnecting set as well.

A *disconnecting edge set* is a subset of the edges of G such that removal of the edges in the set from G disconnects G. This concept is therefore based on a subset of the edges of G. On the other hand one can consider a subset A^* of the vertex set of G. The set of edges joining A^* and its complement, each such edge having one vertex in A^* and the other vertex in $\bar{A}^*$, is called a cut-set of G. Though these concepts produce similar results one is often more convenient to use than the other. A disconnecting edge set having no subset which is also disconnecting is minimal, it is called a *proper cut set*. Though a proper cut set is always a cut set the converse is not true in general. One may readily contemplate cut sets which have disconnecting subsets. In these concepts one must keep in mind that the number of components is not considered but only that the graph G is not connected. The *edge connectivity* (sometimes called cohesion number) of G is defined as the number of elements in a proper cut set (minimal disconnecting edge set) and denoted by $\varepsilon(G)$.

One can construct a collection of independent cut sets which can often be used as an alternative to the cycle basis concept in a graph G. Such a collection of cut sets can be expressed as a fundamental cut set matrix where each row corresponds to a cut set and each column corresponds to an edge of G. A matrix element is one if the corresponding edge lies in the cut set (corresponding to the row) and otherwise is zero. As in the case of a cycle basis, a collection of fundamental cut sets for a connected graph can be specified by means of a spanning subtree T. Each edge of T determines a unique cut set. Removal of such an edge divides T into two components. The edge lies in the cut set as do all edges having one terminal vertex in one component of T and the other terminal vertex in the other component of T. If G has no multiple edges, a column of the fundamental cut set matrix will have a single 1 with all other elements zero if and only if it corresponds to an edge of the spanning subtree T. There are $n - 1$ such edges and hence there are $n - 1$ fundamental cut sets. If one is given a fundamental cut set matrix and one knows the graph G then one can easily determine the spanning subtree T that corresponds to the given matrix.

There exists a number of interesting relations between $\kappa(G)$, $\varepsilon(G)$, and the minimum (local) degree of G. The most basic such relation is the following due to Whitney [34].

Theorem. Let $\kappa(G)$ and $\varepsilon(G)$ denote the vertex connectivity and edge connectivity respectively of a simple linear graph G, then

$$\kappa(G) \le \varepsilon(G) \le \min_{v \in N(G)} d(v),$$

where $N(G)$ is the vertex set of G and $d(v)$ is the local degree at vertex v.

Proof. If G is the null graph or a vertex graph it is clear that $\kappa(G) = \varepsilon(G) = \min d = 0$ so the theorem holds for these (trivial) cases.

Let v^* denote a vertex such that $d(v^*) = \min d(v)$ then removal of all edges at v^* will disconnect G (by isolating v^*) so that $\varepsilon(G) \le \min d(v)$. Of course it may be that less edges will disconnect G in some other way, hence the inequality.

If $\varepsilon(G) = 1$ then removal of a vertex incident on a disconnecting edge will also disconnect G so that $\kappa(G) = 1$ in such a case. When $\varepsilon(G) > 1$ removal of $\varepsilon(G) - 1$ edges leaves an isthmus in the remaining graph. Denote the vertices of the isthmus by v_1 and v_2. Corresponding to each removal edge select a vertex incident on it other than v_1 or v_2 (clearly such an edge cannot be incident on both v_1 and v_2 since G is simple). Removal of all such vertices will necessarily remove the set of $\varepsilon(G) - 1$ edges and possibly additional edges as well. In fact their removal may disconnect G and if so $\kappa(G) < \varepsilon(G)$ (since $\varepsilon(G) - 1$ or less vertices have disconnected G). If G is not disconnected by that process it will certainly be disconnected by the additional removal of v_1 or of v_2 [since (v_1, v_2) is an isthmus in the modified graph]. This establishes the result.   ▲

Along these same lines Chartrand and Harary [35] have shown (by construction) that for a triple of integers (a, b, c) such that $0 < a \le b \le c$ there exists a graph G for which $\kappa(G) = a$, $\varepsilon(G) = b$, and min $d(v) = c$ $v \in G$. From this result one may observe that the above theorem is of the type known in mathematics as a best possible theorem, that is, the inequalities cannot be improved in any way without putting restrictions on the kind of graphs G. For further discussion of these concepts one may consult Harary (Chapter 5 [36]).

The concepts introduced above are illustrated by the graphs in Figure 41. In G_1, $\kappa(G_1) = 3$, $\varepsilon(G_1) = 3$, and min $d(v) = 3$. In G_2, $\kappa(G_2) = 1$, $\varepsilon(G_2) = 2$, and min $d(v) = 3$.

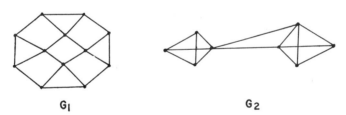

G_1 G_2

Figure 41

BICONNECTED GRAPHS AND MENGER'S THEOREM

There is an important theorem of graph theory that finds application in various parts of the subject. In one sense it is closely related to the cycle structure of a special class of graphs, and for this reason we present it here. A graph G is said to be *biconnected* when it contains no articulation points. The ways (paths using a vertex at most once) of a biconnected graph have special structure as described in the following theorem. The ways α_1, α_2 of the theorem can be used to form simple cycles in G. We denote the ways as sequences of their vertices with the indices indicating that one vertex follows the other as the way is traversed. Thus a way (v_0, v_1, v_2, v_3) from v_0 to v_3 has three edges and as we go from v_0 to v_3 we encounter v_1 before v_2 and so forth.

Theorem (Menger) Let G be a biconnected graph and α a way of length k in G from v_0 to v_k defined by the vertex sequence $(v_0, v_1, \ldots, v_k)$. Two ways α_1 and α_2 can be associated with α such that the following conditions hold:

1. the terminal vertices of α_1 and α_2 are v_0 and v_k and these are the only vertices common to α_1 and α_2,

2. if either of the ways α_1 or α_2 is traversed from v_0 to v_k the indices of the vertices of α which are passed through in the process are in ascending order.

Proof. The proof of this theorem is by cases and though simple is a bit tedious. We proceed by induction and first establish the result for a way of length one.

Consider the way (v_0, v_1). Since v_0 is not an articulation point there must be a way connecting it to v_1 which does not include the one edge way (v_0, v_1). We may take that way as α_1 and the edge (v_0, v_1) as the way α_2 of the theorem. This shows the theorem to be true for ways of length one.

Assume the theorem is true for ways of length k. It then remains to show that utilizing this hypothesis the theorem holds for ways of length $k + 1$. We consider such a way $\alpha^* = (v_0, v_1, \ldots, v_k, v_{k+1})$. The induction hypothesis assures us of the existence of two ways α_1 and α_2 joining v_0 and v_k and satisfying the theorem. We shall denote the way $(v_0, v_1, \ldots, v_k)$ by α so that α^* is α together with the edge (v_k, v_{k+1}). We are to find two ways α_1^* and α_2^* joining v_0 and v_{k+1} and satisfying the theorem.

There is a way β joining v_0 and v_{k+1} which does not include the vertex v_k. Should this not be true every way from v_0 to v_{k+1} would contain v_k so that removal of v_k from G would separate G into components so that v_0 was in one component and v_{k+1} was in another. Thus v_k would be a point of articulation and G not biconnected contrary to hypothesis. We may use the phrase nearest vertex to mean one separated by a way of shortest length. Then we may consider a vertex of β which is nearest to v_{k+1} and also lies on the combination of the three ways α, α_1 and α_2. Call this vertex v and consider the four possible positions v may occupy.

1. One is $v = v_0$. The theorem is satisfied by taking $\alpha_1^* = \alpha^*$ and $\alpha_2^* = \beta$.

2. Consider $v = v_{k+1}$. In this case v does not lie on α hence it must lie on α_1 or α_2. In either case it is simple to define the required ways. For example if v lies on α_2 we define α_1^* and α_2^* as follows

$$\alpha_1^* = \alpha_1 (v_0, v_k) + (v_k, v_{k+1}),$$
$$\alpha_2^* = \alpha_2 (v_0, v),$$

where $\alpha_2(v_0, v)$ means that part of α_2 from v_0 to v. One proceeds similarly when v lies on α_1.

3. The v is not a vertex of α^*, so that as in case (2) it is a vertex of α_1 or α_2. Suppose v belongs to α_1 then we define

$$\alpha_1^* = \alpha_1 (v_0, v) + \beta(v, v_{k+1}),$$
$$\alpha_2^* = \alpha_2 (v_0, v_k) + (v_k, v_{k+1}).$$

This is somewhat harder to see than the previous cases. In particular one may suppose it possible for $\beta(v, v_{k+1})$ and $\alpha_2(v_0, v_k)$ to have a vertex in common. However, this cannot occur since such a situation would allow a vertex nearer than v to v_{k+1} which violates the definition of v.

4. The v lies on α^* but is neither v_0 nor v_{k+1}. Since v lies on β it can not be v_k. Hence $v = v_r$ where $r < k$. Moreover there is a vertex v_s of α with $s \leq r$ which is closest to v_r and belongs to α_1 or α_2. We may define the required ways in terms of these vertices as one may illustrate for the case in which v_s is a vertex of α_2

$$\alpha_1^* = \alpha_1 (v_0, v_k) + (v_k, v_{k+1})$$
$$\alpha_2^* = \alpha_2 (v_0, v_s) + \alpha(v_s, v_r) + \beta(v_r, v_{k+1}).$$

Thus in every case we can define ways having the required properties. ▲

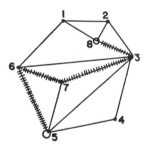

Figure 42.   A biconnected graph.

In Figure 42 the vertices may be denoted by $n_i(i = 1, \ldots 8)$. Let $v_0 \equiv n_8$ and $v_k \equiv n_5$ then the way from n_8 to n_5 marked on the graph is specified by the vertex sequence $(n_8, n_3, n_7, n_6, n_5)$ which for the theorem has the form $(v_0, v_1, v_2, v_3, v_4)$. It may be seen that the ways: $(n_8, n_2, n_3, n_4, n_5)$ and $(n_8, n_1, n_6, n_7, n_5)$ satisfy the theorem. These two ways taken together form a simple cyclic way containing n_8 and n_5 illustrating how such cycles may be formed in a graph G.

ADDITIONAL CONCEPTS IN CONNECTIVITY

Because of the very central and important role connectivity plays in graph theory and in many of its applications a number of additional concepts are presented in this section. These are certainly more specialized and often more complex than the concepts presented above. In most cases they will be defined only without further development, however, some theorems and illustrations will be given. It is felt that the inclusion of these topics here will give the reader a fuller appreciation of the depth and scope of connectivity concepts and will also provide a guide toward his further study of these concepts. In some cases the material may also suggest an area of research or application to the reader. Further details of the material may be found in [11], [21], [37], and particularly in [38].

The most direct way to introduce the concepts of this section is to list them together with a concept number c_k for reference. Anything occurring in a theorem and not listed will be provided as needed. In the list G is a simple graph.

c_1. A *k-isthmus* of G is a complete section graph with k vertices which disconnects G and does not properly contain any complete (non-vertex) section graph of G which disconnects G. A 2-isthmus is an isthmus in the usual sense. Note the articulation points as complete one-graphs are excluded by definition.

c_2. A *k-articulator* G' of G is an incomplete section graph with k vertices which disconnects G, such that each section graph of $G - G'$, (i.e., that part of G remaining after removal of G') except possibly vertex section graphs (isolated vertices) has a neighbor in G of each vertex in G'. By *neighbor* we mean a vertex that is edge connected to a given vertex.

c_3. A connected section graph G' is *dense* if $G' = G$ or if every vertex in $G - G'$ has a neighbor in G'.

c_4. A proper dense section graph which is contained in no other dense section graph (except G) is called *D-maximal* (where the D stands for dense).

c_5. A dense section graph containing no other dense section graph is *D-minimal*.

c_6. Let H be a subgraph of G. A *vertex of attachment* of H in G is a vertex of H that is incident in G with an edge not belonging to $E(H)$ (the set of edges of H). The set of vertices of attachment of H in G is denoted by $W(G, H)$.

Some notation will be presented now that is helpful in discussing the above concepts.

The collection of dense section graphs of G ordered by inclusion together with the empty set Φ is denoted by S_D.

The union of all *D-minimal* section graphs of G is denoted by Γ.

The section subgraph of neighbors of a vertex v in a section graph G' is denoted by $G'(v)$.

If $H_1, H_2, \ldots, H_k$ are graphs their union, denoted by $H = \cup_{i=1}^k H_i$, is defined as a graph with vertex set $N(H) = \cup_{i=1}^k N(H_i)$ and edge collection $E(H) = \cup_{i=1}^k E(H_i)$. The intersection is denoted by $K = \cap_{i=1}^k H_i$ and has vertex set $N(K) = \cap_{i=1}^k N(H_i)$ and edge collection $E(K) = \cap_{i=1}^k E(H_i)$. The fact that H is a subgraph of a graph G may be denoted by $H \subseteq G$.

Some of the concepts above have been generalized but those generalizations are not given here. The concepts presented are now illustrated. Consider Figure 43. In graph G_1 the complete 3-graph determined by vertices a, b, and c

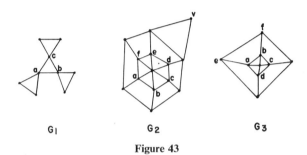

$$G_1 \qquad\qquad G_2 \qquad\qquad G_3$$

Figure 43

is a 3-isthmus. In G_2 the section graph (a cycle) determined by vertices a, b, c, d, e, and f is a 6-articulator which is not dense since vertex v has no neighbor in it. Graph G_3 contains the dense 4-articulator determined by vertices a, b, c, and d. That subgraph is D-minimal but since it is contained in the dense section graph determined by a, b, c, d, e, and f it is not D-maximal. Figure 44 has a subgraph H defined as the section graph specified by vertices a, b, c, and d in the graph G. The set of vertices of attachment is $W(G, H) = \{b, c\}$. The subgraph of neighbors of vertex a, denoted by $H(a)$ is also shown in Figure 44.

The concept of k-connected has already been defined in Chapter 1 as the property that every pair of vertices in G has k distinct paths between them such that the paths have no vertices in common (except of course their

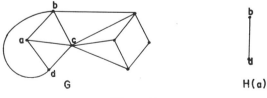

$$G \qquad\qquad\qquad H(a)$$

Figure 44

terminal vertices). Here we shall say that a graph G is (k)-connected if $G - G'$ is connected for every subgraph G' containing fewer than k vertices. Thus, for any integer $m \leq \kappa$ (G) the Graph G is (m)-connected. The two concepts differ from a conceptual point of view in that k-connectedness deals with connecting paths (which may not have vertices in common) while (k)-connectedness is defined in terms of disconnecting sets. However, the latter also implies edge sequence connections in which the sequences have no vertices in common. In fact, the two concepts imply each other so that they simply provide different ways of considering higher order connectivity. Figure 45 shows a graph that is not 2-connected since $\kappa(G) = 1$ due to the articulation point v; it is 1-connected (of course).

G

Figure 45

The following simple theorem illustrates a result that may be obtained from these concepts.

Theorem. If a simple graph G is (m)-connected then $G - G'$ is dense for every section graph G' containing less than m vertices and conversely.

Proof. The converse is a matter of definition. Since the statement that $G - G'$ is dense implies it is connected, it follows that any section graph G' with less than m vertices can be removed from G without disconnecting G, that is, G is (m)-connected.

Let G be (m)-connected and let G' be a section graph on less than m vertices. Suppose $v \in G'$ is a vertex without a neighbor in $G - G'$. Then $G - G' \cup v = G - (G' - v)$ is not connected, contrary to the (m)-connected hypothesis, since $(G' - v)$ has fewer than m vertices.   ▲

Figure 46 illustrates the theorem. Graph G has $\kappa(G) = 4$ since it is the

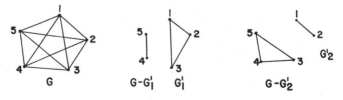

Figure 46

complete 5-graph. Hence G is (4)-connected and an instance of the theorem from that point of view is shown by the section graph G'_1. However G is also (3)-connected and a second instance of the theorem is provided by the section graph G'_2. In each case the remaining graphs are seen to be dense.

The extensive theory based on the above concepts and on their generalizations will not be developed further here. This section will be concluded by an illustration of the vertices of attachment concept following [11], consisting of the development of a calculus for such vertices, (i.e., a method of operating with them). The empty set is Φ and the null graph is Ω. H and K are arbitrary subgraphs of G.

The following properties are obvious:

1. $w(G, G) = w(G, \Omega) = \Phi$.
2. $w(G, H) \subseteq N(H)$, the vertex set of H.

The remaining properties will be established as follows.

3. $w(G, H) = w(K, H \cap K) \cup \{N(H) \cap w(G, H \cup K)\}$.

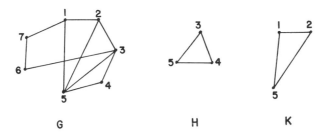

G   H   K

Figure 47

As an illustration consider G in Figure 47 together with the subgraphs H and K shown. For those graphs

$$w(K, H \cap K) = \{5\}, \qquad w(G, H \cup K) = \{1, 2, 3\},$$

$N(H) \cap w(G, H \cup K) = \{3\}$, and the relation property (3) yields the result

$$w(G, H) = \{5\} \cup \{3\} = \{3, 5\}.$$

Proof of (3). The $w(G, H) =$ set of all vertices of H incident with edges of the set $E(G) - E(H)$.

$$E(G) - E(H) = \{E(G) - E(H \cup K)\} \cup \{E(K) - E(H \cap K)\}$$

since the sets on the right consist of the elements of $E(G)$ that are not in H or K and the elements of $E(G)$ that are in K but not in H. These two sets include all edges of G that are not in H. The edges of $E(K) - E(H \cap K)$ are incident on the vertices $w(K, H \cap K)$ of H. On the other hand $N(H) \cap w(G, H \cup K)$ is the

set of vertices of H incident with edges of $E(G) - E(H \cup K)$. The result
follows.   ▲
4. $w(H \cup K, H) = w(K, H \cap K)$.

Proof. Let $G = H \cup K$ in property (3). This results in

$$w(H \cup K, H) = w(K, H \cap K) \cup \{N(H) \cap w(H \cup K, H \cup K)\}$$

but $w(H \cup K, H \cup K) = \Phi$ by property (1) giving the result.   ▲
5. If $H \subseteq K$ then $w(K, H) \subseteq w(G, H)$.

Proof. $w(K, H \cap K) = w(K, H)$ since $H \subseteq K$, hence for this situation
property (3) yields

$$w(G, H) = w(K, H) \cup \{N(H) \cap w(G, K)\}.$$

Thus $w(K, H) \subseteq w(G, H)$.   ▲
6. $w(G, H \cup K) \cup w(G, H \cap K) = w(G, H) \cup w(G, K)$.

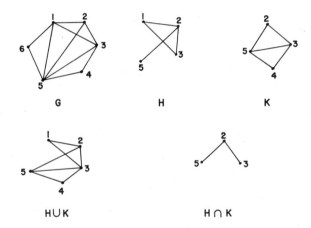

Figure 48

Property 6 is illustrated in Figure 48 by means of the graph G and the two
subgraphs H and K. For those graphs: $w(G, H \cup K) = \{1, 5\}$, $w(G, H \cap K)$
$= \{2, 3, 5\}$, $w(G, H) = \{1, 3, 5)$, and $w(G, K) = \{2, 3, 5\}$.

Proof of Property 6. By property 3: $N(H) \cap w(G, H \cup K) \subseteq w(G, H)$ and
$N(K) \cap w(G, H \cup K) \subseteq w(G, K)$. The union of these results gives

$$[N(H) \cap w(G, H \cup K)] \cup [N(K) \cap w(G, H \cup K)] \subseteq w(G, H) \cup w(G, K).$$

From this expression the calculus of sets yields

$$[N(H) \cup N(K)] \cap w(G, H \cup K) \subseteq w(G, H) \cup w(G, K),$$

but $N(H) \cup N(K) \supseteq w(G, H) \cup w(G, K)$ by property (2). Hence $w(G, H \cup K) \subseteq w(G, H) \cup w(G, K)$. Let us call this result A.

By property (3), using $H \cap K$ for H

$$w(G, H \cap K) = w(K, H \cap K) \cup \{N(H \cap K) \cap w(G, K)\}.$$

In this expression $w(K, H \cap K) = w(H \cup K, H)$ by property (4). Moreover, $N(H \cap K) \cap w(G, K)$ may be enlarged to $w(G, K)$ producing the inclusion

$$w(G, H \cap K) \subseteq w(H \cup K, H) \cup w(G, K).$$

Replacing $w(H \cup K, H)$ by $w(G, H)$ can only enhance the inclusion, according to property (5). Thus one obtains what we shall call result B

$$w(G, H \cap K) \subseteq w(G, H) \cup w(G, K).$$

Also $w(G, H) \subseteq w(G, H \cap K) \cup w(G, H \cup K)$ by properties (3) and (5), and $w(G, K) \subseteq w(G, H \cap K) \cup w(G, H \cup K)$ which we shall call results C and D respectively.

Using A and B gives

$$w(G, H \cup K) \cup w(G, H \cap K) \subseteq w(G, H) \cup w(G, K)$$

and using C and D gives

$$w(G, H \cup K) \cup w(G, H \cap K) \supseteq w(G, H) \cup w(G, K).$$

These results taken together establish property (6).   ▲

A number of other properties may be established in a calculus of articulation point sets. It is felt that the above serves to introduce the reader to these interesting concepts and illustrates how aspects of graph theory may be developed in a formal deductive way.

DISTANCE AND INDEPENDENCE

There are several types of numbers which may be defined for connected graphs and that prove useful in many parts of graph theory and its applications. In this section we shall introduce some of these numbers.

The *coefficient of internal stability* or *independence number* is a measure of the extent to which the vertices of a graph are nonadjacent. A set of vertices in a graph G is *internally stable* or *independent* if no two of its vertices are adjacent, (i.e., edge connected). We may denote the class of all internally stable sets of G by S^* and a typical member of the class by $S \in S^*$. The coefficient of internal stability of a graph G is denoted $si(G)$ and defined by

$$si(G) = \max_{S \in S^*} |S|$$

where $|S|$ is the number of vertices in the set S. This number arises in the study of some classical problems of mathematics such as Kirkman's school-girl problem and the eight queens chess problem. The relation to these problems and to additional theoretical results can be found in Berge [6].

A set of vertices S (of G) is *externally stable* or *dominating* if every vertex not in S is adjacent to at least one vertex in S. We denote the class of all externally stable sets of vertices by S^{**}. The *coefficient of external stability* or *domination number* of a graph G is denoted se(G) and defined by

$$se(G) = \min_{S \in S^{**}} |S|.$$

This quantity measures the extent to which a number of vertices has access to the remainder by means of a single edge. In this sense it is opposite to the coefficient of internal stability. For graph A with 6 vertices and 5 edges in Figure 49 $si(A) = 5$, $se(A) = 1$ while for graph B with 5 vertices and 5 edges $si(B) = 2$ and $se(B) = 2$.

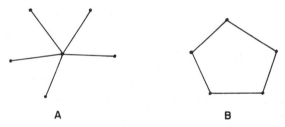

A   B

Figure 49

The concept of the distance between two vertices was introduced in Chapter 1 and utilized in the discussion of centers and bicenters in trees. In a general linear graph G the distance between two vertices u and v is the length of a shortest path between them and is denoted by $d(u, v)$. In a connected graph G $d(u, v) > 0$ for every pair of vertices u and v in G and in a complete graph K_n on n vertices $d(u, v) = 1$ for every pair of vertices in K_n. It is convenient to label the vertices v_i where i is an element of the index set $I = \{1, 2, \ldots, n\}$ and $n = |N(G)|$ denotes the number of vertices of G. The distance between two vertices v_i and v_j can be expressed by the notation $d_{ij} = d(v_i, v_j)$. In terms of this notation two useful concepts of graph theory may be defined as follows.

The radius R of a graph G is defined by the equation

$$R = \min_{i \in I} \max_{j \in I} d_{ij}.$$

A vertex v_r of G is said to be a center (of G) if (Ore [15], p. 29)

$$R = \max_{j \in I} d_{rj}.$$

The *diameter* H of a graph G is defined by the equation (Ore [15], p. 29)

$$H = \max_{i,\, j \in I} d_{ij},$$

so that H is the maximum of the distances between pairs of vertices.

The radius R indicates the degree to which some vertex or vertices of G (the set of centers of G) are close to the other vertices. On the other hand H indicates separation of vertices. A large value of H indicates the existence of vertices that are widely separated in G. In most studies the quantities R and H are only considered for connected graphs. In an unconnected graph each component has a (finite) value for R and for H; these quantities are assigned the value infinity for the (unconnected) graph itself.

In general it is difficult to compute R and H without the help of the distance matrix which will be introduced in Chapter 9. For Graph G in Figure 50 R = 3 and H = 4 while vertex v is the only center.

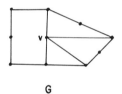

G

Figure 50

EXERCISES

1. Find the vertex connectivity and the edge connectivity of K_5 and $K_{3,3}$. Can you generalize the results to K_n and $K_{r,r}$?
2. Use results of (1) to verify Whitney's inequality
$$\varkappa(G) \le \varepsilon(G) \le \min d(v).$$
3. For any positive integer n form an n-graph in which every edge is an isthmus. Is every vertex an articulation point? (Hint, consider K_2).
4. What are the connectivity numbers for a simple cycle of length n? Consider graphs A and B with the following adjacency matrices:

$$A = \begin{pmatrix} 0 & 1 & 1 & 0 & 0 \\ 1 & 0 & 1 & 0 & 0 \\ 1 & 1 & 0 & 1 & 1 \\ 0 & 0 & 1 & 0 & 1 \\ 0 & 0 & 1 & 1 & 0 \end{pmatrix}, \quad B = \begin{pmatrix} 0 & 1 & 1 & 0 & 0 & 1 \\ 1 & 0 & 0 & 1 & 0 & 1 \\ 1 & 0 & 0 & 1 & 1 & 0 \\ 0 & 1 & 1 & 0 & 1 & 0 \\ 0 & 0 & 1 & 1 & 0 & 1 \\ 1 & 1 & 0 & 0 & 1 & 0 \end{pmatrix}.$$

5. Use graph B to illustrate Menger's theorem.

6. Find a disconnecting edge set for B. Is it a proper cut set? What is the edge connectivity of B?

7. Find the (vertex) connectivity, the independence number, and the coefficient of external stability of B. Give a dominating vertex set for B.

8. Is A biconnected? Find a dominating vertex set for A.

9. Let n be an even positive integer. Find the coefficient of internal stability for a simple cycle with n vertices. Find the coefficient of external stability for such a graph. What are these quantities when n is odd?

10. Carry out Exercise 9 for a wheel and for a star graph with n vertices.

Chapter 3

Planar Graphs

DEFINITIONS AND CONCEPTS OF PLANAR GRAPHS

We recall that a pictorial representation of a graph G is obtained by using lines for the edges of G and points for the vertices of G. Such a representation is drawn in a plane where the points representing vertices of G are clearly specified and the lines representing edges of G are given some convenient form. It may be that in such a representation, lines intersect at other than vertex points of G. In many cases we are not at all concerned with such spurious intersections. In this chapter we are interested in the possibility of avoiding those intersections.

If a graph G can be given a pictorial representation by means of lines and points in such a way that the lines only intersect at vertex points then G is said to be *planar*. One should note that the definition of a planar graph does not require that every pictorial representation have the nonintersection of edge lines property. It only requires that there is some representation having that property. This situation complicates the study of planar graphs since it is often difficult to determine whether a given graph is or is not planar. As a simple illustration the complete four graph is shown in Figure 51; it is planar and is given two representations in the figure, one showing the planar

Figure 51

character while the other is not in proper planar form. A graph drawn in the plane which is of planar form as drawn is often called a *geometric graph*. By definition a planar graph (as an abstract graph) is isomorphic to a geometric graph.

When a graph G has no self-loops or multiple edges and is planar it is said to be a *simple planar graph*. In the general case a planar graph may have self-loops and/or multiple edges.

A planar graph in the planar pictorial representation of a planar graph is seen to have a number of *faces*. A face of a planar graph is a plane area bounded by edges and vertex points of the representation. We may speak of the bounding edges and vertices as the composition of a face. The boundary of a face is formally defined as the sequence of edges and vertices encountered as one moves in a prescribed direction around the face. The concept depends on the representation. For example in Figure 52 the finite face shown has four vertices and four edges lying upon it. Those vertices and edges are considered as part of the boundary of the region forming the face. They are not interior points of the region in the sense of point sets in the plane. Unfortunately they are not really boundary points in that sense either. However, they are boundary points if we define the region of the face as the open set of points that are enclosed within but do not belong to the boundary. Clearly an isomorphic representation of the graph in Figure 52 exists in which the four vertices and edges lie in the infinite face. In that figure the finite face is bounded by a simple cycle but the infinite face is not. By the *infinite face* of a plane graph we mean that portion of the plane surrounding a finite planar graph (representation). An infinite plane graph may have more than one infinite face (the number depending to some extent on how one wishes to define the plane) but we shall not extend that concept here.

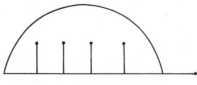

Figure 52

It has been seen that the composition of faces depends on the specific planar representation of a planar graph (in terms of edge and vertex labeling). There are in general many ways of drawing a pictorial representation of a planar graph in what we have called proper planar form. Thus there exists many geometric graphs which are all isomorphic to a particular planar graph (and of course to each other). Such representations differ in the arrangement of the faces and edges of the graph in the plane. Suppose G is a (abstract) planar graph and let G^* denote a plane representation of G. The vertices, edges, and faces of G^* may be labeled. Once that has been done G^* can be used to form alternative (isomorphic) planar representations of G. One method for the generation of alternatives is to project G^* onto a sphere or indeed one may consider G^* as already drawn on a sphere (we shall discuss the concept of linear graphs in non-planar spaces at the end of this section). A new planar representation is then obtained by projecting the sphere graph form of G^* back onto the plane in some specified way. The standard projection mapping used for these alternative formations of representations of G is called a *stereographic projection*. Such a projection is based on placing the plane tangent to the South Pole (any specified point) of the sphere. Then G^* (oriented as desired) is projected onto the plane from the North Pole (the point lying on the opposite end of a diameter to the South Pole). The use of such projections establishes the following class of (isomorphic) different plane representations of a planar graph.

Theorem. Every finite planar graph with connectivity two can be represented in the plane in such a way that any specified face becomes the exterior (infinite) face.

Every finite planar graph can be represented so that any specified edge is part of the boundary of the exterior region (the point set forming the exterior face).

An example of the second result is given in Figure 53.

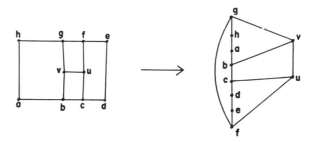

Figure 53

We shall soon establish that the number of faces is a property of the graph G and not of any particular planar representation of G. For this purpose we shall use a special case of euler's formula the general form of which will be given in Chapter 5. As remarked above it is customary to consider the portion of the plane surrounding a finite planar graph as a face of the graph. Thus the complete four graph of Figure 51 has four faces. From one point of view this is a simple convention; however it allows us to define the dual graph and thereby leads to considerable enrichment of the study of planar graph. We shall discuss the dual graph later in this section but at this point we consider euler's formula. Denoting the number of vertices, edges and faces by n, m and f respectively euler's formula states that $n - m + f = 2$ for a planar graph G. It is possible to give a good heuristic proof of this by starting with a single finite face, planar graph (a cycle) and adding faces inductively. Though simple we omit the details in favor of a more rigorous approach which does not depend on construction in so elementary a way. It is possible to challenge the simple proof by pointing out that one can not always construct planar graphs by adding faces as the graph of Figure 52 shows. Our present result will follow from a lemma about planar graphs where we deal with edges rather than faces.

Lemma. The simple cycles that form boundaries of different finite faces of a planar graph G taken together constitute a cyclic basis for G.

Proof. The lemma is true if G has exactly two finite faces. Though this is not completely obvious it becomes clear upon reference to the definition of cyclic basis in Chapter 1.

Assume the lemma is true for every planar graph having $f - 1$ finite faces.

Consider a planar graph G having f finite faces. Remove an edge e on the boundary of a face of G to form a graph G' with $f - 1$ finite faces which, by the induction hypothesis, constitute a cyclic basis for G'. By replacing e we create a new cycle which can not be represented in terms of the existing cyclic basis since e is not any edge of any cycle of that basis. We have seen that introduction of an edge in this way can increase the cyclomatic number by at most one. Thus the cyclomatic number is increased by exactly one in this process and since it is equal to the maximum number of independent cycles in G the f cycles specified form a cyclic basis for G.   ▲

We may now establish *euler's formula*.

For a connected planar graph with n vertices, m edges and f faces the following relation holds

$$n - m + f = 2.$$

Proof. We have seen in the lemma above that the number of finite faces is equal to the maximum number of independent cycles and hence is equal to the cyclomatic number. Thus (including the infinite face in f)

$$f = m - n + 1 + 1$$

since the number of components $p = 1$. This gives the result.   ▲

An extremely useful concept in the study of planar graphs is the *simple incidence graph of faces and edges* associated with a given planar graph G. The simple incidence graph is a bipartite graph, one set of vertices denoted by A corresponds to the faces of G and the other set of vertices denoted by E corresponds to the edges of G. There is an edge joining the vertices $a \in A$ and $e \in E$ if and only if face a has edge e as a boundary edge in the planar graph G. The simple incidence graph associated with a planar graph G can be utilized in many ways as we shall see. Some examples follow. For convenience we denote the simple incidence graph associated with G by $SI(G)$. We present a result which depends on euler's formula.

Theorem. Every simple planar graph G contains a vertex v with degree $d(v) \leq 5$.

Note that a simple planar graph is a planar graph without self loops or multiple edges. The result is certainly not true for non-simple graphs.

Proof. Every face of G is bounded by at least three distinct edges. Form the $SI(G)$ and consider the number of edges k in $SI(G)$. Each vertex of E contributes at most two edges to k so that $k \leq 2m$. On the other hand each vertex of A must contribute at least three edges to k so that $k \geq 3f$. Thus we find that $f \leq 2m/3$.

Suppose every vertex of G is of degree six or greater. Then $2m \geq 6n$ so that $n \leq m/3$. Applying euler's formula to G we find:

$$2 = n - m + f \leq \frac{m}{3} - m + \frac{2m}{3} = 0$$

a contradiction which establishes the result.   ▲

Figure 54 shows a simple planar graph with 12 vertices which is regular of degree 5. The above theorem states that a simple regular graph of degree six or greater can not be planar.

Corresponding to a plane representation of a planar graph G one may define the *dual graph* G^*, or more correctly the geometric dual of the plane representation. Vertices of G^* correspond to faces of G in such a way that for each face of G, including the infinite face there is a vertex of G^*. When two faces of G are contiguous (have a common edge as part of their boundary) the corresponding vertices are adjacent in G^*. Thus the edges of G^* are specified by the contiguity of faces of G. Though rather cumbersome to state the

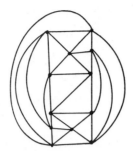

Figure 54. Planar, regular graph.

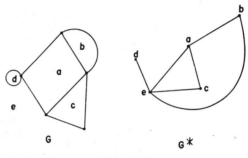

Figure 55

concept of the dual graph is simple enough and is illustrated in Figure 55 where a planar graph G is shown together with its dual graph G^*. The faces of G are given letter labels, e being the infinite face. The corresponding labels are used to denote the vertices of G^*.

The concept of the dual graph to a planar graph given above is too restrictive in that it depends completely on the plane representation for its description. In fact the dual graph concept is a rather powerful one that is useful in advanced studies in graph theory. It must therefore be expressed in terms of the abstract planar graph G itself rather than in terms of specific representations in the plane. A concept for the dual graph that does not depend on the representation is the *combinatorial dual* of a planar graph G. This concept is expressed in terms of the cycle rank and cocycle rank of a simple graph G. It is beyond the scope of this book to enter into a detailed study of these concepts, we wish only to present the resulting formulation, for details one may consult Harary [36] or Ore [22]. We need only observe that the cycle rank is equal to what has been called the cyclomatic number in Chapter 1 (the difference in names arises from the points of view taken for developing the concept). Hence we shall call the cycle rank $c(G) = m - n + p$ (where p is the number of components in G). Thus the cycle rank is the number of elements (cycles) in a cyclic basis for G. In Chapter 2 the concept of a cut set was introduced. That is a set of edges of G that disconnects G. A minimal cut set is called a *cocycle* of G. A vector space based on cocycles can be developed along lines similar to our discussion of the cyclic space in Chapter 1. A basis for such a space is called a cocycle basis and the number of its elements is the cocycle rank of G denoted by $c^*(G) = n - p$. It may be noted that the concepts of cycle rank and cocycle rank can be based on the unifying concept of the spanning subtrees of G.

Consider a simple plane graph G. Let z be any subgraph of G so that $G - z$ denotes the subgraph of G obtained by removing the edges of z from G. Let $\{H\}$ denote the set of edges in any graph H. (We have used $C(H)$ for this in Chapter 1 but the present notation is more useful here, particularly in view of our definition of the cycle and cocycle ranks!)

A simple graph G^d is a *combinatorial dual* of a simple graph G if there exists a one to one corresponding between $\{G\}$ and $\{G^d\}$ such that for any sets $\{z\} \subseteq \{G\}$ and $\{z^d\} \subseteq \{G^d\}$ that are in correspondence, (i.e., $\{z\} \leftrightarrow \{z^d\}$):

$$c^*(G - z) = c^*(G) - c(z^d)$$

where z^d is the subgraph of G^d determined by the set $\{z^d\}$ of edges. Ore calls G and G^d, *w*-dual after Whitney who developed the concept of the combinatorial dual in detail [it is interesting that Ore also uses the following alternative criteria equation: $c^*(G^d - z^d) = c^*(G^d) - c(z)$]. Figure 56 illustrates the concept of the combinatorial dual criteria equation based on the subgraph $Z = \{e_7, e_8, e_9, e_{10}\}$. For the graphs of Figure 56: $c^*(G - z) = 5$, $c^*(G) = 6$, and $c(z^d) = 1$. To establish that G^d is a combinatorial dual of G one would have to show the result for all subgraphs $z \subseteq G$. Thus the concept is used primarily in theoretical investigations.

We have remarked that the number of faces in a planar graph does not depend on the particular planar representation of such a graph. That number is a simple example of a quantity that does not depend on the representation, any such quantity is called a (topological) *invariant* of G over the set of isomorphic representations of G. Invariants are defined for graphs in many ways. In this section we will give a brief description of two invariants that are closely related to the concepts of planar graphs.

The *thickness* of a graph G, denoted by $\tau(G)$ is the minimum number of planar subgraphs which together constitute a partition of G (i.e., the union of all vertices and edges forms the vertex set and edge collection of G). In this notation G is planar if and only if $\tau(G) = 1$. Direct evaluation shows that $\tau(K_j) = 2$ for $j = 5, 6, 7, 8$ (the complete graphs K_j for $j = 1, 2, 3, 4$ are planar) and for K_9, $\tau(K_9) = 3$. In general $\tau(K_j) \geq [(j + 7)/6]$, (note that $[x]$ denotes the greatest integer not greater than x). There are some results for $\tau(G)$ when G is a complete graph, a complete bipartite graph and some other

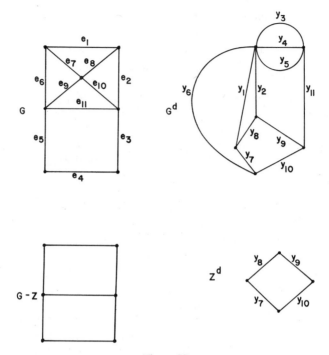

Figure 56

special cases. These results involve detail, special cases and so forth; there is very little in the way of general theory. Reference may be made to Beineke [39] or Harary [36].

When a graph is given a pictorial representation in the plane there will be some such representation in which the number of non-vertex points at which edges are required to cross will be minimum. The number of such edge crossing points is called the *crossing number* of G, denoted by $h(G)$. It seems very difficult to obtain values for $h(G)$ beyond the trivial fact that $h(G) = 0$ if and only if G is planar. It is clear that $h(K_5) = 1$ and that $h(K_{3,3}) = 1$ and it is known that $h(K_j) = 1/4\,[j/2]\,[(j-1)/2]\,[(j-2)/2]\,[(j-3)/2]$ for $j \le 10$ (Saaty [40]) and some results are known for $K_{m,n}$. Beyond these results, discussed for example by Harary [36], very little is known about crossing numbers. However, these numbers seem important to the relationship between graph theory as an abstract study and the concepts of graphs as subsets of points in a plane or other (topological) space. It has been conjectured that the above formula for $h(K_j)$ holds for all j and that a similar formula may be given for $h(K_{m,n})$. A discussion of such conjectures is given by Guy [41]. A fundamental approach to crossing number studies is given by Tutte [42].

Considerations based on the idea of a graph as a point set in topological space lead to the concept of imbedding a graph in a space. Equivalence of graphs is then given in terms of homomorphisms. The invariant properties under such mappings and the possibility of particular forms of imbedding are part of topology and the transition from (abstract) graph theory to usual topological analysis has been effected. The ideas of planar graphs illustrate the imbedding concept as does the remark that any finite linear graph can be expressed in the interior of a sphere (i.e., in a bounded 3-space) without edges crossing except at vertices.

STRAIGHT LINE REPRESENTATION OF PLANAR GRAPHS

A simple planar graph has no self loops or multiple edges. For such graphs we state the following interesting theorem due to Fary [43] in which a *straight graph* is one whose edges are all straight lines. The theorem clearly does not hold for non-simple planar graphs.

Theorem (Fary). Every simple finite planar graph can be represented as a straight graph.

Proof. The proof is based on the concept of triangulated graphs. A planar graph in which each finite face is bounded by a cycle of exactly three edges is called a triangulated graph. Such graphs are always simple.

Every simple planar graph G is a subgraph of a triangulated graph based on the same set of vertices as G.

This result is trivially true when G has three vertices. Thus we suppose G has more than three vertices. We construct a triangulated graph G^* of which G is a subgraph by adding edges to G according to the following scheme.

Select two vertices on the boundary of a face that are not connected by an edge. Introduce an edge between these two vertices and lying interior to the face. Continue in this way until no such pair of vertices can be found. The process will always stop since G is finite. The resulting graph is G^*. Certainly G^* is connected for if it has separated components there would be vertices on the boundary of a face, (i.e., the infinite face) which should be connected by an edge according to the above scheme. Such a situation can not exist since by definition of G^* the process has terminated.

Graph G^* is triangulated. If there was a finite face of G^* with more than three edges the above process could not have terminated which contradicts the definition of G^*. Hence the stated result is established.

It is clear that every subgraph of a straight graph is straight. Since the straight aspect is one of representation only it follows that if a planar graph has a representation as a straight graph so does any of its subgraphs. The representation of the subgraphs is in fact directly generated by the representation of the graph itself.

The two results stated above show that the theorem will be established if every triangulated graph is a straight graph. We shall now establish that fact. To do so requires two results.

Result 1.   Let T be a triangulated graph with at least four vertices. Let v be a vertex of T and $(v, u_1), (v, u_2), \ldots, (v, u_k)$ be all the edges with v as one terminal vertex. The edges $(u_1, u_2), \ldots, (u_k, u_1)$ are in T and form a cycle $\eta(v)$ which separates v from every other vertex of T.

Suppose v is not on the boundary of the infinite face. Then there is an innermost cycle $C(v)$ which has v inside (i.e., there is no cycle inside $C(v)$ such that v is inside it). Denote the vertices of $C(v)$ by $w_1, \ldots, w_m$. Since T is triangulated the edge (w_i, w_{i+1}) of $C(v)$ is adjacent to a face lying within $C(v)$ and having a third vertex x on its boundary. Suppose $x \neq v$ then the cycle consisting of edges $(w_1, w_2), \ldots, (w_i, x), (x, w_{i+1}), \ldots, (w_m, w_1)$ has v in its interior and lies within $C(v)$ which contradicts the definition of $C(v)$ as the innermost cycle. Hence $x = v$ and this is the situation for all the edges of $C(v)$. Thus $C(v)$ contains only v and is the cycle required to establish Result 1.

When v is on the boundary of the infinite face one uses the outermost cycle having v in its exterior for the cycle $C(v)$. The same kind of argument used above shows that $C(v)$ can be used for the cycle $\eta(v)$ and the result is fully established.

The theorem is to be established by induction and to this end we introduce another graph. Let G be the triangulated graph in question (we wish to show it is straight) and form a related graph G' having one less vertex. A vertex v of G which does not lie on the boundary of the infinite face is removed from G. All edges incident on v are also removed. That yields G^*, which by Result 1 contains a cycle $\eta(v)$. Graph G^* is identical with G outside $\eta(v)$ and is empty inside $\eta(v)$. We connect vertex u_1 of $\eta(v)$ with each of the vertices $u_3, u_4, \ldots, u_{k-1}$ of $\eta(v)$ by edges lying within $\eta(v)$ that do not intersect each other. The resulting graph is denoted by G'; it has one less vertex than G. Having defined G' we must establish our second result.

Result 2.   If G' is not simple, then G has a cycle of three edges which separates two vertices.

We remark that the character of G' may depend on how the vertices of $\eta(v)$ are labeled since they are connected in an order which depends on this labeling in forming G'. Nevertheless the possibility of forming a non-simple graph for G' must be taken into account. The situation is illustrated in Figure 57 where G' is not simple.

The result is established by detailing the situation when G' is not simple. Since G is triangulated, G' can fail to be simple only when a new edge connects

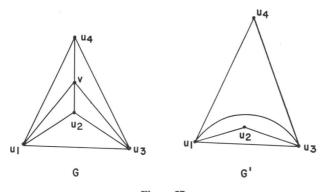

Figure 57

two vertices which are also connected by an edge lying in the exterior of $\eta(v)$. Since all new edges are incident on u_1 this means there is a vertex $u_i (2 < i \le k - 1)$ which is joined by an edge in the exterior of $\eta(v)$ to u_1. Call this edge e. The cycle (u_1, v), (v, u_i), e separates u_2 from u_k, hence proves the result. In Figure 57, e is the edge (u_1, u_3) of G.

We now utilize induction to establish the theorem. Certainly a triangulated graph with less than four vertices is straight. Assume a triangulated graph with v vertices $3 \le v \le n$ is straight. We consider a triangulated graph G with $n + 1$ vertices. Let v be a vertex of G not on the boundary of the infinite face and construct G' described above. Two cases occur.

CASE 1. G' is simple. In this case G' is straight by the induction hypothesis so we put it into a straight representation. The edges $(u_1, u_2), \ldots, (u_1, u_k)$ are removed leaving the interior of $\eta(v)$ empty. We may now place v into this region and connect it to each of the vertices $u_1, \ldots, u_k$ with edges. The result is G in a straight representation.

CASE 2. G' is not simple. Thus by our second result there is a cycle having three edges in G which separates two vertices of G. Denote this cycle by C. The cycle C and that part of G which is exterior to C forms a triangulated graph G_1. The cycle C and its interior in G also forms a triangulated graph G_2. Since G_1 and G_2 have at most n vertices they are straight graphs by the induction hypothesis. Putting these graphs into a straight representation shows that G is straight and the theorem is established. ▲

In Figure 58 we show at A a straight representation of the complete four graph which is a straight graph since it is simple planar. The reader may wish to draw a straight representation of the simple planar graph at B in the figure.

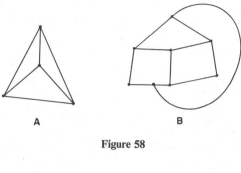

Figure 58

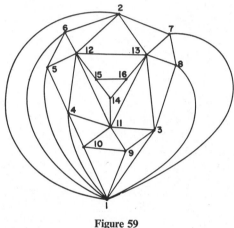

Figure 59

A nontrivial illustration of Fary's theorem is given by the straight graph representation in Figure 60 of the simple planar graph shown in Figure 59.

CRITERIA FOR PLANAR GRAPHS

There are essentially three approaches to the characterization of the planar property for graphs. One is in terms of the existence or not of special classes of subgraphs, a second is in terms of the concept of the dual graph, and the third is in terms of cycle bases for the graph. The first will be treated in some detail in this section and the other two will be stated as criteria without any detailed discussion.

Two graphs play a key role in the study of planar graphs. They are shown in Figure 61 as K_5, the complete 5-graph and as $K_{3,3}$, the complete bipartite graph with three vertices in each disjoint vertex set. Graph $K_{3,3}$ is a pictorial representation of the problem of connecting three homes with three utilities

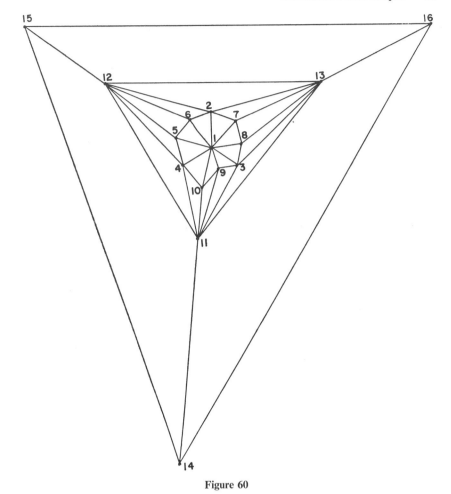

Figure 60

without having utility lines cross (it is often called the utilities graph). A solution to this problem would require $K_{3,3}$ to be planar. We will find that it is not planar so that the utilities problem has no solution.

The following two results establish that neither K_5 nor $K_{3,3}$ is planar.

Result 1.   Graph $K_{3,3}$ is not planar.

Proof. In graph $K_{3,3}$, $n = 6$, $m = 9$ so if $K_{3,3}$ is planar euler's formula yields $f = 5$ for the number of faces. Since $K_{3,3}$ is bipartite its vertices fall into two sets such that vertices of the same set are not adjacent. Thus every face must have at least four edges in its boundary. Consider $SI(K_{3,3})$ and count the number of edges k in $SI(K_{3,3})$. By considering the vertex set of $SI(K_{3,3})$ which

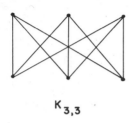
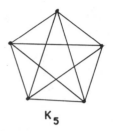

$K_{3,3}$ K_5

Figure 61

corresponds to the edges of $K_{3,3}$ we find $k \leq 2m$. From the point of view of the set of vertices of $SI(K_{3,3})$ corresponding to the faces of $K_{3,3}$ we find $k \geq 4f$. Hence $f \leq m/2$. But we have seen that $f = 5$ so that $5 \leq 9/2$ which is a contradiction that establishes the result.   ▲

Result 2. Graph K_5 is not planar.

Proof. If K_5 was planar euler's formula would yield $f = 7$ since $n = 5$ and $m = 10$. Each face must have at least three edges in its boundary. Consider $SI(K_5)$ with k edges and we see that $k \leq 2m$ and also that $k \geq 3f$ so that $f \leq 2m/3 = 20/3$. But $f = 7$ hence $7 \leq 20/3$ which is false. Hence K_5 is not planar.   ▲

The graphs $K_{3,3}$ and K_5 are used to formulate two fundamental classes of non-planar graphs which because of their utilization in the Kuratowski criteria for planar graphs are often called Kuratowski type 1 and Kuratowski type 2 graphs respectively. Graphs of type 1 are formed by introducing vertices along the edges of graph $K_{3,3}$ and graphs of type 2 are formed by introducing vertices along edges of graph K_5. Thus $K_{3,3}$ and K_5 are of types 1 and 2 respectively. Examples of graphs of type 1 are shown in Figure 62 while Figure 63 shows some examples of type 2 graphs.

An alternative description of type 1 and type 2 graphs is obtained by using the concept of contracting a graph as described in Chapter 1. One contracts a graph by removing each vertex that is incident to only two edges and replacing the two edges by a single edge. Any graph that can be contracted to form graph $K_{3,3}$ of Figure 61 is a type 1 graph. Similarly any graph that can be contracted to form graph K_5 of Figure 61 is a type 2 graph. Graphs that can be related by means of this simple type of contraction are said to be *conformal* (as described in Chapter 1).

It is clear from Results 1 and 2 above that all type 1 and 2 graphs are non-planar.

We have introduced the concept of simple cycles before. Such cycles use an edge at most once. When a cycle uses a vertex at most once, the cycle may be

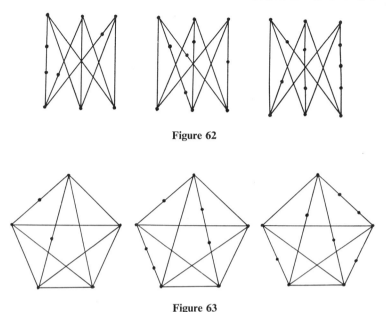

Figure 62

Figure 63

called elementary. In the following, elementary cycles will be denoted by Greek letters such as α. If u and v are vertices of an elementary cycle α then $\alpha(u, v)$ denotes the sequence of vertices of α including u and v which are met in moving along the cycle from u to v and $\overline{\alpha(u, v)}$ denotes the vertex sequence excluding u and v. There is ambiguity as to which of the two possible vertex sequences are to be taken. This is resolved by specifying that we move from u to v in the prescribed sense (or positive sense). Prescribed sense means that the direction of travel around the cycle is specified in some (usually arbitrary) way. For example a direction in which the inside of the cycle lies on the left as one moves along the cycle could be utilized. This is a convention which, though not rigorously defined (e.g., we require careful designation of inside of a cycle) would be satisfactory for our present purpose.

If a graph G has a set of articulation points A then $G - A$ has several components. A connected component C of $G - A$ together with all edges incident to C and their terminal vertices is called a *piece of G* relative to the articulation points.

Sufficient concepts are now available to establish a very basic theorem on planar graphs due to Kuratowski.

Theorem (Kuratowski) (For example see Berge [6] and Busacher and Saaty [23]. A finite graph G is planar if and only if it contains no subgraphs which are of either Kuratowski type 1 or type 2.

Proof. Results 2 and 3 above have established that a graph must not contain any subgraphs of type 1 or type 2 if it is to be planar. Thus the condition of having no such subgraphs is seen to be necessary if G is to be planar. The difficult part of this theorem is to establish the sufficiency of the condition.

Proof of sufficiency is established by induction. Certainly the theorem is true for graphs having less than four edges. Assume the condition is sufficient for all graphs having less than m edges. We are required to show on the basis of this assumption that the condition is sufficient for all graphs having m edges.

Consider a graph G having m edges which satisfies the condition, that is, has no subgraphs of type 1 or of type 2. Assume G is not planar; we shall call this the major assumption. The establishment of a contradiction will therefore prove the theorem.

The graph G is connected for if it was not all of its components would have less than m edges and by the induction hypothesis would be planar (we omit trivial cases where isolated vertices are parts of G as such vertices will not relate to the planar character of G). Moreover G is biconnected. If G contained an articulation point each piece of G relative to such a point would be planar by the induction hypothesis. These pieces can be connected at the articulation point without introducing crossing of the connecting edges. For example the articulation point can be in the infinite face of each piece. Thus we are dealing with a connected, biconnected graph G.

An edge may be denoted by (v, w) where v and w are its terminal vertices. If the edge (v, w) is removed from G there exists an elementary cycle α passing through v and w. Denote the graph obtained from G by the removal of (v, w) by $\bar{G}$, which is a connected graph since G is biconnected, moreover, it is planar by the induction hypothesis. Assume there is no elementary cycle in $\bar{G}$ passing through v and w. Since $\bar{G}$ is connected, Menger's theorem assures us that $\bar{G}$ has an articulation point based on this assumption. Furthermore, since there is assumed to be no elementary cycle containing v and w these vertices must lie on two different pieces of $\bar{G}$ relative to an articulation point u (by Menger's theorem on each piece). These pieces are denoted $\bar{G}_v$ and $\bar{G}_w$. When edge (v, u) is added to $\bar{G}_v$ we denote the result by G_v and when edge (w, u) is added to $\bar{G}_w$ we denote the result by G_w.

The graph G_v is planar since it has less than m edges and, being a subgraph of G, satisfies the induction hypothesis. Since we have already stated as a basic result of planar graphs that there is a planar representation in which any specific edge lies on the infinite face, G_v may be represented in such a way that the edge (v, u) is on the infinite face. In the same way G_w is planar and may be represented with edge (w, u) on its infinite face. Now if we introduce the edge (v, w) the result is a planar graph containing G which contradicts our major assumption that G is non-planar. Hence we have established that an elementary cycle α exists in $\bar{G}$ which contains v and w.

There may be several cycles such as α. Let us denote the elementary cycle in the planar graph $\bar{G}$ containing v and w which encloses the greatest possible number of faces in its interior by η. We impose an orientation on η which serves to specify a region which is inside η and a region which is outside. The pieces of $\bar{G}$ having vertices inside η are called internal pieces and those having vertices outside η are called external pieces relative to the elementary cycle η. Recall that $\eta(v, w)$ denotes the sequence of vertices of η which are encountered in moving in the prescribed orientation from v to w along η.

An external piece cannot contain more than one vertex of the sequence $\eta(v, w)$. If it did so we could find another elementary cycle enclosing more faces than η. In the same way an external piece can contain at most one vertex of the sequence $\eta(w, v)$. Thus an external piece meets η in one or two vertices only. There is at least one external piece and one internal piece which meet both $\eta(\overline{v, w})$ and $\eta(\overline{w, v})$. If this were not so we could introduce the edge (v, w) in such a way that G would be planar contrary to the major assumption.

It is necessary to establish that an internal piece I and an external piece E both exist which meet both paths $\eta(\overline{v, w})$ and $\eta(\overline{w, v})$ in such a way that the points of contact are in alternating order. Thus if we denote the points of contact by a and b for I and c and d for E then when one traverses the cycle η the points are encountered in the order $acbd$.

Suppose such pieces do not exist. Let I' be an internal piece which meets $\eta(\overline{v, w})$ and $\eta(\overline{w, v})$. Denote two consecutive points of contact between I' and η by x_1 and y_1. Since we have assumed that pieces I and E with the alternating intersection property do not exist, we can connect x_1 and y_1 by means of a continuous line σ which lies outside η. The statement that we could not do so would indicate that an external piece E of the desired type existed, contrary to our assumption. Reference to Figure 64 may clarify the situation.

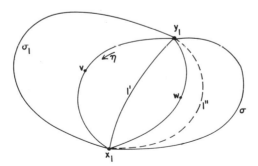

Figure 64

Every internal piece which only meets $\eta(x_1, y_1)$ can be transferred to the exterior region lying within the curves η and σ. Thus I' can be so transferred to become I''. If every internal piece could be so transferred we could connect v and w so that the result was planar contrary to our major assumption. Thus there exists an internal piece I_1' which cannot be transferred in this way. If I_1' only meets $\eta(y_1, x_1)$ it could be transferred outside in a similar way to I' by connecting y_1 and x_1 with another continuous curve σ_1 (refer to Figure 64). If I_1' meets $\eta(x_1, y_1)$ in only one point it would serve together with I'' (which is now an external piece) to form the pieces I and E violating our assumption that such pieces do not exist. Hence I_1' meets $\eta(y_1, x_1)$ at some point and meets $\eta(x_1, y_1)$ in two consecutive points x_2 and y_2 at least one of which is in $\eta(\overline{x_1, y_1})$ (otherwise I_1' could be transferred into an external piece in the region bounded by σ_1 and η).

Now the transformations described above can be carried out using I_1' as a starting piece rather than I'. If the process were to terminate we would violate either the present assumption that pieces I and E don't exist or the assumption that G is not planar. Thus the process can not terminate. Since G is finite we are forced into a contradiction which establishes the existence of I and E (since at this stage we wish to maintain the major assumption that G is not planar).

There are four points at which I meets η that are of particular interest. We denote these points by e, f, g, h and specify them to lie on the following vertex sequences:

$$e \in \eta(\overline{c, d}), f \in \eta(\overline{d, c}), g \in \eta(\overline{v, w}), h \in \eta(\overline{w, v}).$$

The existence of these four points is assumed by the existence of I and E established above. There are four general situations which may arise relative to the four points e, f, g and h. Certainly $e \neq f$ and $g \neq h$. The possible situations are dealt with by studying a typical example of each case.

In studying the four cases it is helpful to utilize diagrams. These are pictorial and a single line is used to represent paths or even pieces of a graph (such representation is common in graph theory). For simplicity the edge (v, w) is omitted in each figure but as subgraphs of G each would contain that edge. Its presence is necessary for the arguments and should be kept in mind by the reader. The cases which lead to type 1 graphs make use of square and circular enclosure of vertices to show the two sets of vertices present in the basic (bipartite) type 1 graph.

CASE 1.   One of the points e, f is on $\eta(\overline{v, w})$ and the other is on $\eta(\overline{w, v})$. One such example is $e = g$ and $f = h$. This is illustrated in Figure 65 where we observe that G contains a type 1 subgraph (recall that (v, w) is an edge of G). Thus cases of this type cannot arise by our assumption on G.

CASE 2. Both points e and f are on $\eta(\overline{v, w})$ so that we may associate f and g. Point h may equal d or not; in any case, we observe that G must contain a subgraph of type 1 as illustrated by Figure 66. Related cases are dealt with in the same way.

CASE 3. When exactly one of the points e, f lies in such a way as to equal w or v respectively we get another class of situations. As a typical example we may suppose $e = w$ and $f \neq v$. A typical situation is shown in Figure 67 and we see that G must contain a subgraph of type 1.

We have therefore established that none of the situations included in the above three cases can occur. There is only one case left. Demonstration that this type of situation is also impossible will establish the theorem for it will mean that the major assumption must be false.

CASE 4. In this case $e = w$ and $f = v$. The only case that is not eliminated at once by our previous arguments is when $g = c$ and $h = d$. Figure 68

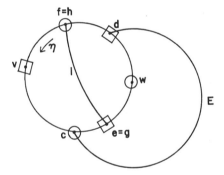

Figure 65

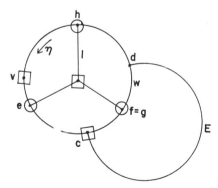

Figure 66

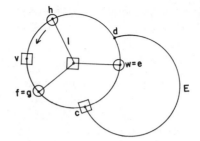

Figure 67

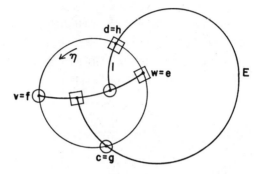

Figure 68

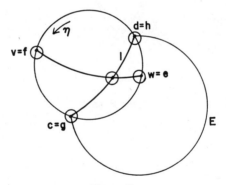

Figure 69

shows a situation in which the paths of I which join vertices c and d and join vertices v and w have more than one vertex in common. We see that in this situation G contains a subgraph of type 1. Hence the case cannot arise.

In case the paths described above have just one vertex in common the situation is as shown by Figure 69. Here we see that G has a subgraph of type 2 (as indicated by the five enclosed vertices). Since this is the last possible case we have a contradiction of the major assumption. This contradiction establishes the theorem.   ▲

In terms of the special notion of a conformal homomorphism (the special contraction which removes all vertices of degree 2), we can state the following form of Kuratowski's theorem:

Theorem—Alternative Form. A finite graph G is planar if and only if it contains no subgraph conformal to K_5 or to $K_{3,3}$.

A more subtle alternative is also possible. One may define a special type of contraction. An *elementary contraction* consists of the coalescence of two adjacent vertices, eliminating the edge between them and resulting in a single vertex upon which all edges of the original pair remain incident. One may now state the following alternative of Kuratowski's theorem:

Theorem—Second Alternative Form. A finite graph G is planar if and only if it contains no subgraph contractable to K_5 or to $K_{3,3}$ by means of a sequence of elementary contractions.

It is convenient to compare the forms of Kuratowski's theorem by reference to the famous Petersen graph (used as a starting point for several areas of research in graph theory, for example, the study of so-called permutation graphs) shown in Figure 70.

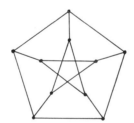

Figure 70.   Petersen graph.

Since every point of the Petersen graph is of degree 3 no subgraph can be made conformal to K_5. It does have subgraphs conformal to $K_{3,3}$ (in the trivial sense that they are directly isomorphic to $K_{3,3}$) and so is nonplanar by the first alternative form of the theorem. However, a sequence of elementary contractions can be applied to show that the Petersen graph is contractable by them to K_5, thus showing it satisfies the property in the second alternative form.

Now we shall give two other types of criteria for planar graphs.

In terms of the combinatorial dual, Whitney has provided the following criteria.

Theorem (Whitney). A graph G is planar if and only if G has a combinatorial dual.

As we have seen in our discussion of the combinatorial dual, it is very difficult to test a given graph for the property.

In terms of the set of independent cycles of a graph, MacLane [44] gives the following criteria.

Theorem (MacLane). A graph G is planar if and only if G has a cycle basis and one other cycle such that the set of cycles contains each edge of G exactly twice.

It will be seen that none of the three characterizations of planar graphs given in this section lend themselves to direct testing for planarity. An algorithm approach is needed by means of which a given graph G can be tested to determine whether or not it is planar. The next section will discuss the problem of determination of planarity.

DETERMINATION OF PLANARITY

In some cases one can find a planar representation of a planar graph by simply trying various drawings. However, for more complicated graphs such a procedure is difficult to apply and may fail to yield a planar representation. Therefore it is interesting to consider methods for testing a graph for planarity in some systematic way. The criteria theorems of Kuratowski and Whitney do not seem to lend themselves to such procedures. Of course one might use Kuratowski's conditions to show a graph is not planar by finding a subgraph conformal to or contractable to K_5 or $K_{3,3}$. But because of the need to include these homomorphism considerations such a procedure is hardly more systematic than the pictorial trial and error method. This leaves MacLane's criteria as a possibility and indeed one can develop a systematic procedure to test for planarity based on the cycles of a graph G. Some aspects associated with the determination of planarity were discussed by MacLane in [44]. We shall show how the question of planarity can be cast into a question about a cycle matrix of G (essentially MacLane's matrix). An algorithm for testing that matrix for the planar graph property may be based on a method of Tutte which he presents in terms of matroids as discussed in [45]. A detailed discussion of a planarity algorithm was made by L. J. Osterweil for a student project at Princeton University (1965). The details of such an algorithm are too involved for inclusion here but we shall discuss what must be done by any such algorithm if it is to test for planarity.

The property of planarity is not affected by removal of all pendant ways so we need only consider graphs in which each edge lies on at least one cycle.

A cycle or set of disjoint cycles can be represented as an m vector $C_i(c_{i1}, \ldots, c_{im})$ where m is the number of edges in G. If edge e_j is in C_k (where C_k is a cycle or set of disjoint cycles) then $c_{kj} = 1$ otherwise $c_{kj} = 0$ for $j = 1, \ldots, m$. It can be shown that any m vector $\bar{C}$ with elements zero or one satisfying the condition $\bar{C} I_g{}^T = 0$ corresponds to a cycle or set of cycles and conversely. Hence the set C^* of all vectors representing one or more cycles is the null space of the matrix $I_g{}^T$ (the transpose of the incidence matrix of G) over the field $\{0, 1\}$ (so that modulo 2 arithmetic is used in all calculations). Any basis B for C^* can be used to form a cycle matrix M_c whose rows correspond to the elements of B (equal in number to the dimension of the null space of $I_g{}^T$) and whose m columns correspond to the edges of G. The graph formed from the matrix M_c by construction of the basic cycles will be identical to the graph G since in each case every edge lies on cycles and all cycles are generated from the elements of B. It may be remarked that it is possible for elements of a basis to represent more than one cycle. This gives no difficulty and one may simply think of the single vector as a sum of the vectors representing the several disjoint cycles that may be involved. Thus the planarity of G may be studied by considering the planarity of the (same) graph represented in terms of the cycle matrix M_c.

Form the quantities $r_j = \sum_{i=1}^{R} c_{ij}$ where c_{ij} is the (i, j) element of M_c and R is the number of rows in M_c. The sum is take modulo 2. Add the vector $r = (r_1, \ldots, r_m)$ as a row to M_c and call the resulting matrix $M(G)$. One can now discuss $M(G)$ from the point of view of Tutte's algorithm which utilizes the concept of a matrix in graphic form. His method does not deal directly with graphs and matrices but rather with matroids. However, for our purpose one may think of a matrix as having graphic form if each column contains at most two ones. For such a form one can interpret the matrix as an incidence matrix for some graph H since the columns give allowable edge incidence on vertices corresponding to the rows. One has the result as follows.

Theorem. The matrix $M(G)$ may be put into graphic form if and only if G is planar.

In fact this is MacLane's theorem because of the definition of graphic form. It means that there is a basis for the null space of $I_g{}^T$ which can be used to form a matrix $M(G)$ in graphic form. Of course, the problem is to find the proper basis. In general, one selects a basis and forms $M(G)$ which may not be in graphic form. Then the question is can $M(G)$ be modified by changing the basic cycles so that a matrix in graphic form results? Tutte's algorithm gives a systematic procedure for answering that question.

In [44] MacLane relates his cycle basis criteria to the possibility of representing a graph as a map on a sphere (or plane). Such a mapping is possible if and only if G is planar and the resulting map consists of $N(I_g) + 1$ connected regions, where $N(I_g)$ is the nullity of I_g (i.e., the dimension of the null space associated with the matrix I_g). This result follows from MacLane's demonstration that the cycles of G that constitute region boundaries, under a mapping, form a cycle basis for G. Of course there is also the extra region "outside" the map itself.

Many people have investigated the planarity property of graphs and topics closely related to that property. In particular L. Weinberg has done considerable research on planar graphs, some of which may be found in [46] and the references given there (further aspects of that work were presented by L. Weinberg, "Two new characterizations of planar graphs," presented to Fifth Annual Allerton Conference on Circuit and System Theory, 1967).

The cycle matrix concept developed above can be illustrated by considering a graph such as (1).

$$
I_g = \begin{pmatrix}
1 & 0 & 1 & 1 & 0 & 0 & 0 & 0 & 0 & 1 \\
1 & 1 & 0 & 0 & 0 & 0 & 0 & 0 & 0 & 0 \\
0 & 1 & 1 & 0 & 1 & 1 & 0 & 0 & 0 & 0 \\
0 & 0 & 0 & 1 & 1 & 0 & 1 & 1 & 0 & 0 \\
0 & 0 & 0 & 0 & 0 & 1 & 1 & 0 & 1 & 1 \\
0 & 0 & 0 & 0 & 0 & 0 & 0 & 1 & 1 & 0
\end{pmatrix}
\tag{1}
$$

A typical vector in the null space of I_g (for $I_g{}^T$ one uses V^T) is shown in (2).

$$
V = \begin{pmatrix}
a \\
a \\
a+b+c+d+e \\
b+c+d \\
b \\
c+d+e \\
c \\
d \\
d \\
e
\end{pmatrix}
\tag{2}
$$

The null space has dimension 5 and a basis is given by the rows of the cycle matrix (3).

$$
M_c = \begin{pmatrix}
1 & 1 & 1 & 0 & 0 & 0 & 0 & 0 & 0 & 0 \\
0 & 0 & 1 & 1 & 1 & 0 & 0 & 0 & 0 & 0 \\
0 & 0 & 1 & 1 & 0 & 1 & 1 & 0 & 0 & 0 \\
0 & 0 & 1 & 1 & 0 & 1 & 0 & 1 & 1 & 0 \\
0 & 0 & 1 & 0 & 0 & 1 & 0 & 0 & 0 & 1
\end{pmatrix} = \begin{pmatrix}
n_1 \\
n_2 \\
n_3 \\
n_4 \\
n_5
\end{pmatrix}.
\tag{3}
$$

The basic cycles consist of the edges given in Table 2.

Table 2

Basic Cycle	Edges
n_1	1, 2, 3
n_2	3, 4, 5
n_3	3, 4, 7, 6
n_4	3, 4, 8, 9, 6
n_5	3, 10, 6

As an example of how other cycles are expressed in terms of the basic cycles note that $(5, 7, 6) = (3, 4, 5) + (3, 4, 7, 6)$, a result obtained by translating the cycles into vector notation; adding the vectors (modulo 2) and translating back into direct edge membership notation.

The matrix $M(G)$ formed from M_c is not in graphic form because several columns have more than two ones. However the graph with incidence matrix I_g is planar (the reader is invited to sketch it). Thus another set of basic cycles exists which would lead to an $M(G)$ in graphic form.

EXERCISES

1. Draw the following graph in straight line only, planar representation.

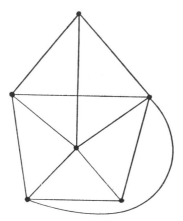

2. Draw the dual of the graph in Exercise 1. Verify euler's formula for both graphs.

3. Find the simple incidence graph for the graph in Exercise 1 and also for its dual. Are these graphs planar?

4. What is the crossing number of $K_{3,3}$. Construct a cycle matrix for $K_{3,3}$ and use MacLane's criterion to prove $K_{3,3}$ is not planar.

5. What is the crossing number of Petersen's graph?

6. Consider a graph with N cycles whose cycle matrix $C = (c_{ij})$ has the form $c_{i1} = 1$ for $i = 1, \ldots, N$ and for $i > 1$ $c_{ij} = 0$ for $i \neq j$ and $c_{ii} = 1$. Apply the criteria of MacLane's theorem to such a graph. In particular exhibit a cycle basis for the case $N = 3$.

7. By finding a basis for the null space of I_g illustrate the procedure for determination of planarity for the following graph

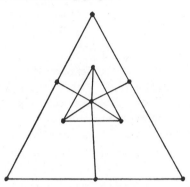

8. Find the dual and simple incidence graph for the graph of Exercise 7. Draw that graph in straight line planar representation.

9. Find the thickness of K_5, K_6 and $K_{3,3}$.

10. Use Kuratowski's theorem to test the following for planarity.

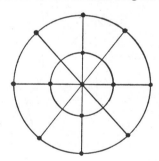

Chapter 4

Hamilton Graphs

DEFINITIONS AND BASIC CONCEPTS

An edge sequence in which vertices are given specific identity as opposed to being simply carried along with the edges of which they are terminal vertices can be useful in discussing paths in a graph which deal with vertices in some prescribed way. Of particular interest is a simple path in a finite graph G having distinct vertices which we have called a way in G. A way consists of distinct vertices and edges in an alternating sequence with the first and last elements being (terminal) vertices. Each pair of successive elements consists of one edge and one vertex which are incident. Recalling our previous notation a way is seen to be a simple path using a vertex only once. When a way has more than two vertices and terminal vertices which are adjacent (i.e., are terminal vertices of an edge) in the graph G one forms a circuit by adding the joining edge to the terminals of the way. This agrees with our earlier notation but provides us with an additional type of special sequence in a graph where vertices and edges play an equal role.

A *hamilton graph* is a finite graph G with $n > 2$ vertices in which there exists a circuit of length n. Any such circuit is called a *hamilton circuit*.

A hamilton circuit covers the vertices of a finite graph G exactly once using whatever edges it may require. The similarity to a cyclic euler path which covers the edges of a graph G exactly once is clear. Though we have a simple criterion for euler paths there is no known criterion of this type for hamilton circuits. In this sense the hamilton circuits are of richer theoretical interest than euler paths on finite graphs.

Since there is no simple criterion to specify hamilton graphs it is of interest to determine whether or not a given graph or class of graphs has hamilton circuits. A method by which this can be done is described in the last section of the Chapter. In studying hamilton graphs multiple edges and self-loops can be excluded from consideration since they do not affect the construction of any hamilton circuit. Self-loops may not occur in a hamilton circuit since such a loop would isolate its incident vertex (a vertex must be used exactly once). With multiple edges at most one is used when $n > 2$ or the pair of vertices terminal to the multiple edge would be isolated. Thus for studying hamilton circuits one may assume without loss of generality that G has no loops or multiple edges. Furthermore G is assumed to be finite with $n > 2$ vertices. It is clear that G must be connected if it is to have a hamilton circuit. However, in the sufficient condition theorem to be presented later in this section, we shall not make connectedness part of the hypothesis. Let us now consider some graphs with respect to the property of possessing a hamilton circuit.

We have made an analogy between cyclic euler paths and hamilton circuits with vertices of one corresponding to edges of the other. However, it is a very weak analogy at best since in euler paths any vertex may occur several times whereas the edges of a hamilton circuit can occur only once. This follows from the definition of a way as a sequence of distinct edges and vertices. Thus any graphs with at least one vertex of degree one cannot contain a hamilton path.

A graph which is a single circuit certainly is a hamilton graph. Its hamilton circuit is the graph itself. One is tempted to phrase this result as "any regular graph of degree two is a hamilton graph" but this is only true if one requires the graph to be connected. A connected regular graph of degree two is a single circuit provided it has more than one vertex and the statements become equivalent in this case. If we allow the graph to have a single vertex with a self loop so it is of degree two, it becomes a matter of definition as to whether or not this constitutes a circuit and hence a hamilton graph.

A way of length $n - 1$ in a finite connected graph G with n vertices is a hamilton way. Such a way contains every vertex of G exactly once and $n - 1$ distinct edges of G. As is the case for hamilton circuits there are no known complete criteria for the property of possessing a hamilton way. Certainly one sufficient condition is that the graph be hamiltonian. However, many non-hamiltonian graphs possess hamilton ways. Figure 71 shows a hamilton graph A, a nonhamiltonian graph with a hamilton way at B and a graph with no hamilton way at C.

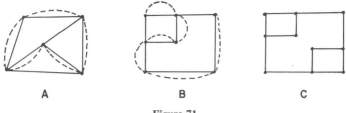

Figure 71

The graph *P* shown in Figure 72 which is a linear graph representation of a pentagonal prism is a hamilton graph, a hamilton circuit is (6, *i*, 7, *j*, 8, *k*, 9, *m*, 10, *f*, 5, *d*, 4, *c*, 3, *b*, 2, *a*, 1, *o*, 6). It is interesting and useful for further investigations to notice that the edges marked *o* and *p* cannot both occur in any hamilton circuit. We shall establish this result by means of a rather long consideration of cases. The reader is encouraged to develop alternative arguments if he desires. The argument is by contradiction and we suppose throughout that both *o* and *p* are members of a hamilton circuit which we wish to construct. We shall show that in every case it is impossible to construct a hamilton circuit. Denote any hamilton circuit for graph *P* by *H*.

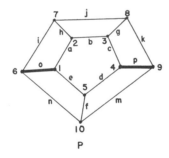

Figure 72

If *H* contains *a* and *c* vertex 5 is isolated, hence *H* can contain *a* but not *c* or *c* but not *a* or neither *a* nor *c*.

In case *H* contains neither *a* nor *c* it must contain *d* and *e* hence it may not contain *f* and therefore contains *m* and *n*. This isolates vertices 1, 5, 4, 9, 10, and 6 into a component and therefore cannot yield a hamilton circuit.

In case *H* contains *a* but not *c* it must contain *d* but not *e*. Hence it contains *f*. *H* must not contain *m* for to do so would isolate vertices 4, 5, 9, 10 into a component. Hence *H* must contain *n* and *k*. Since *c* is absent *H* must also contain *b* and *g*. Thus *H* may not contain *j* nor *h* nor *i*. This means vertex 7 is isolated and therefore we may not develop a hamilton circuit of this form.

The only remaining possibility is when *c* is present in *H* and *a* is not. In this case *d* must be absent so that *e* and *f* must be contained in *H*. Edge *n* may not be present for its presence would cause vertices 1, 5, 6, 10 to form an isolated component. Thus m and i must be in H so that k may not be an edge of H. Now we have two possibilities:

1. If h is present and j is not then b must be present and vertex 8 is isolated.
2. If j is present and h is not then g must be present and vertex 2 is isolated.

In either case we may not form a hamilton circuit.

Since all possible cases have been considered we have established the fact that any hamilton circuit of the pentagonal prism graph contains at most one of the edges o and p.

We shall now characterize a class of hamilton graphs by presenting sufficient conditions for hamilton circuits. These are contained in a theorem of Pośa as discussed by Nash-Williams [47], whose method of proof will be followed below. An interesting feature of the theorem is the absence of connectedness from the hypothesis. As above we are considering any finite graph G with more than two vertices with no self loops or multiple edges.

Theorem (Pośa). If G satisfies the conditions:

1. For every positive integer k less than $\frac{1}{2}(n-1)$, the number of vertices with degree not exceeding k is less than k.
2. The number of vertices with degree not exceeding $\frac{1}{2}(n-1)$ is less than or equal to $\frac{1}{2}(n-1)$,

then G is a hamilton graph.

Proof (Nash-Williams). Assume G satisfies conditions 1 and 2 then G is connected. To show this consider a connected component of G having r vertices. The degree of these r vertices cannot exceed $r-1$, hence we have r vertices with degrees all less than or equal to $r-1$ so that $r-1 \geq \frac{1}{2}(n-1)$ by condition (1) hence $r > \frac{1}{2}n$. Thus every connected component of G has more than $\frac{1}{2}n$ vertices which is impossible for more than one component (i.e., connected) graphs.

Consider all ways in G and let m denote the maximum length of these ways. Among the ways of length m select a way such that the sum of the degrees of its terminal vertices is maximal. Denote the vertices of the way by $w_1, w_2, \ldots, w_{m+1}$ with w_1 and w_{m+1} being the terminal vertices. We shall call the selected way W. Let A be the set of all vertices w_i from W such that w_1 is adjacent to w_{i+1} in G, (i.e., w_1 and w_{i+1} are terminal vertices of an edge in G). If w_1 were adjacent to any vertex not in W we could construct a way of increased length which is impossible by our selection of W. Thus w_1 is adjacent to $d(w_1)$ vertices in W and A has $d(w_1)$ elements. For any vertex $w_i \in A$ we can form the way $(w_i, w_{i-1}, \ldots, w_1, w_{i+1}, \ldots, w_{m+1})$ of length m which must have $d(w_i) \leq d(w_1)$ by the selection criterion for W. This results because w_{m+1} is one terminal of both W and the way starting with w_i. The degrees of the

$d(w_1)$ elements of A are therefore seen to never exceed $d(w_1)$ so by the same reasoning used to show connectedness $d(w_1) \geq \frac{1}{2}(n - 1)$ by condition (1). Similarly, $d(w_{m+1}) \geq \frac{1}{2}(n - 1)$. Not both $d(w_1)$ and $d(w_{m+1})$ can equal $\frac{1}{2}(n - 1)$. Suppose they did, then if we include w_{m+1} with the set A we obtain a set of $\frac{1}{2}(n + 1)$ vertices (by the addition of w_{m+1}) with degrees not exceeding $\frac{1}{2}(n - 1)$ which contradicts condition (2). In this argument we use the fact that w_{m+1} is not an element of A, if it were we could form a way of greater length than W which is impossible by definition of W. Thus $d(w_1) + d(w_{m+1}) \geq n$. Vertex w_{m+1} is adjacent to $d(w_{m+1})$ vertices. The number of vertices to which it is nonadjacent is $n - d(w_{m+1})$ which cannot exceed $d(w_1)$ by the last result. Since there are no self loops w_{m+1} is itself one of the nonadjacent vertices. It does not belong to A hence there is at least one vertex in A which is adjacent to w_{m+1}. We shall call it w_a. If this were not the case one could construct a set of more than $d(w_1)$ vertices which were nonadjacent to w_{m+1}. Let X be the way $(w_a, w_{a-1}, \ldots, w_1, w_{a+1}, \ldots, w_{m+1})$ and denote its vertices by $(x_1, \ldots, x_{m+1})$. Note that w_{a+1} is adjacent to w_1 since w_a is in A. By definition of w_a we have x_1 adjacent to x_{m+1}. Suppose $m < n - 1$, then there is a vertex v of G not in X. Since G is connected there is a path connecting v to some vertex of X. Let the vertices of this path be $y_1, \ldots, y_p$ where $p \geq 1$ and let $y_p \equiv x_j$ be the vertex of X to which v is connected. The way

$$v \equiv y_1, y_2, \ldots, y_p \equiv x_j, x_{j+1}, \ldots, x_{m+1}, x_1, x_2, \ldots, x_{j-1}$$

has length at least $m + 1$ which is impossible by selection of W. Note that x_{m+1} is adjacent to x_1 by construction. Thus $m = n - 1$ and the way X together with the edge (x_1, x_{m+1}) constitutes a hamilton circuit in G.   ▲

A graph G which consists of a single circuit is hamiltonian but does not satisfy the conditions of Pośa's theorem. This shows the conditions of the theorem to be sufficient but not necessary. As an additional example of this fact we may consider the cubical graphs. A regular graph of degree three is called a cubical graph. If G is cubical with n vertices the conditions of the theorem may or may not be satisfied. The complete four graph is cubical and satisfies Pośa's theorem while the pentagonal prism graph of Figure 72 does not satisfy the theorem. In the latter $n = 10$ and for $k = 3$ we have $k < \frac{1}{2}(9)$ or $k \leq 4$. But there are 10 vertices of degree 3 so the conditions of the theorem do not hold. Of course, this graph is hamiltonian. In Figure 73 we show a graph which does satisfy Pośa's theorem and for which the reader may readily construct a hamilton circuit.

Additional considerations of sufficient conditions for hamilton circuits and hamilton ways exist as discussed by Ore [15] (Section 3.4). These are closely related to the concepts developed in our study of Pośa's theorem, the employment of maximal length ways being of great value. In fact one may remark (with apology to the reader) that where there is a will to investigate existence of hamilton circuits there is a way.

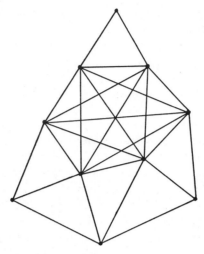

Figure 73

A class of problems that has enjoyed considerable attention from mathematicians over the past two hundred years deals with the knight's tour. In such a problem one wishes to determine a succession of moves of a knight over a rectangular chess board of size $n \times n$ in which the knight occupies every square exactly once. The relation to hamilton graphs is obtained by letting each square of the board be represented by a vertex of a graph G. Then G is given what may be called knight's connection, that is, any vertex which corresponds to a square reachable by the knight in one move from the square corresponding to a vertex v of G is connected by an edge to v. A solution to the knight's tour is equivalent to a hamilton way on G.

A method for obtaining such a tour on a regular 8×8 board is to divide the board into a central 6×6 board with a border of width 2 squares. Move the knight along in a fixed direction keeping within the border as much as possible. When the border has been completely visited by the knight, fill in the central portion. In this manner one can obtain solutions such as De Moivre's (1), where the numbers indicate the order of moves in the tour.

$$
\begin{array}{llllllll}
34 & 49 & 22 & 11 & 36 & 39 & 24 & 1 \\
21 & 10 & 35 & 50 & 23 & 12 & 37 & 40 \\
48 & 33 & 62 & 57 & 38 & 25 & 2 & 13 \\
9 & 20 & 51 & 54 & 63 & 60 & 41 & 26 \\
32 & 47 & 58 & 61 & 56 & 53 & 14 & 3 \\
19 & 8 & 55 & 52 & 59 & 64 & 27 & 42 \\
46 & 31 & 6 & 17 & 44 & 29 & 4 & 15 \\
7 & 18 & 45 & 30 & 5 & 16 & 43 & 28
\end{array}
\tag{1}
$$

The tour is said to be reentrant when the final position is such that with an additional move the knight may return to the initial position. This is equivalent to G being hamiltonian.

There can be no reentrant tour on an $n \times n$ board if n is odd. On a single move a knight always goes from one color square to one of a different color (the board is colored in two alternating colors, say red and green). Thus in any tour the colors must alternate and if the tour is reentrant it must have an even number of steps, that is, squares. If n is odd the board has an odd number of squares and therefore possesses no reentrant knight's tour.

The reader can easily construct a tour on the 4×4 board and a tour on the 6×6 board based on the method outlined above, that is, filling the edges first. One tour due to Euler for the 6×6 board is shown in (2).

$$
\begin{array}{cccccc}
30 & 21 & 6 & 15 & 28 & 19 \\
7 & 16 & 29 & 20 & 5 & 14 \\
22 & 31 & 8 & 35 & 18 & 27 \\
9 & 36 & 17 & 26 & 13 & 4 \\
32 & 23 & 2 & 11 & 34 & 25 \\
1 & 10 & 33 & 24 & 3 & 12
\end{array}
\tag{2}
$$

A method used by Euler and suggested by Bertrand starts with a careful movement of the knight over the board filling as many squares as possible. The resulting configuration should consist of only a few unvisited squares and a path that can be made reentrant by reordering it appropriately. The unvisited squares are then added with renumbering (reordering the tour) of squares at various steps. The method requires skill and its success depends to some extent on good luck as well. A further description of this method and of the knight's tour in general can be found in Ball and Coxeter [48]. An 8×8 reentrant solution due to Euler appears in Ore ([15], page 53). Another obtained by the method indicated above is shown in (3).

$$
\begin{array}{cccccccc}
22 & 25 & 50 & 30 & 52 & 35 & 60 & 57 \\
27 & 40 & 23 & 36 & 49 & 58 & 53 & 34 \\
24 & 21 & 26 & 51 & 38 & 61 & 56 & 59 \\
41 & 28 & 37 & 48 & 3 & 54 & 33 & 62 \\
20 & 47 & 42 & 13 & 32 & 63 & 4 & 55 \\
29 & 16 & 19 & 46 & 43 & 2 & 7 & 10 \\
18 & 45 & 14 & 31 & 12 & 9 & 64 & 5 \\
15 & 30 & 17 & 44 & 1 & 6 & 11 & 8
\end{array}
\tag{3}
$$

Though the knight's tour problem has been extensively studied, we shall mention only two more methods for developing knight tour paths. The reader may find some history of the subject and an indication of more recent consideration in Ball and Coxeter [48]. The more recent studies deal with generalizations to nonrectangular boards and paths with restrictions such as specifying the terminal square (some of these restricted problems were considered from the beginning of the subject by Euler and others).

A method due to Warmsdorff developed in 1823 does not yield reentrant paths but is simple to state and apply. It has not been proven to always yield a knight's tour but always seems to work even when false moves, (i.e., minor infractions of the rule) are made so long as one does not make them in one of the last few moves. The rule is that a knight is moved to one of the cells from which it will command the fewest squares not already visited. A (nonreentrant) knight's tour obtained by Warmsdorff's method is shown in (4).

$$
\begin{array}{cccccccc}
1 & 14 & 63 & 40 & 11 & 16 & 19 & 38 \\
62 & 41 & 12 & 15 & 64 & 39 & 10 & 17 \\
13 & 2 & 57 & 60 & 51 & 18 & 37 & 20 \\
42 & 61 & 52 & 45 & 56 & 59 & 54 & 9 \\
3 & 32 & 43 & 58 & 53 & 50 & 21 & 36 \\
26 & 29 & 46 & 33 & 44 & 55 & 8 & 49 \\
31 & 4 & 27 & 24 & 47 & 6 & 35 & 22 \\
28 & 25 & 30 & 5 & 34 & 23 & 48 & 7
\end{array}
\qquad (4)
$$

One of the most interesting methods for developing a reentrant tour is due to De Lavernede and was developed by Roget in 1840. The entire tour is divided into four circuits which are combined in such a way that one may begin on any cell and terminate on any other cell of a different color. Thus the necessary condition for a reentrant tour exists and one can be constructed by properly imposing the terminal square with respect to the initial square. The method applies to square boards with $4n$ cells on a side and not to other cases (e.g., it does not apply to a 10×10 board).

Roget's method begins by dividing an 8×8 board into four equal quarters of 16 squares each. Within each quarter the 16 squares are divided into four sets each consisting of four distinct squares which form a closed knight's path. All squares of one set are given the same label hence the four letters a, b, c, d can be used to specify the sets and one can observe the closed knight's path nature of the sets in Figure 74, where we see that the sets marked a and b form diamond-shaped closed paths and those marked c and d form square-shaped paths.

$$
\begin{array}{cccc}
a & d & c & b \\
c & b & a & d \\
d & a & b & c \\
b & c & d & a
\end{array}
$$

Figure 74

This figure indicates some relation to Latin squares (of which Figure 74 is one) and points out that the knight's tour is also a problem in structural combinatorial analysis. A Latin square of size n consists of n columns and n rows of an $n \times n$ square with each space occupied by one of n Latin letters such that each letter occurs exactly once in each row and in each column.

The method of Roget now combines the four marked quarters into a complete chess board. One then transverses all squares having the same label till the circuit is completed then goes to the next label circuit and so forth. The only caution one must exercise is to terminate on the proper square of opposite color from the starting square so as to obtain a reentrant path. In the knight's tour (5) the first circuit uses the b squares and first develops the circuit 1 through 16 then the c squares for circuit 17 through 32 and so forth.

$$
\begin{array}{cccccccc}
34 & 51 & 32 & 15 & 38 & 53 & 18 & 3 \\
31 & 14 & 35 & 52 & 17 & 2 & 39 & 54 \\
50 & 33 & 16 & 29 & 56 & 37 & 4 & 19 \\
13 & 30 & 49 & 36 & 1 & 20 & 55 & 40 \\
48 & 63 & 28 & 9 & 44 & 57 & 22 & 5 \\
27 & 12 & 45 & 64 & 21 & 8 & 41 & 58 \\
62 & 47 & 10 & 25 & 60 & 43 & 6 & 23 \\
11 & 26 & 61 & 46 & 7 & 24 & 59 & 42
\end{array}
\tag{5}
$$

The various solutions to the knight's tour problem show how to find hamilton circuits and ways in some forms of hamilton graph (i.e., those having knight's connection). The general problem of constructing a hamilton circuit is related to an algorithmic approach to testing a graph for the hamilton property. This is discussed in the last section of this chapter. Another algorithmic method for the construction of a hamilton circuit is given by Berge [6], (Chapter 11).

CUBICAL GRAPHS

A cubical graph is a regular graph of degree three. Cubical graphs have received considerable study, particularly in connection with coloring of graphs, a topic to be discussed in the next Chapter. A result of Tutte and Smith [49], of special interest to this section deals with the number of hamilton circuits in a cubical graph. We remark that finite connected graphs without self loops or multiple edges are being considered. We may observe that since the number of vertices of odd degree must be even in a finite connected graph every cubical graph has an even number of vertices. Any hamilton circuit in a cubical graph has an even number of edges.

For a cubical graph C we require two concepts which are defined as follows. An S-subset of C is any union of disjoint circuits having even length such that every vertex of C is contained in some circuit of S. The circuits of S are called its components and their number is denoted by $\sigma(S) + 1$ so that $\sigma(S) \geq 0$. Since hamilton circuits in cubical graphs have even length they may be considered as elements of S-subsets. If an S-subset contains a hamilton circuit it contains no other component and if an S-subset has exactly one component this component must be a hamilton circuit in cubical graphs. Thus the hamilton circuits of C are those S-subsets of C for which $\sigma(S) = 0$. Let us suppose that the edges of C are numbered $1, 2, \ldots, n$. We may then associate the linear form: $\sum_{k=1}^{n} s_k x_k$ with S by setting $s_k = 0$ if edge k is not in S and $s_k = 1$ if edge k is in S. Such a form shall be denoted by $X(S)$. These forms may be added together by taking the coefficients as residues modulo 2.

The second concept we shall require is a T-coloring of a cubical graph C. A T-coloring is an unordered set of three disjoint sets of edges of C, called T-classes, which together contain all edges of C, and are such as to cause each vertex of C to be incident with exactly one member of each class.

Theorem (Tutte, Smith). In a cubical graph C

$$\sum_H X(H) = 0$$

where H runs through the hamilton circuits of C.

Proof. The union of any two of the T-classes of a T-coloring of C forms an S-subset of C. Thus to each T-coloring we may associate three S-subsets which may be denoted S_1, S_2 and S_3. We observe that

$$X(S_1) + X(S_2) + X(S_3) = 0, \tag{6}$$

where the addition is modulo 2 as previously defined. This follows from the fact that every edge of C occurs exactly twice in the three sets S_1, S_2 and S_3 so that all coefficients are zero modulo 2 in (6).

A particular S-subset S will determine a number of T-colorings of C. This is carried out as follows.

The edges of C which do not belong to S form one class of T and the other two classes are determined by the ordering of their respective membership on each component of S. In any component there are two possible orderings of the edges into the two T-classes since every component has even length. This would result in $2^{\sigma(S)+1}$ cases but this number must be divided by 2 since we do not count cases as different in which the two classes are interchanged throughout the set. This is the significance of defining a T-coloring as an unordered triple of T-classes. It is this that allows us to simply treat one T-class as the edges not in S as well. Thus it is only the ordered structure of the three classes on C that counts as distinct cases and not specific interchanges of classes among themselves. The number of distinct T-colorings associated with an S-subset is therefore $2^{\sigma(S)}$. Now for each T-coloring of C we form the sum of the three linear forms of the associated S-subsets as in (6). If we add all of these for all T-colorings each $X(S)$ will occur $2^{\sigma(S)}$ times and the total is:

$$\sum_S 2^{\sigma(S)} X(S) = 0.$$

In this expression every coefficient for which $\sigma(S) > 0$ is zero modulo 2 so that we get the expression of the theorem, since only the terms corresponding to hamilton circuits explicitly remain.   ▲

Figure 75

Consider the cubical graph shown in Figure 75. Three Hamilton circuits H_1, H_2, H_3 have the linear forms

$$X(H)_1 = x_1 + x_2 + x_4 + x_6, \, X(H_2) = x_2 + x_3 + x_5 + x_6,$$

and

$$X(H_3) = x_1 + x_3 + x_4 + x_5.$$

Clearly $\sum_{i=1}^{3} X(H_i) = 0$ as stated by the theorem (using modulo 2 arithmetic).

Corollary. A cubical hamilton graph C contains at least three hamilton circuits.

Proof. If there were at most two different hamilton circuits in C the sum of their associated linear forms could not be zero as required by the above theorem since they cannot contain exactly the same edges. The smallest number of different hamilton circuits which can satisfy the theorem is three. ▲

A particular class of cubical graphs has been rather extensively studied both in connection with the question of the hamilton property and for other reasons (e.g., in the theory of structures of molecular compounds). That class consists of all planar cubical graphs that correspond (e.g., by projection) to trivalent convex polyhedra, that is, a convex polyhedron with exactly three edges touching at each vertex. One may think of the vertices and edges of such a graph as isomorphic to the simplicial skeleton (formed by the bounding one-simplices and their points of incidence) of a polyhedron. The isomorphism takes the skeleton into the plane to produce the planar linear graph.

A result from the theory of convex bodies states that a linear graph is 3-connected and planar if and only if it is isomorphic to the simplicial skeleton of a convex polyhedron. This result is known as Steinitz's Theorem as discussed by Grunbaum [50]. In particular this implies that cubical graphs of the type under consideration (i.e., corresponding to convex trivalent polyhedra) are 3-connected.

In 1884 Tait [51], formulated a conjecture concerning hamilton circuits in the class of cubical graphs described above. The conjecture was felt to be likely to be correct for a period of 62 years until a counterexample was provided by Tutte in 1946 [49]. Since the conjecture received rather widespread interest, it came to exist in (at least) two forms:

Conjecture A.   Every planar cubical graph corresponding to a trivalent polyhedron, is hamilton.

Conjecture B.   Every planar three connected graph is hamilton.

The reader should notice that conjectures *A* and *B* differ. A planar three connected graph may not be cubical. However, every graph considered in conjecture *A* is included in conjecture *B* (by the result given above for convex polyhedra). In conjecture *A* it is not enough to speak of planar cubical graphs. The fact that the graphs correspond to polyhedra is basic to the statement. For example, the graph of Figure 76 is planar and cubical but it is clearly not

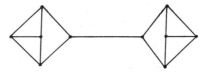

Figure 76

hamilton. It does not correspond to a polyhedron. On the other hand, the graph *P* of Figure 72 is a member of the class of planar cubical graphs under study and it is hamilton giving support to the conjecture.

If conjecture *A* were true then conjecture *B* would still be open. Tutte's counter-example is for conjecture *A* and therefore shows that conjecture *B* is also false.

Any hamilton circuit in the graph *N* shown in Figure 77 will determine a hamilton circuit in the pentagonal prism graph *P* of Figure 72. This is established by noting that *N* can be converted to *P* by shrinking vertices 9, 11, 12 to one vertex called 9 and shrinking 6, 13, 14 to one called 6. Such conversion

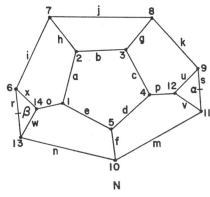

Figure 77

would automatically establish a hamilton circuit in *P* corresponding to any such circuit in *N*. Thus by the argument previously given for *P* any hamilton circuit of *N* cannot contain both *o* and *p* as edges. Let *H* denote any hamilton circuit in *N*. If *H* contains edge *s* it cannot contain both *u* and *v* for this would isolate vertices 9, 11, 12 in a component, hence *H* must contain *p*. In the same way if *H* contains *r* it must contain *o*. Since *H* cannot contain both *o* and *p* it cannot contain both *r* and *s*.

Let α be a point on edge *s*, say the midpoint of *s* and β be the midpoint of *r*. Form a new graph *N** by treating α and β as additional vertices and introducing new edges $(9, \alpha)$, $(\alpha, 11)$, $(13, \beta)$ and $(\beta, 6)$. The edge (α, β) is also added to form *N**. Any hamilton circuit in *N** must contain (α, β) otherwise it would contain the other four added edges and determine a hamilton circuit in *N* containing *s* and *r* which is impossible.

Of course *N* is a hamilton graph as is *N** but *N** allows us to establish a counterexample to conjecture *A*. That is a cubical graph of the class under study that is nonhamiltonian as first shown by Tutte [49]. The graph is denoted by *N*** and shown in Figure 78. The graph *N** has the same structure as the part *ABC* of *N*** with its interior and three additional edges joining vertices *A*, *B*, *C* to an extra vertex *v*. The reader can verify this by drawing *N** and the graph just described and distorting the latter until it is evident to him that the two are the same. Each graph has three quadrilateral faces which may be brought to correspondence as may the edges (α, β) and (B, v). Let *N'* denote the graph which corresponds to *N**. Then any hamilton circuit in *N'* must contain edge (B, v).

Any hamilton circuit in *N*** defines one in *N'* by identifying everything in *N*** except the interior of *ABC* with the added vertex *v*. Thus any hamilton circuit of *N*** must contain *BR*. By symmetry any such circuit must also contain *DR* and *ER*. This is impossible since *R* would then have three incident

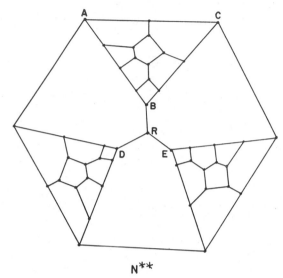

Figure 78.   Tutte's graph.

edges in a hamilton circuit thus N^{**} is nonhamiltonian. The reader may wish to construct a hamilton way in N^{**}. Such ways exist.

Though Tutte's graph from Figure 78 showed Tait's conjecture to be false, it gave rise to further questions concerning hamilton graphs. One asks if

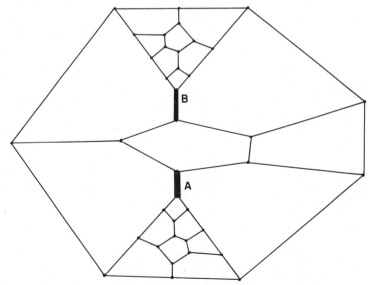

Figure 79.   Barnette's graph.

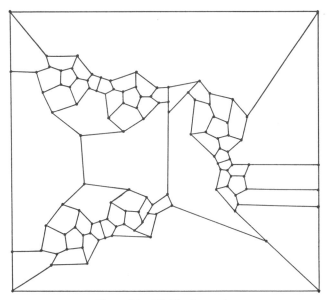

Figure 80. Walther's graph.

there is a graph belonging to the class under study that has less than 46 vertices and is not hamiltonian. Barnette (as reported by Lederberg in [52]) produced the graph in Figure 79 having 38 vertices which is of the desired class and is nonhamilton. (This example was also found independently by Tutte.) Using studies by Grace [53], Lederberg showed that every graph of the class under discussion, having eighteen or less vertices was Hamilton [52]. Thus the situation for such graphs between 18 and 38 vertices remains more or less open.

One can look more deeply into the question of the hamilton property for planar cubical graphs of the above type (i.e., polyhedral). A graph is said to be cyclically k-edge connected if, by removing less than k edges, two disjoint subgraphs cannot be formed with each containing a circuit. Tutt's graph is cyclically 3-edge connected and Barnette's graph is cyclically 4-edge connected. In 1965, Walther [54], gave a 5-edge connected, nonhamilton graph of the desired class. That 114 vertex graph is shown in Figure 80 and is discussed in some detail by Sachs [55].

ADDITIONAL RESULTS ON HAMILTON GRAPHS

Lattice graphs have been investigated with regard to the hamilton property by Nash-Williams [56], Kotzig [57], and others. Since the points of an n-dimensional euclidean lattice form a natural model for various physical structures, some of the results for lattice graphs will be included here. The type of graph under consideration has an n-dimensional representation (and of course an isomorphic representation in the plane which, however, is not useful since the n-dimensional representation is the more easy to think about). The vertices are the lattice points of an n-dimensional euclidean space. Two vertices u and v are joined by an edge if and only if the edge is of unit length. Here we specialize the general idea of lattice graphs (from Chapter 1) and include all edges and vertices that are bounded by n positive integers greater than one. The notation for such a graph is $L^n = G\,[a_1, a_2, \ldots, a_n]$, $a_i > 1$ for $i = 1, \ldots, n$, and $n > 1$. The coordinate axes are ruled out as vertex points for the graph which is therefore considered to start at the lattice points with lowest value equal to unity. The graph L^n has $\prod_{i=1}^{n} a_i$ vertices and the local degree values range from n to $2n$, where vertices of degree $2n$ are interior to the graph and other degree vertices lie on the boundary of the n-dimensional representation. Vertices of degree n are the extreme points of the n-dimensional convex figure determined by L^n. A major result on the hamilton property for lattice graphs is the following [57].

Theorem. A hamilton circuit exists in a lattice graph $G\,[a_1, \ldots, a_n]$ if, and only if, the number of vertices in G is even. That is $\prod_{i=1}^{n} a_i \equiv 0$ (mod. 2).

The theorem is illustrated in Figure 81 for $n = 2$ and in Figure 82 for $n = 3$. In each case a lattice graph having a hamilton circuit is shown (with a circuit specified) and one failing to have a hamilton circuit is also shown. The reader may check the condition of the theorem for each of these illustrations.

Returning to the general problem of characterizing hamilton graphs we can give a sufficient condition other than the (more involved) conditions of Poša's Theorem given previously. This result is due to Ore [15].

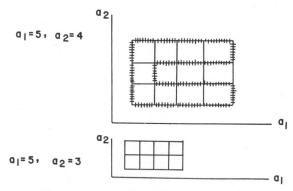

Figure 81.   Two-dimensional lattice graphs.

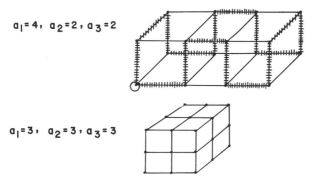

$a_1 = 4, \; a_2 = 2, \; a_3 = 2$

$a_1 = 3, \; a_2 = 3, \; a_3 = 3$

Figure 82.   Three-dimensional lattice graphs.

Theorem. Let G be a graph with m vertices such that for any pair of non-edge connected vertices u and v, the local degrees $d(u)$, $d(v)$ satisfy the condition $d(u) + d(v) \geq m$. Then G has a hamilton circuit.

This theorem establishes a result of Dirac that a graph with m vertices is hamilton if the local degree of each vertex is not less than $m/2$.

Of course the conditions above are sufficient only and Figure 83 shows a

Figure 83

graph which does not satisfy either type of condition but is hamilton as indicated by the hamilton circuit shown.

To conclude our discussion of hamilton graphs we shall give a method for determining whether or not any given graph is hamilton. It is a direct numerical method based on concepts of integer valued linear programming.

Let I_g denote the incidence matrix of a graph G having n vertices and $m \geq n$ edges (to reduce the size of I_g, self loops and multiple edges are assumed removed from G). Let $x = (x_1, \ldots, x_m)^t$ denote a (column) vector consisting of m components. Consider the following zero-one programming problem:
Problem P_H.

$$\text{minimize } z = \sum_{i=1}^{m} x_i$$
$$\text{subject to } A x \geq 2$$
$$x_i = 0, 1 \qquad \text{for} \quad i = 1, \ldots, m.$$

Where the notation $y \geq a$ for vectors y and a means that the inequality holds for each and every component of those vectors. The constraint conditions in problem P_H require at least two edges (specified by those x_i that assume the value one in a solution) at each vertex. There are two major possibilities for the solutions to P_H. The minimum value of the objective function z may equal n or it may not. If min z is not equal to n there can be no hamilton circuit in G. If min z is equal to n then it is possible to select a set of n edges that form a set of circuits including all vertices of G. In fact the solution values specify such a set of edges; those for which $x_i = 1$ being selected to form the edges. The result min $z = n$ does not however insure a hamilton circuit since several disjoint circuits may be involved. Thus when min $z = n$ the solution set of edges must be tested to see whether it forms a single circuit or several circuits (on disjoint subsets of vertices of G). If the solution gives a single circuit it is a hamilton circuit, G is hamilton, and the solution provides a hamilton circuit.

The above procedure may be difficult to apply, particularly when n and m are large. Difficulties arise from the requirement that x_i be zero or one (thus making the optimization problem essentially nonlinear) so that the relatively easy methods of linear programming, (e.g., the simplex method) do not apply. In addition, the requirement that a single circuit exists is an additional computational effort. Of course there is some compensation for this effort in that the procedure not only determines whether or not G is hamilton but gives a hamilton circuit in the affirmative case. Moreover, for all its computational difficulty the method does allow a direct investigation of the existence of hamilton circuits in a graph G.

In general, problem P_H must be solved by methods of zero-one programming (briefly discussed in the Appendix). However, it may be that in special cases the solutions obtained by using normal linear programming methods, (e.g., the simplex algorithm) will yield solutions in which $x_i = 0$ or 1. In such cases the relatively simple simplex method solves P_H. In any case, the single circuit nature of the solution is easily checked by following the solution edge sequence along using the incidence matrix I_g. Either isolated circuits will be found or a complete (hamilton) circuit of G will be established. Thus as a practical procedure one might solve P_H by the simplex method without requiring that $x_i = 0$ or 1 (but only that $x_i \geq 0$). If the solution is indeed 0 or 1 for each x_i it is a useful solution. Conditions on I_g that will insure this result are difficult to obtain. Certainly a necessary condition is the so-called integrity property for problem P_H as described in the Appendix. That property insures integral solutions but those may not be zero or unity. Moreover, the integrity property itself is difficult to characterize. As an example of the considerations described above, consider the graph of Figure 84. That graph has no circuits of odd length and hence its incidence matrix has the unimodular property (see

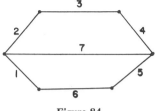

Figure 84

Chapter 1). Thus in this case problem P_H has integrity (Appendix). This does not insure a zero-one solution but does encourage us to apply the simplex method for a trial solution. A solution to P_H formulated for this graph yields the optimal solution: min $z = 6$ with $x_1 = x_2 = x_3 = x_4 = x_5 = x_6 = 1$. This edge sequence can easily be seen to yield a hamilton circuit for the graph.

EXERCISES

Let the linear graph G have the following adjacency matrix:

$$G = \begin{pmatrix} 0 & 1 & 1 & 0 & 0 & 1 \\ 1 & 0 & 1 & 2 & 0 & 0 \\ 1 & 1 & 0 & 0 & 0 & 0 \\ 0 & 2 & 0 & 0 & 1 & 1 \\ 0 & 0 & 0 & 1 & 0 & 1 \\ 1 & 0 & 0 & 1 & 1 & 0 \end{pmatrix}$$

1. Does G satisfy Poša's Theorem? Is G a hamilton graph?
2. Use G to illustrate Ore's theorem on vertex degree conditions for the hamilton property.
3. Set up the linear programming formulation of the hamilton graph problem for G. Solve to illustrate the method.
4. Consider a plane rectangular lattice graph of size m by n in which each lattice point is considered to be a vertex and all grid lines are edges. For what values of m and n is the graph hamiltonian? For what values of m and n does a hamilton path exist? Do these graphs satisfy Poša's theorem?
5. Remove the four vertices of degree two in each of the lattice graphs of Exercise 4 by means of a contracting homomorphism. The number of vertices in the resulting graph is $(nm - 4)$. Do these graphs satisfy Poša's theorem? Do they satisfy Ore's condition? Are they hamilton graphs?
6. Express the knight's tour problem as a hamilton graph problem. Find the hamilton ways on the resulting graphs that correspond to the various tour solutions given in the text. Can you find any other hamilton ways and hence tour solutions? Find all hamilton ways on the graph corresponding to the 4×4 board.

7. Consider two simple cycle graphs C and C' having n vertices each. Label the vertices of C as $1, 2, \ldots, n$ and those of C' as $1', 2' \ldots, n'$. Form a graph H from C and C' by introducing n additional edges of the form (i, i') for $i = 1, \ldots, n$. Observe that H is cubical. Is H hamilton? Is H three connected? Find an S-subset of H. Find $\sigma(S)$ and state what this value tells you about H.

8. The wheel W on four vertices is hamilton. Find all hamilton circuits of W. Show that $\sum_R X(R) = 0$ where R runs through the hamilton circuits of W. Show that W satisfies Ore's conditions for a hamilton graph.

9. How is the general structure of the linear programming formulation for hamilton graphs affected by restricting attention to cubical graphs? Use that formulation to prove that Tutte's graph is not hamilton.

10. Is the Petersen graph hamilton?

Exercises 11, 12, and 13 are due to the referee.

11. A group of men, all recent fathers, had congregated in a hospital waiting room. Unfortunately, there was but one cigar among them. Its owner was able to organize the group so that every man was able to congratulate every other man exactly once (by passing the cigar). Naturally, the original owner finally left the room puffing on it. How did he do it? (There are many ways to organize the ritual, but one exceptionally simple one).

12. Show that every tournament (complete simple digraph) has a hamilton way (following the arrows).

13. Show that the graph consisting of the edges and vertices of the n-dimensional hypercube has a hamilton circuit, but that it has a hamilton way with its ends at antipodal (diametrically opposite) vertices if and only if n is odd.

Chapter 5

Graph Coloring

CHROMATIC NUMBER

Using k different colors we may color each of the n vertices of a graph G. For example, the same color could be assigned to each vertex or if $n < k$ a different color could be given to each vertex. When it is possible to color the vertices of G with k colors (not necessarily using all colors) in such a way that adjacent vertices are assigned different colors we say G is k-*colorable*. This concept can be expressed more abstractly by saying that G is k-colorable implies that the set of vertices of G can be decomposed into k disjoint internally stable subsets. Thus adjacent vertices always belong to different subsets. Let $K(G)$ denote the set of integers k for which G is k-colorable. Of course K is infinite for any finite graph G. Consider the complete n-graph G_n in which every vertex is adjacent to every other vertex. At least n colors are required to color G_n and it is k-colorable only for $k \geq n$. Thus as $n \to \infty$ the set $K(G_n)$ tends to a one element set, the single element being aleph nought (the cardinal number of the set of all integers) but this is a very special case indeed. We may remark that graphs with self loops are not colorable at all. Our interest here is with non-self loop graphs. Multiple edges have no effect on questions of colorability we may or may not consider G to have multiple edges.

119

The *chromatic number* $\gamma(G)$ of a graph G is defined as follows

$$\gamma(G) = \min_{k \in K(G)} k \quad .$$

The vertex (only) n-graph is one-chromatic while the complete n-graph is n chromatic. Any star graph with more than one vertex is two chromatic. If n is even the wheel on n vertices is 4 chromatic while if n is odd the wheel is 3 chromatic.

The questions concerning chromatic numbers refer to classes of graphs on the one hand and to specific graphs on the other. One is interested in characterizing those graphs for which the chromatic number has a specific value γ. For a given graph the chromatic number may be desired or when known a specific coloring using the chromatic number of colors may be required. Such questions are in general very difficult to resolve; we shall present some of the more important results.

When the chromatic number of a graph G is two we say that G is a *bichromatic graph*. Such graphs have interesting properties and occur widely in graph theory. The vertices of a bichromatic graph G divide into two disjoint sets such that every edge of G has exactly one terminal in each set. A cycle graph of even length is bichromatic as is every tree. An odd length cycle graph is trichromatic (requires three colors). Every bichromatic graph can be colored in just two distinct ways for if we call the colors red and green then one set of vertices may be red or green while the other set must be assigned the opposite value. This makes clear that we are not concerned with specific colors but with the way a given number of colors are utilized in dividing the vertices of a graph into independent sets. Figure 85 shows a bichromatic graph (the utilities graph) and a trichromatic graph with a coloring specified on each by means of letters representing colors.

An early theorem of graph theory deals with bichromatic graphs:

Theorem (König). A graph G is bichromatic if and only if it contains no cycles of odd length.

Proof. Suppose G has no cycles of odd length. A coloring in at most two colors is possible as follows. We shall use the colors red and green. Color an arbitrary vertex v_0 red. If a vertex is colored red color the vertices adjacent to

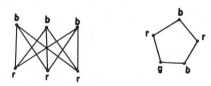

Figure 85

it green. If a vertex is colored green color vertices adjacent to it red. By this process every vertex in the same connected component as the starting vertex v_0 will be colored. If the coloring rule requires a vertex v to receive both colors then v and v_0 are on a cycle of odd length which is impossible. Each connected component of G can be so treated in turn. Hence G is bichromatic.

If G is bichromatic it can contain no cycles of odd length since such cycles require three colors. ▲

It should be observed that a bichromatic graph is bipartite and in fact a k-chromatic graph is k-partite. The two concepts describe the same graph structure property but differ in terms of the point of view being taken.

Construction of a coloring on a graph G is equivalent to the formation of a special function on the vertices of G. This function is called a *Grundy function* of G. As previously defined $d(v)$ is the number of edges of G incident to a vertex v. When G has no self loop or multiple edges as we shall assume in our studies of graph coloring, $d(v)$ is also the number of vertices adjacent to v. Let $A(v)$ denote the set of adjacent vertices to v. The $d(v)$ elements of $A(v)$ may be written $v_i, i = 1, \ldots, d(v)$. We shall denote the Grundy function of G by $f(v)$ and define it as follows.

The $f(v)$ is the smallest positive integer such that $f(v) \neq f(v_i)$ for $i = 1, \ldots,$ $d(v)$. One can always assign a value $f(v)$ not greater than $d(v) + 1$ and the values $1, 2, \ldots, f(v) - 1$ are assumed by $f(\cdot)$ at neighboring vertices to v. These concepts are illustrated by the graph of Figure 86. One Grundy function is shown and it is clear that the Grundy function is not unique. From the Grundy function we get a coloration of a graph by assigning the vertices with the same Grundy function value the same color. The connection between Grundy functions and chromatic number is given by the following result.

Theorem. A graph G is k-chromatic if and only if G possesses a Grundy function $f(v)$ such that

$$\max_{v} f(v) \leq k.$$

Proof. Should a Grundy function satisfying the condition of the theorem exist the k-coloring induced by $f(v)$ shows G to be k-chromatic.

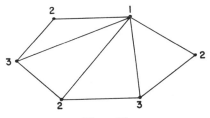

Figure 86

If G is k-chromatic its vertices can be divided into k disjoint independent sets $C_1, \ldots, C_k$. A Grundy function with the desired property is constructed as follows. Let $f(v) = a$ when $v \in C_a^*$ where C_a^* is defined inductively. C_1^* is C_1 together with all vertices not in C_1 which are not adjacent to vertices of C_1 or to each other. Form C_2^* from $C_2 - (C_2 \cap C_1^*)$ and adding any vertex not in C_1^* or C_2 which is not adjacent to $C_2 - (C_2 \cap C_1^*)$ or to each other. Continue until all vertices are utilized.   ▲

The determination of the chromatic number γ of a graph G and a γ-coloration of G can be carried out by computational means. First one constructs a set of algebraic conditions which are equivalent to a k-coloration of G then tests for solutions to these conditions. If the smallest value of k for which solutions can be found is k_0 then $k_0 = \gamma$ and G is γ chromatic. Any solution will yield a coloration. The transformation to a computational problem for the integer k situation is as follows: denote the vertices of G by $v_i, i = 1, \ldots, n$ the edges by $u_j, j = 1, \ldots, m$ and let $s = 1, \ldots, k$ be the possible colors.

Let
$$x_{is} = 1 \quad \text{if vertex } i \text{ is assigned color } s$$
$$= 0 \quad \text{otherwise}$$

and using the notation of the incidence matrix of G let

$$g_{ij} = 1 \quad \text{if edge } u_j \text{ is incident on vertex } v_i$$
$$= 0 \quad \text{otherwise.}$$

A k-coloration is a set of kn values for the variables x_{is} satisfying the conditions

$$\sum_{s=1}^{k} x_{is} = 1 \qquad i = 1, \ldots, n, \tag{1}$$

$$\sum_{r=1}^{n} g_{rj} x_{rs} \leq 1 \qquad j = 1, \ldots, m \text{ and } s = 1, \ldots, k, \tag{2}$$

$$x_{is} = 0 \text{ or } 1 \qquad i = 1, \ldots, n \text{ and } s = 1, \ldots, k. \tag{3}$$

Condition (1) results from the requirement that each vertex be assigned exactly one of the k colors. Condition (2) ensures that adjacent vertices receive different colors for if two adjacent vertices had the same colors and the edge they are both incident to is u_a then $\sum_{r=1}^{n} g_{ra} x_{rs}$ is at least equal to 2. The final condition is clearly required if the algebraic representation of the problem is to be meaningful. One may search for solutions to the set of conditions above. If solutions exist then $\gamma \leq k$ and k may be decremented and solutions sought for the new problem. If no solutions exist then $\gamma > k$ and k is incremented. Though rather different in their main areas of application the algorithms of integer linear programming or zero-one programming can be used as a tool to study this problem. In particular it is the so-called Phase I technique which is of specific utility here. That technique consists of forming an objective function (called the infeasibility form) consisting of artificial variables that have been added to the problem. If that form can be minimized to zero all artificial variables will be driven out (assume zero values) and the result will provide a (feasible) solution to the programming problem. To formulate the problem in a standard form one removes all inequalities by means of auxiliary variables. Let y_{js} be an auxiliary variable for $j = 1, \ldots,$ m and $s = 1, \ldots, k$ and let w_i, $i = 1, \ldots, n$ denote artificial variables. The following problem treats the k-colorability of G with incidence matrix I_g.

Minimize

$$w = w_1 + w_2 + \cdots + w_n$$

subject to

$$\sum_{s=1}^{k} x_{is} + w_i = 1, i = 1, \ldots, n,$$

$$\sum_{r=1}^{n} g_{rj} x_{rs} + y_{js} = 1, j = 1, \ldots, m \quad \text{and} \quad s = 1, \ldots, k,$$

$$w_i, y_{js}, x_{is} = 0, 1 \quad \text{for all} \quad i, j, s \text{ values.}$$

If min $w = 0$ in the above problem G is k-colorable. In such a case one tries $k - 1$ and so forth until min $w > 0$. When this occurs it will occur in a problem with some number k_0 of colors. Then G is $k_0 + 1$ chromatic. If min $w > 0$ for the first k value used then one tries $k + 1$ and so forth. The procedure is clear but it does have all the practical difficulties associated with zero-one (or integer) programming. As usual one can try to solve the problem by dropping the zero-one constraint. To do this the simplex method may be applied. If the solution comes out with zero or one values then it is a useful solution. If not one must resort to zero-one techniques (as discussed in the Appendix).

An interesting relation exists between the coefficient of internal stability and the chromatic numbers of a graph G. The former is denoted by $s_i(G)$ and was defined as the number of vertices in the largest independent set of vertices (no two vertices of such a set are adjacent). If G is γ chromatic its vertices can be partitioned into γ independent sets. Let $h_1, \ldots, h_\gamma$ denote the number of vertices in each set, then $h_i \leq si(G)$ for $i = 1, \ldots, \gamma$. Hence we have the result:

$$n = h_1 + \cdots + h_\gamma \leq si(G) + \cdots + si(G)$$
$$n \leq \gamma(G) \, si(G).$$

In addition to relating two of the numbers associated with a graph G this relation shows that a graph with specific chromatic number and coefficient of internal stability can not be arbitrarily large.

The concept of *critical graphs* was developed by G. Dirac and occurs in a number of studies of chromatic graphs. We shall introduce the concept here

Figure 87.   Critical 4-chromatic graph.

without going into studies employing it. A graph G is called (vertex) critical when the removal of any vertex and the edges on which that vertex is incident reduces the chromatic number $\gamma(G)$. Very simple examples are the single vertex graph and the single edge graph (K_2) which are critical 1-chromatic and critical 2-chromatic respectively. A circuit of odd length is a critical 3-chromatic graph. Removal of any vertex (and its two associated edges) results in a bichromatic graph (a path). Any complete graph is critical since removal of any vertix of K_n results in K_{n-1} which is $n-1$ chromatic. The graph in Figure 87 is 4-chromatic and is critical. Removal of any vertex results in a 3-chromatic graph. A typical theorem on critical graphs is the following:

Theorem. A critical graph G has the properties:

1. G is finite and connected.
2. For every vertex v of G, $d(v) \geq \gamma(G) - 1$.
3. G has no separating vertices.

Classification of graphs in terms of their chromatic number is an important area of study. One of the most interesting partial results is known as Hadwiger's conjecture.

Hadwiger's Conjecture

Every finite connected k-chromatic graph has a subgraph which has a complete k-graph as its image under a connected homomorphism.

In Chapter 1 a connected homomorphism was called a contraction so that Hadwiger's conjecture really deals with the images of subgraphs of G under contractions. There is a close connection between the conjecture and map coloring studies. That connection will be indicated in the next section; it is extensively discussed by Ore in [22].

The conjecture is clearly true for $k = 1$ and $k = 2$. For $k = 3$ any simple cycle has a homomorphism to a triangle as shown by the homomorphism η for the graphs of Figure 88 where the contraction is determined by the mapping (4).

$$
\begin{aligned}
&\eta \\
1 &\to 1' \\
2 &\to 2' \\
3 &\to 2' \\
4 &\to 3' \\
5 &\to 3'
\end{aligned}
\qquad (4)
$$

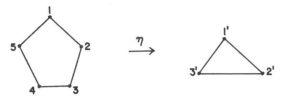

$$\eta$$
$$1 - 1'$$
$$2 - 2'$$
$$3 - 2'$$
$$4 - 3'$$
$$5 - 3'$$

Figure 88

This example also illustrates the need for stating the conjecture in terms of homomorphisms. The conjecture is certainly not true when stated without the use of homomorphisms. Thus the statement that a k-chromatic graph has a complete k-graph as subgraph is false as Figure 88 shows. Of course if a graph contains a complete k-graph it must be at least k chromatic.

The question of existence of k-chromatic graphs without cycles of various lengths has received some attention. We present here one result of this type.

Theorem (Mycielski 1955 [58]). For each positive integer n there exists a finite graph without triangles (cycles of length 3) which can not be colored by n colors.

Proof. We consider graphs without self loops or multiple edges. The proof is by induction.

$n = 1$ *case*. Two vertices joined by a single edge constitutes a finite graph without triangles which can not be colored with one color. This establishes the result for $n = 1$.

Assume the theorem is true for n and show that this implies its truth for $n + 1$. Let $a_1, \ldots, a_m$ be the vertices of a graph A without triangles which can not be colored with n colors.

Construct a new graph A^* as follows:

Denote the vertices of A which are adjacent to vertex a_i by $b_{i1}, b_{i2}, \ldots, b_{ir_i}$ where $r_i = d(a_i)$ for $i = 1, \ldots, m$. Introduce $m + 1$ new vertices a_0', $a_1', \ldots, a_m'$. Introduce new edges by connecting an edge each of the vertices $b_{i1}, \ldots, b_{ir_i}$ to a_i' for $i = 1, \ldots, m$ and also connect by an edge each of the vertices $a_1', \ldots, a_m'$ to a_0'. Thus A^* has the vertices $a_1, \ldots, a_m$ and $a_0', a_1', \ldots, a_m'$.

We shall show that A^* contains no triangles and can not be colored in $n + 1$ colors.

Since A contains no triangles the vertices $b_{i1}, \ldots, b_{ir_i}$ have no adjacent pairs (since all are connected to a_i by an edge and the existence of such a pair would yield a triangle in A). Any new vertex such as a_i' is adjacent to two points of a set like $b_{i1}, \ldots, b_{ir_i}$ and hence does not lie on a triangle. Since the $a_1', \ldots, a_m'$ are not adjacent in pairs a_0' also does not lie on a triangle. Thus A^* has no triangles.

Assume that A^* can be colored with $n + 1$ colors. The graph A can not be colored with n colors. Since A^* is assumed to be colorable with $n + 1$ colors A must be colorable with $n + 1$ colors. For each such coloring there is a set of $n + 1$ vertices of A which all have different colors. We may denote these by $a_{s_1}, \ldots a_{s_{n+1}}$. For each s_i the vertex a_{s_i}' is the same color as a_{s_i} for $i = 1, \ldots, n + 1$. This follows from the fact that a_{s_i}' is adjacent to all the vertices to which a_{s_i} is adjacent and these adjacent vertices must contain all other colors (if this was not the case for $i = 1, \ldots, n + 1$ one could obtain a coloring in less than $n + 1$ colors). Thus the vertices $a_1', \ldots, a_m'$ contain among them vertices using each of the $n + 1$ colors. Then a_0' can not be colored since it is connected to each of the vertices $a_1', \ldots, a_m'$. This is a contradiction hence A^* can not be colored in $n + 1$ colors.   ▲

A closely related result is the following.

Theorem. For each positive integer k there are k-chromatic graphs without triangles (cycles of length 3).

The theorem is established by means of a specific, inductive construction procedure that produces graphs having the desired properties. Suppose G is a k-chromatic graph without triangles which has n vertices. Form the graph H by taking G as a (section) subgraph and call its vertices $v_1, \ldots, v_n$; add n additional vertices $v_1', \ldots, v_n'$, and another vertex w. If v_i is connected to a vertex s in G connect v_i' to s in H then connect w to v_i' for $i = 1, \ldots n$. The reader will observe that the resulting graph H is $k + 1$ chromatic, has no triangles, and has $2n + 1$ vertices. Figure 89 shows the first few graphs of the desired type. The $k = 5$ graph may be constructed from the $k = 4$ graph shown but it is rather involved. Colorings are indicated by representative letters on the graphs of the figure. It should be remarked that the result is not obvious in that the most common illustrations of higher chromatic number graphs do contain triangles, for example, the complete graphs.

Various investigations have been made regarding k-chromatic graphs without certain cycles and with certain properties. Some of that work is discussed by Grünbaum [59] in which he indicates the rather sparse nature of specific results. The main interest in [59] is to graphs with requirements on the

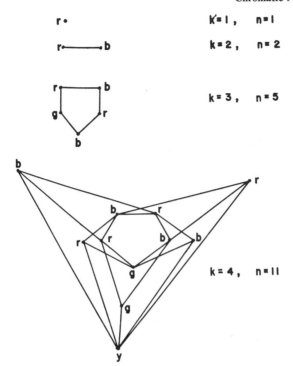

Figure 89.   Triangle free chromatic graphs.

valence (degree) of the vertices. For that topic Grünbaum states the following result of Brooks:

If all vertices of a connected graph G have degree at most $k \geq 3$, then either G is k-colorable or else G is the complete graph with $k + 1$ vertices (K_{k+1}).

The following conjecture (of Grünbaum) is also stated.

If $k \geq 3$ and $n \geq 3$, there exists, k-chromatic, k-valent graphs $G(k, n)$ that contain no cycle of length $\leq n$. A graph is k-valent if each vertex has degree k (and is, therefore a regular graph).

Graphs $G(3, n)$ are described in [59] and a 25-vertex graph $G(4, 4)$ is given. Mention is also made of a 12-vertex graph $G(4, 3)$ due to Chvátal. Except for these results the conjecture seems to be unresolved.

We conclude our present discussion of k-coloration and chromatic number by introducing another method for studying these concepts. Though combinatorial in nature we place the topic here rather than in the combinatorial chapter because it is so very closely related to coloration studies. When a graph G is k-colorable we may ask how many different k-colorings are possible. This number may be denoted by $C(G, k)$ and for a given graph can be shown to be a polynomial in k called the *chromatic polynomial* or coloration polynomia of G. If we do not restrict the coloring of vertices of G so as to require a coloration, (i.e., adjacent vertices having different colors) then there are k^n colorings where n is the number of vertices of G since each vertex can be colored in k ways independently of the coloring of all other vertices. Let $\beta(e_{i_1}, e_{i_2}, \ldots e_{i_r})$ denote the number of colorings of G with k colors in which the terminal vertices of the r edges $e_{i_1}, \ldots, e_{i_r}$ have the same color. The principle of inclusion and exclusion yields:

$$C(G, k) = k^n - \sum \beta(e_{i_1}) + \sum \beta(e_{i_1}, e_{i_2}) - \cdots,$$

where the sums are taken over all edges, all pairs of edges, all triples of edges and so forth. Now any set of r edges such that the terminal vertices have the same color defines a subgraph H of G consisting of all vertices of G and the specified edges. Any connected component of H must have all vertices colored the same thus the ways of coloring this set of edges is $k^{p(H)}$ where $p(H)$ is the number of connected components of H. Thus the number of colorings of j-tuples of edges is $\sum_p h(j, p) k^p$ where $h(j, p)$ is the number of subgraphs of G having p connected components and j edges. One then obtains the polynomial representation of $C(G, k)$ due to G. D. Birkhoff:

$$C(G, k) = \sum_{j, p} (-1)^j h(j, p) k^p.$$

It may clarify the *Birkhoff formula* by observing that for $j = 0$ there are n components and $p(0, n) = 1$ yielding the first term of the polynomial. For $j = 1$ each subgraph has $n - 1$ components, $n - 2$ of which are isolated vertices and one a single edge, this case yields the second term and so forth.

The chromatic number γ of a graph G can be obtained from the roots of the chromatic polynomial $C(G, k)$. Let k_0 be the greatest positive root of $C(G, k)$ then γ is the smallest integer which is greater than k_0.

The graph of the tetrahedron shown in Figure 90 is the complete 4-graph and therefore is 4-chromatic. It may be used to illustrate the application of Birkoff's formula.

For $c = 1$ we get the results: $p(6, 1) = 1$, $p(5, 1) = 6$ since any edge can be removed from the full graph, $p(4, 1) = \binom{6}{2} = 15$ since any 2 edges can be

Figure 90. Tetrahedron graph.

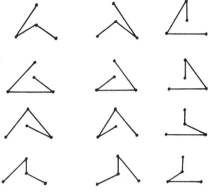

Figure 91

removed, and $P(3, 1) = 4 + 3 \cdot 4 = 16$ as discussed below. For $c = 2$ we get: $p(3, 2) = 4$ by removing all edges at any vertex, and $p(2, 2) = \binom{6}{2} = 15$ by retaining any two edges. For $c = 3$, $p(1, 3) = 6$ since any edge can be retained and for $c = 4$, $p(0, 4) = 1$ by using no edges.

The value $p(3, 1)$ is the only result that is not immediate. Four cases arise by retaining all edges at any vertex; the additional 12 cases represent all chains of length 3 shown in Figure 91.

Using the above results one obtains the chromatic polynomial for the tetrahedron graph as

$$C(G; x) = x^4 - 6x^3 + 11x^2 - 6x.$$

We find that $C(G; 3) = 0$ and $C(G; 4) = 4! = 24$, hence (as we know) the graph is 4-chromatic and there are (clearly) 24 different ways to color the graph in 4 colors.

Some special results concerning chromatic polynomials will be presented here. There are many more and also some alternative approaches to forming the polynomials, for example, methods due to Whitney as described in Ore [15], Chapter 14.

When G has no edges (is a vertex graph) and has n vertices: $C(G; x) = x^n$, the chromatic number of such a graph is one. For a graph on n vertices with m edges it is clear that $p(1, n - 1) = m$ since $n - 1$ components result from the selection of any one of the m edges. In the same situation $p(1, j) = 0$ for $j \neq n - 1$ since every one edge subgraph has $n - 1$ components (in these studies the vertex set of any subgraph of a graph G is taken to be the vertex set of G).

When G has a disjoint decomposition $G = \sum G_i$, $i = 1, \ldots, k$ into k parts, the definition of the coloration polynomial yields:

$$C(G; x) = C(G_i; x) \cdots C(G_k; x).$$

This is because the number of ways of coloring G with x colors represents the value of the polynomial so that the basic product axiom of combinations yields the result.

Let $G = G_1 + G_2$ be a decomposition in which G_1 and G_2 have a single vertex in common. All colorings of G are obtained from colorings of G_1 and G_2 in which the common vertex has the same color; thus:

$$C(G; x) = \frac{1}{x} C(G_1; x)\, C(G_2; x).$$

This follows since having colored G_1 there is only one way to color the common vertex with G_2 rather than x ways in the completely unrestricted case.

A more advanced result than the simple situations given above is stated in the following.

Theorem. When E is a separating edge for G then

$$C(G; x) = \left(1 - \frac{1}{x}\right) C(G - E; x)$$

We may now consider the coloration polynomials for trees. In a tree T any edge E may be thought of as a connecting edge so that the above theorem yields:

$$C(T; x) = \left(1 - \frac{1}{x}\right) C(T - E; x)$$

On the other hand if we let T' denote a tree having one vertex less than T as shown in Figure 92 it follows that $C(T - E; x) = xC(T'; x)$.
Applying these two results yields:

$$C(T; x) = (x - 1)\, C(T'; x).$$

If T_0 denotes a single vertex so that $C(T_0; x) = x$ the above result may be applied $n - 1$ times to yield the coloration polynomial for a tree on n vertices:

$$C(T; x) = (x - 1)^{n-1} x.$$

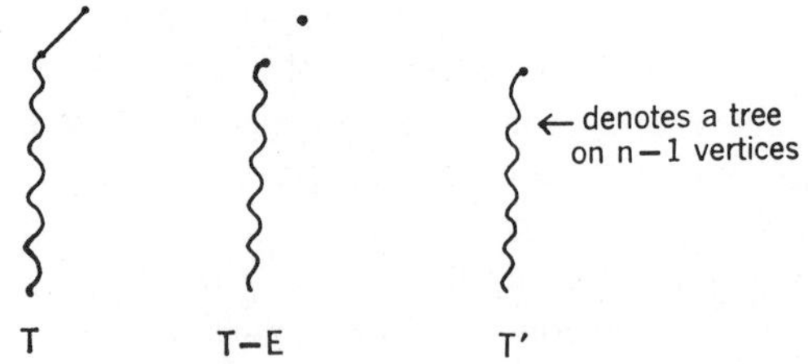

← denotes a tree on $n-1$ vertices

T   T−E   T'

Figure 92

The roots of such a polynomial are zero and one so that the chromatic number of any tree is two as is obvious from the structure of a tree, (i.e., the absence of cycles). Of course any tree is bichromatic as has already been observed. The calculation of the coloration polynomial for a tree concludes our discussion of these polynomials.

One may define another number associated with a graph G by considering the coloration of the edges of G in a way similar to the coloration of vertices just discussed. This number is called the *chromatic index* of G and is the smallest positive integer r such that all edges of G can be colored with r colors in such a way that adjacent edges are of different colors. It is clear that the study of chromatic index and edge colorings in general can be dealt with by forming a new graph G^* from G in which the vertices represent the edges of G and are adjacent if and only if the edges are adjacent in G. The chromatic number of G^* is the chromatic index of G. Thus it is unnecessary to develop a separate study of edge colorings.

MAP COLORING (Much of the material in this section is based on Ball [48]).

For the purpose of this section a map may be considered to be a subdivision of some type of surface into contiguous regions. Two regions are considered to be contiguous if and only if they have an arc in common; a single point in common is not enough. One may think of the regions of the map as countries and for convenience we exclude islands and other non-simple region countries, (e.g., a two part country such as Pakistan is excluded). These restrictions have no effect on the generality of the problems considered but merely serve to simplify the discussion. A map coloring consists of an assignment of a color to each region so that contiguous regions all have different colors. Any map in the plane can be used to construct a corresponding planar graph as follows. Let R_i denote the ith region of a planar map M. Let v_i be a vertex of a planar graph $G(M)$ corresponding to the region R_i. Then (v_i, v_j) is an edge of $G(M)$ if and only if R_i and R_j are contiguous regions in M. The index i runs over the number of regions in M, which can be at most denumerable. When the number of regions is finite one (outside) region is taken as infinite in extent. By considering projections the maps on the plane may be related to the maps on a spherical surface. Hence problems of map coloring for spherical or planar maps are directly related to the chromatic number of planar graphs. In this section we will discuss the planar map coloring problem and give some of the historical results for that problem. Most advanced modern studies make use of the corresponding planar graph representation and the methods of graph theory (such as critical graphs, Hadwiger's conjecture). Map coloring problems in other surfaces such as the Torus and the Möbius strip are solved directly. These will be included because of their close connection with planar map colorings. We shall not discuss map coloring in other types of surfaces though that is an interesting topic in topology (it is not part of linear graph theory).

Before considering the map coloring problems some background material will be developed.

We consider unbounded surfaces (without any periphery or edge such as sphere, torus, plane, etc.) which are orientable. A surface is orientable if one can assign a positive sense of rotation consistently at all points, that is, about each point one can draw a small circle with direction indicated by an arrow and this direction is the same for every point. For example, this will not be possible if a closed path exists on the surface such that the point encirclement sense is reversed as the path is transversed. This occurs on the Möbius strip which therefore is not orientable.

Divide an unbounded, orientable surface into F polygonal districts (some may be infinite on the plane, e.g., but we treat this by using the sphere as a model then projecting it onto the plane as needed.) That division is carried out by utilizing E arcs joining (pairs of) V points. When such a surface is crumpled or stretched (homeomorphicly mapped onto itself) these numbers F, E, and V are unaltered since the districts are not intrinsically changed thereby. The interesting thing is that the quantity

$$F - E + V$$

is a numerical property of the surface and does not depend on the way the districting was carried out. This fact is the generalization of the Euler theorem which follows from it at once. We shall call $F - E + V$ the surface characteristic, then we have:

Theorem on the Characteristic. The quantity $F - E + V$ is independent of the particular formation of districts (defining the F, E, and V) on an unbounded, orientable surface. It is, therefore a property of the surface.

Proof. (J. W. Alexander following Ball and Coxeter [48]). Consider two districtings formed by two maps upon the surface. One yields F, E and V while the other yields F', E' and V'.

Superpose the 2 maps with lines of one breaking up districts of the other. The result is a third map with f districts, e lines and v points. For convenience we suppose $V \cap V' = \Phi$, otherwise simple displacement of some parts will yield this. Then v consists of $V \cup V'$ plus the points where the lines of the 2 maps cross.

Modify the first map to include the crossing points as vertices. This breaks up the lines on which they lie increasing E by the same amount v is increased and keeping $F - E + V$ unchanged.

We now add the remaining lines and vertices of the third map in chains, each joining two given vertices (already present) and dividing a given district into 2 parts. Such a chain has one less new vertex than new lines. Its insertion increases F by 1 and $-E + V$ by -1 which leaves $F - E + V$ unchanged.

By completing the third map in this way we find:

$$f - e + v = F - E + V,$$

similarly

$$f - e + v = F' - E' + V',$$

hence

$$F' - E' + V' = F - E + V,$$

but the two maps were arbitrary, hence the theorem is established.   ▲

Corollary—Euler's Theorem. Every polygonal map on the sphere (plane) has $F - E + V = 2$ [i.e., the sphere has characteristic 2].

Proof. By the above theorem we need only exhibit a polygonal map on the sphere with $F - E + V = 2$. Consider a tetrahedron on the sphere $E = 6$, $F = 4$, $V = 4$ so $F - E + V = 2$ as required.   ▲

Remark 1.   On a torus $F - E + V = 0$ consider the two circles A and B on the torus in Figure 93. They cross at one point so $V = 1$; there is just one face so $F = 1$, and there are 2 edges so $E = 2$, yielding the result.

Remark 2.   Euler's theorem on the plane yields Euler's Formula for finite plane polygons to the effect that for any finite plane polygon, (i.e., one whose edges do not cross), $F - E + V = 1$. Consider a finite plane polygon together with the infinite face as a division of the plane into domains so that $F - E$

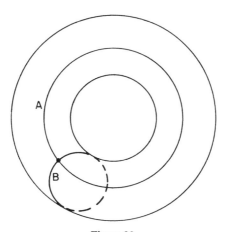

Figure 93

$+ V = 2$. Remove the infinite face to obtain the relation for the polygon. We have given an alternative proof of Euler's Formula in Chapter 3.

Remark 3. In *three space* a simple polyhedron, (i.e., has no holes) is a solid whose faces are polygonal. Such a polyhedron can always be placed so that its vertices lie on a sphere and hence it can be mapped onto a sphere in such a way that its faces map into disjoint region. Thus for such a polyhedron $V - E + F = 2$ as we have seen this for a sphere.

The "regular solids" have all polygonal faces congruent and all angles at vertices equal.

We shall now use Euler's formula to prove that there are no more than 5 regular polyhedra. Since we can construct 5 of them, this is therefore their total number.

Suppose a regular polyhedron has F faces, each of which is an n-sided regular polygon, and that r edges meet at each vertex. Let E be the number of edges then $2E = nF$ (by counting edges by means of faces) each edge belongs to two faces and there are n-edges per face.

Counting edges by vertices we have $2E = rV$ where V is the number of vertices; each edge is incident on 2 vertices and there are r edges per vertex. Hence by Euler's formula:

$$\frac{2E}{n} + \frac{2E}{r} - E = 2 \Rightarrow \frac{1}{n} + \frac{1}{r} = \frac{1}{2} + \frac{1}{E} \tag{5}$$

A polygon must have at least 3 sides so $n \geq 3$ and in a polyhedron at least 3 sides must meet at a vertex so that $r \geq 3$. But if both n and r exceed 3 the left side of (5) could not exceed $1/2$, this is impossible since the right side of (5) always exceeds $1/2$. Hence either n or r or both must be equal to 3. This will enumerate all possible regular polyhedra.

$n = 3$ CASE. Formula (5) becomes $1/r - 1/6 = 1/E$ the only possible values are $r = 3, 4, 5$ (other values yield negative or zero E which is impossible). These give $E = 6, 12, 30$ respectively (Tetrahedron, octahedron, icosahedron).

$r = 3$ CASE. Formula (5) becomes $1/n - 1/6 = 1/E$ so as before we have $r = 3, 4, 5$ and $E = 6, 12$ or 30 giving tetrahedron (again) and adding the cube and dodecahedron.   ▲

General attention was called to the problem of coloring a map in such a way that contiguous districts have different colors by De Morgan about 1852 (who heard it from Francis Guthrie). Cayley discussed it in 1878 and remarked that he could not prove the 4-color conjecture (stated below).

A false proof was given by A. B. Kempe in 1879 and Tait gave a solution in 1880. These stood for some time but were found to be incomplete. At present, though much effort has been put into the 4-color problem, it remains unsolved.

There are two types of map coloring theorems.

Type I. On a surface S (defined), k colors are sufficient to color any map such that contiguous districts have different colors. These are called k-color theorems for surface S.

Type II. On a surface S, k colors are necessary and sufficient to color any map such that contiguous districts have different colors. These are called the k-color theorem for surface S.

The major surfaces S studied are the sphere (which is the "same" as the plane), the torus, and surfaces of higher genus, and the Möbius strip.

A question concerning the truth of any theorem of the above form is called a map coloring problem. We shall now discuss planar map coloring. In such studies considerable use is made of reduction techniques.

MAP REDUCTION PROCEDURE

The formation of a standard map will now be described.

1. If the map does not cover the entire surface regard the unused part as one more district.

2. Reduce all vertices to degree three. At a vertex of degree greater than three, there must be a pair of non-contiguous districts we may open out the vertex and merge these two into one district as shown in Figure 94.

3. Get rid of districts having one, two or three sides. Remove one side and merge into a contiguous district (this will not change the coloring number which is taken to be greater than three).

4. Get rid of four-sided districts. Of the four surrounding districts two must be non-contiguous; this pair can be merged with the four-sided district. If the map so formed can be colored in four or more colors so can the original (proven by reconstructing the original map).

5. Get rid of ring-shaped districts. Then each district is bounded by a single continuous line (which will include many edges, however) and no district encloses one or more other districts.

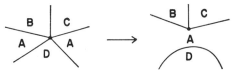

Figure 94

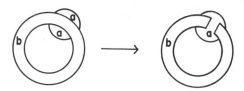

Figure 95

An example is shown in Figure 95. This step is carried out by cutting a corridor in the ring-shaped district and merging the corridor with the non-contiguous regions on either side (as illustrated in Figure 95.) This procedure will not change the coloring number.

We will now give the major historical results on map coloring.

Theorem. *A* 6-*color theorem* for the plane (or sphere).

Six colors are sufficient to color any map on a sphere (or plane).

Proof. Reduce the map to standard form. Then no district has less than five sides (edges).

Such a map may be drawn as a polygon with F faces (districts), E edges and V vertices.

Since each edge meets two vertices and there are 3 edges at each vertex, we have $3V = 2E$.

Denote the number of edges in face i by e_i then the total number of edges is $(e_1 + e_2 + \cdots + e_F)/2$ since each edge is associated with two faces. Thus

$$2E = \frac{e_1 + e_2 + \cdots + e_F}{F} F.$$

We let $a = (e_1 + \cdots + e_F)/F$ denote the average number of edges per face. Then $2E = aF$.

Euler's formula states $F - E + V = 2$ which, using the above results yields:

$$a = 6 - \frac{12}{F}.$$

Thus the average number of edges is less than 6 so there must be at least one pentagon. Now an induction procedure establishes the theorem.

Consider a district having five sides and merge it with one of its contiguous districts by removing one edge (side).

If the new map (which may not be in standard form) can be colored in six colors so can the original by reconstruction.

One can reduce step by step in this way and the theorem is proven.   ▲

Note that the new map each time may require a transformation to standard form before the next reduction step is taken. This proof was done before 1890 and the procedure used may be called "the crude induction argument".

In 1890 Heawood used a finer form of the induction argument to establish the five-color theorem:

Five colors are sufficient to color any map on the plane (or sphere).

The argument may be found in Ball, [48].

The five and the six color theorems are of Type I. We now give a theorem of Type II for the Torus.

The 7-Color theorem on the Torus

Seven colors are necessary and sufficient to color all maps on a torus so that contiguous districts have different colors.

Proof (Necessity). The map in Figure 96 requires 7 colors since every one of the 7 hexagonal districts is contiguous to all the others. Note the torus is formed by bending the diagram so that the points having the same letters come together.

(Sufficiency). By using standard map representation we have (as in the 6-color sphere theorem): $2E = 3V$ and $2E = aF$ where a is the average number of edges per face.

On the torus Euler's formula is $F - E + V = 0$ which yields $F - (a/2)F + (aF/3) = 0 \Rightarrow a = 6$ so there is a district having not more than 6 sides (if every district had more than 6 sides the average could not be 6). We now apply the

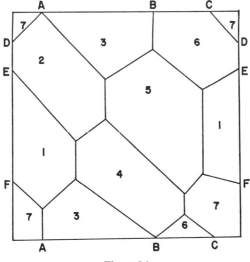

Figure 96

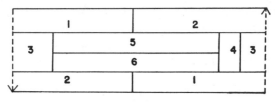

Figure 97

crude induction argument starting with the face having no more than 6 sides (7 colors suffice for the map if it suffices for the reduced map, etc.)

As remarked before the induction steps are subtle, one may have to start over completely on the entire reduction to standard form, etc., but note we are always on the torus (or other surface as the case may be). ▼

Consider a Möbius strip which is an unorientable surface. The six regions on the Möbius strip of Figure 97 are such that each region is adjacent to all five others (i.e., has an edge in common with all others). Note that the surface is formed by bending the figure so that the arrows coincide. This division into regions shows that 6 colors are necessary for maps on a Möbius strip. Results of Franklin [60], show that 6 colors are also sufficient. Thus we have a 6-color theorem of Type II for the Möbius strip.

Returning to our consideration of the plane (sphere) the 5 color theorem insures us that at most 5 colors will suffice to color any map on the plane. The plane map *A* in Figure 98 requires 4 colors because each region is continuous to each of the 3 other regions. At *B* there is a map requiring 4 colors even though it does not contain a set of 4 neighboring regions.

Thus on the one hand 5 colors are sufficient and on the other hand 4 colors are necessary. The 4-color conjecture states that 4 colors are sufficient. No example exists requiring 5 colors and a number of theorems have been established indicating that any such example must have very involved structure. Some computer search of maps has been made to help study the structure of examples but these yield limited results (computations get complicated rather fast as the number of vertices increase). As an example of such studies one can consult Yamabe and Pope [61]. However, a proof that 4 colors are

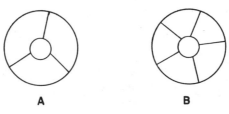

A   B

Figure 98

sufficient also seems to be very elusive despite considerable effort by a number of investigators over a period of many years. The following chronology provides some of the highlights in the long history of the 4-color problem. For a deep study of that problem and the important and extensive theory that has been developed as an outgrowth of studying it, the reader should consult Ore [22]. Of course it is that body of related theory that makes the problem important in mathematics; though the difficulty in resolving what seems to be a simple problem attracts the interest of almost anyone who hears about it.

A CHRONOLOGY OF THE 4-COLOR PROBLEM

1840   Problem related to the 4-color problem but not directly. The problem was stated by A. F. Möbius in lectures. This seems to have had no effect on the problem itself. *Leipzig Trans.* (*Math. -phys. Classe*), **37** (1885), 1–6.

1852   Francis Guthrie transmitted the problem to this brother Frederick who made it known to Frederick's teacher, Augustus De Morgan. De Morgan tried with little success to interest others, such as W. R. Hamilton. Francis Guthrie, *Proc. Roy. Soc.* (*Edinburgh*), **10** (1880), 728.

1878   Arthur Cayley stated that he could not prove the conjecture. *Proc. London Math. Soc.*, **9** (1878), p. 148. *Proc. Roy. Geographical Soc., London*, n.s. **1** (1879), 259–261.

1879   A. B. Kempe gave incorrect solution. *Amer. J. Math.*, **2** (1879), 193–200. *Trans. London Math. Soc.*, **10** (1879), pp. 229–231. *Nature* **21** (Feb. 26, 1880), 399–400.

1880   Peter Guthrie Tait gave a proof depending on a proposition that had not been established. *Proc. Roy. Soc. Edinburgh* **10** (1880), 729. *Philosophi. Mag.* Series 5, **17** (1884), 41.

1890   P. J. Heawood showed Kempe's solution to be incorrect. He established the 5-color theorem. *Quart. Jour. Math.* (*London*), **24** (1890), 332–338 and **29** (1897), 270–285. Heawood established the 6- and 7-color theorems for the Möbius strip and Torus and treated other surfaces.

This concludes the classical phase of the problem. The "modern" phase has two (at least) major parts. One is the formulation of conditions for non-4-colorability; the other is a deep study of properties of graphs relative to their chromatic number. The latter are indicated by the work of G. Dirac.

The other series of studies are indicated by the following.

1922   P. Franklin. Map requiring 5 colors must have at least 26 districts. *Amer. J. Math.*, **44** (1922), 225–236.

1924   A. Errera showed such a map must include at least 13 pentagons (that it must contain at least 12 is an "elementary" result). *Comp. Rendu. Assoc. Francaise Avan. Sci., Liege* (1924), 96.

1926   C. N. Reynolds. Must have 28 districts.

1936   P. Franklin. Must have 32 districts and 15 pentagons.

1937   C. E. Winn. More complicated "structural" requirements on non-4-colorable maps. *Amer. Jour. Math.*, **59** (1937), 515–528.

1968   O. Ore and J. Stemple showed a non-4-colorable map must have at least 40 countries. *J. Comb. Theory* (1970), **8**

The reader will find a good account of the origin of the 4-color problem in May [62]. The evidence seems to establish that Francis Guthrie conceived of the problem in essentially its present form and had a good feeling for the difficulties involved in its complete solution. Though early workers (e.g., Cayley, Kempe, and Tait) attributed the problem to De Morgan, it seems more accurate to consider it as Guthrie's problem. This was May's conclusion on the basis of his detailed historical research.

EXERCISES

1. Let G be a graph with n vertices. Write the chromatic number of G as a function of n when: G is a tree, G is a wheel, G is a simple circuit, G is the complete graph K_n, and G is the plane square lattice graph of side n.

2. Consider the graph corresponding to the cube with 4 principle diagonals. Prove this graph is non-planar and bichromatic.

3. Consider the following graph G:

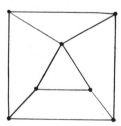

Set up a Grundy function for G and determine the chromatic number. How many different 5-colorings are possible for G? Construct the chromatic polynomial for G.

4. Formulate the determination of the chromatic number for G, Exercise (3), in terms of a linear programming problem. Solve to verify your previous result.

5. Show that any complete graph is a vertex critical graph. Is the wheel on n vertices critical?

6. Illustrate Hadwiger's conjecture for $k = 4$.

7. Construct a 5-chromatic graph without any cycles of length 3. Is the graph you constructed critical?

8. Show that every critical graph on more than two vertices contains cycles of odd length.

9. Prove that the vertices of a planar graph can always be partitioned into five disjoint sets such that any pair of vertices in the same set have no common edge. Do you think a similar situation is possible using four sets? What is the situation when only three sets are used? Illustrate your conclusions.

10. What is the chromatic number of Petersen's graph?

Chapter 6

Combinatorial Theory

BASIC CONCEPTS

Combinatorial theory (or combinatorial analysis) is concerned with problems of enumeration and structure of mathematical objects. The objects may represent physical situations or things in applications or may be purely abstract and under study for theoretical reasons. It is common practice to refer to the subject matter of combinatorial theory as combinatorics. Counting the number of objects of a certain type or the number of ways a particular operation can be carried out forms the central problem of enumerative combinatorial theory. However, the structural features of the subject are of equal importance in theory and applications. These features are concerned with such things as how to specify when two objects (or processes) are the same (i.e., definitions of equivalence) and the existence of specified structure in a given class of objects (theory of structural synthesis). Clearly these latter aspects of combinatorial theory are also vital to its enumerative aspect and in addition cover a wide range of nonenumerative studies.

Discrete probability problems are essentially combinatorial since they require the enumeration of various cases leading to the desired events. Many difficulties in discrete probability are due to their connection with combinatorics. In this section we introduce and survey some of combinatorial theory.

The material presented should serve to introduce the reader to the kinds of problems and to some of the methods of solution encountered in combinatorial theory. Our main interest in this chapter of course is to present some topics from graph theory that are of a combinatorial nature and conversely to present some combinatorial topics that are very close to graph theory. To the extent that graphs deal with structure, the whole subject of graph theory is very close to combinatorics. Our meaning here however will be restricted to questions related to the existence of certain types of graphs and to the enumeration of types of graphs. Some examples of such questions have already been treated, illustrating how close graph theory really is to combinatorics. One can hardly talk about graphs at all without giving some combinatorial type results. An example of enumerations is the number of spanning arborescences given in Chapter 1 while Mycielski's theorem (Chapter 5) on the existence of triangle free k-chromatic graphs gives an example of a structural result.

Much of the material in this chapter is based on Riordan [13] and Ryser [9] who give particularly important accounts of basic combinatorial theory. In addition to the specific reference citations made to these works, their influence pervades the entire chapter.

Basic to many combinatorial problems is the utilization of sets in the mathematical sense. In enumerative combinatorics the objects we wish to enumerate can be taken as elements of some appropriate set of objects and the identification problem becomes one of setting up equivalent classes of subsets in which all elements of a given subset are equivalent (i.e., indisguisable from one another).

If A is a finite set of elements the collection $S(A)$ of all its subsets, including A itself and the empty set Φ, is a set. The set $S(A)$ has the subsets of A as its elements. The number of elements in a set is often called the power of the set. From these ideas we can describe one of the most basic enumerative problems: "If the power of A is n what is the power of $S(A)$?" This is equivalent to asking "How many elements are there in $S(A)$, (i.e., How many subsets are there for a set with n elements)?" The answer is $\sum_{i=0}^{n} \binom{n}{i} = 2^n$ gives the power of $S(A)$. Verification will be left to the reader as an exercise. Note that there are $\binom{n}{r}$ subsets of r elements for $r = 0, 1, \ldots, n$.

A *partition* of a set A is a collection of subsets of A denoted by $\{A_i\}$ where $i \in I$ and I is an index set (i.e., I is a subset of the set $\{k | k = 1, \ldots, n\}$ when the power of A is n) and furthermore the A_i satisfy the conditions $A_i \cap A_j = \Phi$ for $i \neq j$ and

$$\bigcup_{i \in I} A_i = A.$$

One may well ask "How many partitions of A are there?" where partitions are different if they contain different subsets. This is a difficult question which we will not answer. However, a special case of this problem plays an important role in combinatorics. If the set A is taken to be a collection of n units then each subset contains $k \le n$ units. If a subset contains k units we may identify it with the integer k. A partition of the set A as defined above is thus equivalent to the *partition of an integer n* into a sum of positive integers less than or equal to n. Though the details of solutions to the integer partition enumeration problem are beyond our scope some of the important results in this area will be given as illustrations of combinatorial results and utilized as solutions to still other combinatorial problems.

Basic to the development of enumerative combinatorial theory are the following fundamental rules of combination:

Logical Union rule

If an object H can be selected in n ways and another object K can be selected in m ways then either H or K can be selected in $n + m$ ways.

Logical Intersection rule

If an object H can be selected in n ways and another object K can be selected in m ways then both H and K can be selected, in the stated order, in $n\,m$ ways.

We now consider some basic problems and methods in enumerative and structural combinatorics. Objects being enumerated or considered for structural properties are often called designs.

Consider n distinct objects. The total number of ways of arranging r of these objects in a line with respect to order is $P_n(r) = n(n - 1) \cdots (n - r + 1)$ and is called "the number of r permutations of n distinct objects". One derivation of this illustrates a technique of enumerative combinatorics, that is, direct evaluation (by logical argument based on the *structural form* of the objects being enumerated.)

There are r positions in the line to be filled. The first may be filled in n ways, then independently the second may be filled in $n - 1$ ways and so forth till, finally, the rth location may be filled with any of the $n - r + 1$ objects remaining. This gives the result which may also be written $P_n(r) = n!/(n - r)!$.

If the order of objects in the line is not to be distinguished but only the set of r objects making up the line, the design is called a *combination of n objects taken r at a time* and the number of these is denoted by $\binom{n}{r}$, the so-called binomial coefficient. This value can be determined in a direct way similar to that used to obtain $P_n(r)$. However, an alternative deviation will indicate an even more important method of enumerative combinatorics. The method of difference equations will be used to obtain $\binom{n}{r}$ which for convenience will be denoted by $C_n(r)$. The difference equation is the following:

$$C_{(n+1)}(r) \text{ can arise in two ways:}$$

1. The r elements contain the new element.
2. The r elements do not contain the new element. This leads to:

$$C_{(n+1)}(r) = C_n(r-1) + C_n(r)$$

The first term on the right arises from case (1) where we select $(r-1)$ elements from the original set of n elements. The second term arises from case (2) in which all r elements are selected from the original elements. The reader may show that $C_n(r) \equiv \binom{n}{r} = n!/r!(n-r)!$ satisfies this equation so that the usual expression for the number of combinations is obtained. This equation can of course be solved directly but it is rather involved to do so as it is a double difference equation involving two integral valued variables.

In the case of permutations with $r = n$ we speak of $P_n = n!$ as the number of permutations of n distinct objects. Note that if the objects are not distinct a more involved development is required which we do not consider here. This is a typical way in which structure affects enumeration.

PERFECT PERMUTATIONS

A perfect permutation S_n of n distinct objects is a permutation for which the kth object is not in the kth position for $k = 1, \ldots, n$. This may also be called a complete permutation or *derangement*. Following Euler we develop a recurrence relation (i.e., difference equation) for this quantity by observing that $S_{(n+1)}$ is formed in two distinct ways:

1. The new object goes into some location and the element that was there (in the standard form) goes at the end; then the remaining $n-1$ elements undergo a perfect permutation. This occurs in $nS_{(n-1)}$ ways.
2. The new object goes into some location and the element that was there does not go at the end. Allow it to go to end and then force all objects but the new one to undergo a perfect permutation: this will insure that the replaced element does not go at the end. This occurs in nS_n ways.

The recurrence is then:

$$S_{(n+1)} = n(S_{(n-1)} + S_n).$$

This is not difficult to solve (see Kemeny, Snell and Thomson [63]). The equation implies that $S_n = n\,S_{n-1} + (-1)^n$. Iteration of this expression yields the result obtained below by another method. A different approach to the problem will show another important tool of combinatorics, the *principle of inclusion and exclusion*. This principle may be given the following formulation:

Consider n properties $a_1, a_2, \ldots, a_n$ and let $N_{i_1, i_2 \ldots i_r}$ denote the number of ways properties $a_{i_1}, a_{i_2}, \ldots a_{i_r}$ can occur. Let N denote the totality of occurrences. Then the number of occurrences for which none of the properties occur is given by the following: (algebraic formulation of the principle of inclusion and exclusion. For an extended discussion, see Riordan [13]).

$$N(0) = N - \sum_{i_1 = 1}^{n} N_{i_1} + \sum_{i_1 < i_2} N_{i_1\, i_2} + \cdots + (-1)^r \sum_{i_1 < i_2 < \cdots < i_r} N_{i_1 i_2 \cdots\, i_r} + \cdots .$$

To apply this formulation to the problem of perfect permutations let a_j denote the property that element j is in position j. Then $N_{i_1,\, i_2 \cdots i_r} = (n - r)!$ since the r elements are in fixed position and all permutations of the remaining $n - r$ elements enumerate the case. The total number of such cases is $\binom{n}{r}$ which enumerates how to select the r elements to hold fixed for the evaluation of the combinatorial (symbolic) sums. Note $N = n!$ total number of permutations and $N(0) = S_n$ the number of permutations where elements j is not in location j for $j = 1, \ldots, n$.

Then

$$S_n = n! - \binom{n}{1}(n - 1)! + \binom{n}{2}(n - 2)! + \cdots + (-1)^r \binom{n}{r}(n - r)! + \cdots$$

$$+ (-1)^n \binom{n}{n} 0!$$

$$S_n = n! - n! + \frac{n!}{2!} + \cdots + (-1)^r \frac{n!}{r!} + \cdots + (-1)^n \frac{n!}{n!}$$

$$S_n = n!\left[\frac{1}{2!} + \cdots + (-1)^r \frac{1}{r!} + \cdots + (-1)^n \frac{1}{n!}\right]$$

Since $e^{-1} = 1/2! + \cdots + (-1)^r/r! + \cdots$ the number S_n is the integer nearest to $n!/e$, that is, it is given by $[n!/e]$. Here $[x]$ is the "greatest integer in" symbol, meaning the greatest integer not greater than x.

Moreover,

$$\lim_{n \to \infty} \frac{\sqrt{2\pi n}\ n^n}{e^{n+1}\ S_n} = 1$$

so that

$$S_n \sim \frac{\sqrt{2\pi n}}{e}\left(\frac{n}{e}\right)^n$$

is an asymptotic expression for the perfect permutations.

Note. Since the numbers in combinatorics grow so rapidly with n it is often of interest to develop such asymptotic formulas which are useful approximations for n "large."

As a final example of enumerative combinatorics we may consider *Terquem's problem* (Riordan [13]).

For combinations of n numbered things in natural rising order, and with $f(n, r)$ the number of r-combinations with odd elements in odd positions and even elements in even positions, find $f(n, r)$. Also study the total number of such combinations $f(n) = \sum_r f(n, r)$.

1. Recurrence relation. Note $f(n, 0) = 1$. Consider the case n even. If r is odd we proceed as follows: Note that in this case the last integer cannot be n. However, the last may be $n - 1$ or it may not be. This gives two cases.

(a) The last element is $n - 1$ so we must select $r - 1$ other elements. The last of these is even and $n - 1$ is odd so this can be done in $f(n - 1, r - 1)$ ways. (The element $n - 1$ will not be counted in this among the elements contributing to the $r - 1$).

(b) The last element is not $n - 1$ so we must select r other elements. Since we may not use n or $n - 1$ the number of cases is $f(n - 2, r)$. Thus for n even and r odd

$$f(n, r) = f(n - 1, r - 1) + f(n - 2, r)$$

Now suppose r is even. The last number may be n or not giving:

$$f(n, r) = f(n - 1, r - 1) + f(n - 2, r)$$

The first term on the right arises when n is last. Since the last is n, we must select the remaining $r - 1$ ($r - 1$ odd) from the other $n - 1$ ($n - 1$ odd). The second term on the right arises when n is not last. Since the last is not n we must select all r from the other $n - 2$ (even). By such reasoning one arrives at the recurrence for all cases:

$$f(n, r) = f(n - 1, r - 1) + f(n - 2, r)$$

It should be remarked that this is a more difficult relation than the others we have given. It is also hard to solve. One may wish to show the following:

2. Solution to the recurrence:

$$f(n, r) = \binom{q}{r},$$

where $q = [(n + r)/2]$ using the greatest integer notation:

3. Formula for $f(n)$. Consider the recurrence for $f(n, r)$ and sum from $r = 1$ to $r = n$.

$$\sum_{r=1}^{n} f(n, r) = \sum_{r=1}^{n} f(n - 1, r - 1) + \sum_{r=1}^{n} f(n - 2, r)$$

By definition $f(n) = \sum_{r=0}^{n} f(n, r)$ and $f(n, 0) = 1$. Moreover, it is clear that $f(n, k) = 0$ for $k > n$. Hence the sum expression gives:

$$f(n) - f(n, 0) = f(n - 1) + f(n - 2) - f(n - 2, 0) + f(n - 2, n - 1)$$
$$+ f(n - 2, n).$$

Finally, evaluating the known terms above yields:

$$f(n) = f(n - 1) + f(n - 2), \qquad f(0) = 1, f(1) = 1.$$

This is the recurrence defining the *Fibonacci numbers*. These are of great interest and wide occurrence throughout combinatorics, number theory and elsewhere (e.g., some optimal binary search algorithms can be based on them). We shall not go into them further here.

As we have remarked, *structural combinatorics* deals with problems of the existence of certain types of objects, for example, geometrical designs, special linear graphs, etc. It also deals with identification problems and for both these reasons is closely related to enumerative combinatorics.

A simple but elegant example is the following "covering" problem. A domino is an object consisting of two squares joined along an edge. Consider a chess board having 8 rows and 8 columns, composed of 64 squares. If the two squares on opposite ends of a diagonal are removed can the resulting board be covered by 31 dominoes? Note in such a covering each domino covers two squares. Before reading the solution the reader may wish to solve this problem. It is typical of combinatorial problems, simple to state and understand but far from obvious in solution. By thinking in the proper way the solution becomes simple but the reader must understand that direct mathematical techniques are extremely few in this subject and hard to apply. Here is a solution. Each domino must cover one white and one black square and since there are 62 squares the number of dominos required is 31. However, the two squares removed were of the same color hence among the 62 squares 30 are of one color and 32 are of another, hence it is not possible to cover the modified board with dominos. We remark that it is clearly possible to cover a complete 64 square board with dominos in many ways. How many ways (an enumerative problem) is a difficult problem only recently solved. It is related (and in fact is an instance of) the dimer covering problem in statistical mechanics (or that branch of the subject which may be called combinatorial statistical mechanics) which will be discussed in Chapter 10.

We give some further examples in structural combinatorial analysis.

LATIN SQUARES

An interesting type of combinatorial design which has application to statistical design of experiments is the Latin square and some of its generalizations. Example of 3×3 latin square:

$$
\begin{array}{ccc}
A & C & B \\
B & A & C \\
C & B & A
\end{array}
$$

These are used for three factor experiments, in the above case each at three levels. Orthogonal latin squares or greco-latin squares example of 3×3:

$$
\begin{array}{ccc}
A\alpha & C\beta & B\gamma \\
B\beta & A\gamma & C\alpha \\
C\gamma & B\alpha & A\beta
\end{array}
$$

These are useful in four factor experiments, in the above case each is at three levels. Greco-Latin squares exist for 1, 2, 3, 4 and 5 levels (i.e., an $n \times n$ square is of level n).

Euler (1779) considered the following famous problem of 36 officers (Ball [48]).

Arrange 36 officers from six regiments; the officers being in six groups, each consisting of six officers of equal rank one drawn from each regiment in a square so that each row and each file (column) contains one and only one officer of each rank and one and only one from each regiment.

For the case of nine officers a solution is given by the above Greco-Latin square. Let the ranks be A, B and C and the regiments be α, β, γ. Then the first row has an officer of rank A from regiment α, one of rank B from regiment γ and one of rank C from regiment β, etc.

For 36 officers there is no such arrangement. This result was a conjecture of Euler and was established by G. Tarry in 1900 by an extensive enumeration method of possibilities. In fact Euler made a more general conjecture: Euler's conjecture (in our words).

There exists no Greco-Latin square of size n with $n \equiv 2 \pmod 4$. The first instance of this is $n = 6$ which as we said was established by Terry. The next is ten, then fourteen, etc. Since Terry's result, an attempt to establish the Euler conjecture has been a major research problem in combinatorics; it was generally believed to be true. However, in 1960 R. C. Bose, I. S. Shrikhande and E. T. Parker disproved the Euler conjecture and showed how to construct Greco-Latin squares of size n for all $n \equiv 2 \pmod 4$, $n \geq 10$.

GENERATING FUNCTIONS (An excellent introduction to these concepts is given in Riordan [13], Chapter 2).

One of the most useful concepts in enumerative combinatorics is the generating function. If we want to enumerate some type of objects that depend on a collection of n objects then we can call the desired enumerated value $r(n)$ for example. The generating function $Z(z)$ for $r(n)$ is defined to be a power series in a (often assumed complex but may often be considered only as a formal variable with no special properties.) variable z such that $Z(z) = \sum_{n=0}^{\infty} r(n) z^n$. For some problems the sum is over a modified range. In other problems the quantities being enumerated may depend on two or more "store" sizes; then the generating function becomes a function of two or more variables, for example

$$H(z, t) = \sum_k \sum_n r_n(k)\, z^k\, t^n.$$

Such functions arise in two different ways in enumeration problems. When difference equations can be written for the quantity desired they can be converted into algebraic or differential equations for the appropriate generating function. This application is technical and straightforward though it can lead to complicated mathematical problems (note for multivariable generating functions one most often must deal with partial differential equations). The other method of using generating functions is to employ them directly as counting devices, different terms being specifically included to account for specific quantities one wishes to count. This technique is very fundamental to combinatorics and has been widely used; it is however, very difficult since it depends to a considerable extent on the intuition, etc. of the particular person making the study.

In some applications generating functions are used as purely formal algebraic objects and manipulated as such. A formalism of such manipulation has been developed. In this kind of application convergence and other analytic aspects of the function are of no concern. Other applications utilize analytic techniques such as contour integration in the complex plane, solution methods for differential equations, etc. When such methods are to be utilized the generating function must have appropriate analytic character such as a nonzero radius of convergence.

PARTITIONS

One of the most important areas of enumeration problems is the enumeration of partitions of integers. This is so because many other problems can be put into the form of a partition problem. We give some of the basic aspects of partition enumeration. There exists a number of results in this general topic. Much of the classical work was developed by MacMahon as extended and discussed by Riordan in [13].

The partitions of an integer n may be unrestricted or have certain restrictions placed on them such as no repetitions of any part (integer), etc. If we write a partition by writing each part with repeated parts indicated by powers we have a convenient representation for partitions. For example $3 + 2 + 2 + 1 = 8$. This partition of 8 is written $3\, 2^2\, 1$. The partitions may also be classified by the number of parts, the one part partition being the number n itself and the n part partition being represented by 1^n. Table 3 shows partitions of

Table 3

Integer	Number of Parts					
n	1	2	3	4	5	6
1	1					
2	2	1^2				
3	3	2 1	1^3			
4	4	3 1	2 1^2	1^4		
		2^2				
5	5	4 1	3 1^2	2 1^3	1^5	
		3 2	$2^2 1$			
6	6	5 1	4 1^2	3 1^3	2 1^4	1^6
		4 2	3 2 1	$2^2 1^2$		
		3^2	2^3			

n for $n = 1, \ldots, 6$ by number of parts, showing each partition. It may be that the order of a partition is important. For example, for 3 we can have $2 + 1$ or $1 + 2$, then we call the quantity a composition thus there are more compositions of an integer n than there are partitions.

To study partitions more completely we denote the number of unrestricted partitions of an integer n by P_n, restricted partitions with no part exceeding a value k are denoted by P_{nk} and the number of partitions with exactly k parts is denoted by $P_n(k)$. It is clear that

$$P_n = \sum_{k=1}^{n} P_n(k).$$

The enumerations of these quantities are most often carried out by generating functions combined in the fundamental method we have spoken of. For example, we have the generating function $P_k(t)$ for the quantities P_{nk} which by definition of a generating function is:

$$P_k(t) = \sum P_{nk} t^n.$$

This is found to be

$$P_k(t) = \prod_{i=1}^{k} \frac{1}{(1 - t^i)}.$$

One can show that $P_{n+k}(k) = P_{nk}$ so that the P_{nk} can be used to enumerate the $P_n(k)$ and hence the P_n as well.

Some interesting results on partitions are as follows.

1. The number of partitions with unequal parts, (i.e., no repeated parts) is equal to the number of partitions having all parts odd.

2. The number of partitions of n with at most k parts is equal to the number of partitions of n with no part greater than k.

3. The number of partitions of n with exactly m parts is equal to the number of partitions of n with the largest part equal to m.

A useful quantity in the study of partitions is the Ferrers graph or diagram. These are not linear graphs but are diagrams of points. A Ferrers diagram corresponding to a partition of n consists of n points separated in k rows if the partition has k parts. The rows all start at the same vertical position to the left (are left verified) and contain as many points as the part represents. For example, the partition 4 3 1 of 8 has the Ferrers diagram shown in Figure 99. These diagrams may be operated on to obtain other partitions and indeed

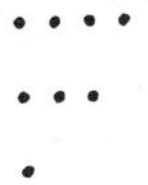

```
. . . .
. . .
.
```

Figure 99

may be utilized in achieving general results. One example is the *conjugate* of the partition obtained by reading the Ferrers diagram column wise (or alternatively forming a new diagram by interchanging rows and columns). The conjugate of the above is shown in Figure 100. Which is the partition

```
. . .
. .
. .
.
```

Figure 100

3 2^2 1 of 8. It may be that a Ferrers diagram is self-conjugate and yields the same partition reading by rows or by column. As an example consider the partition 4^2 3 2 of the integer 1 3 which has the diagram in Figure 101. In any self-conjugate diagram there are two important parts one is a corner square of k^2 dots. This is called a Durfee square and when it is removed the two like tails represent partitions of $(n - k^2)/2$ into at most k parts. For example, in the above case we have $n = 13$, $k = 3$ the Durfee square and the remaining tail

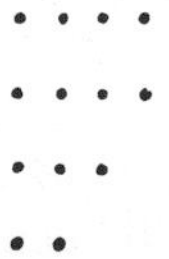

```
. . . .
. . . .
. . .
. .
```

Figure 101

square   tails

Figure 102

diagrams are as shown in Figure 102. The tails represent partitions of 2 into at most 3 parts, there being of course two such partitions, namely 2 and 1^2 represented by the two tails shown above, each treated separately as a Ferrers diagrams.

Many more results are possible but we limit our discussion of partitions to the above, including a table of useful partitions under our discussion of distributions to follow.

DISTRIBUTIONS (We survey some of the results from this topic. Our material is based on Riordan [13], particularly Chapter 5).

Many combinatorial problems can be expressed as the number of ways n objects can be distributed among m cells. There are four cases; we give the major results for each case.

1. The number of ways of putting n *different objects* into m *different cells* is m^n. With no cell empty there are $m!\ S(n, m)$ ways where $S(n, m)$ is a Stirling number of the second kind.

2. The number of ways of putting n *like objects* into m *different cells* is

$$\binom{n + m - 1}{n}$$

With no cell empty

$$\binom{n - 1}{m - 1}.$$

3. The number of ways of putting n *different objects* into m *like cells* is $S(n, 1) + S(n, 2) + \cdots + S(n, m)$.

With no cell empty $S(n, m)$.

4. The number of ways of putting n *like things* into m *like cells* is $P_n(1) + P_n(2) + \cdots + P_n(m)$.

With no cell empty $P_n(m)$.

(Note) $P_n(m)$ is the number of partitions of n into m parts.

Stirling numbers of the second kind are given by the recurrence:

$$S(n + 1, k) = S(n, k - 1) + k\ S(n, k)$$

where $S(n, 1) = 1$ all n and $S(n, n) = 1$ all n.

$$S(n, k) = 0 \text{ for } k > n.$$

Table 4 shows $S(n, k)$

Table 4

n/k	1	2	3	4	5	6	7	8	9
1	1								
2	1	1							
3	1	3	1						
4	1	7	6	1					
5	1	15	25	10	1				
6	1	31	90	65	15	1			
7	1	63	301	350	140	21	1		
8	1	127	966	1701	1050	266	28	1	
9	1	255	3025	7770	6951	2646	462	36	1

As discussed above $P_n(m)$ the number of partitions of n into m parts is a complicated combinatorial quantity. Small values are shown in Table 5, of course, $P_n(m) = 0$ for $m > n$.

Table 5

n/m	1	2	3	4	5	6	7	8
1	1							
2	1	1						
3	1	1	1					
4	1	2	1	1				
5	1	2	2	1	1			
6	1	3	3	2	1	1		
7	1	3	4	3	2	1	1	
8	1	4	5	5	3	2	1	1

Also $P_n(1) = 1$ and, $P_n(n) = 1$ for all n.

Example: $n = 6$, $m = 2$ the four possible cases yield:

$$2^6 = 6 4 \tag{1}$$

$$\binom{7}{6} = 7 \tag{2}$$

$$S(6, 1) + S(6, 2) = 1 + 31 = 32 \tag{3}$$

$$P_6(1) + P_6(2) = 1 + 3 = 4 \tag{4}$$

Generating functions may be combined according to the basic rules of combinatorics. Thus a standard technique is to formulate the simple generating functions for cases that contribute to a total result and then to form the total generating function by composition. That technique is illustrated by finding the (exponential) generating function for the enumeration of m distinct objects into n distinct cells with at most two objects per cell.

The enumerating generating function for single cell occupancy has three terms in this case. These terms correspond to no object, exactly one object, and exactly two objects per cell respectively. The function is: $1 + x_i t + x_i^2$ $t^2/2!$ for cell i where $i = 1, \ldots, n$. We use exponential generating functions which have the factors $1/k!$ as coefficients for ease in computation (such factors play no fundamental role in the calculations). Let $F(m; n)$ denote the number of restricted distributions using m objects and n cells. The generating function of these numbers is:

$$F(t) = \sum_{m=0} F(m; n) \frac{t^m}{m!} = \prod_{i=1}^{n} \left(1 + x_i t + x_i^2 \frac{t^2}{2!} \right)$$

by the composition rule of combinatorics. This expression is difficult to simplify. We shall illustrate it for the case $n = 4$ and $m = 6$. In the expansion for $n = 4$ the coefficient of t /6! is the following:

$$90(x_1^2 x_2^2 x_3^2 + x_1^2 x_2^2 x_4^2 + x_1^2 x_3^2 x_4^2 + x_2^2 x_3^2 x_4^2$$
$$+ 2 x_1 x_2 x_3^2 x_4^2 + 2 x_1 x_3 x_2^2 x_4^2 + 2 x_1 x_4 x_2^2 x_3^2$$
$$+ 2 x_2 x_3 x_1^2 x_4^2 + 2 x_2 x_4 x_1^2 x_3^2 + 2 x_3 x_4 x_1^2 x_2^2).$$

Looking at this result we see that the total number of distributions is obtained by setting $x_i = 1$ for $i = 1, 2, 3, 4$. This gives $F(6 ; 4) = 1440$. Special distributions satisfying the general requirement may also be observed. For example the $x_1^2 x_2^2 x_3^2$ term shows that there are 90 distinct cases which have exactly 2 objects in each of the cells 1, 2, and 3, and no objects in cell 4. Of course we may also obtain this result directly by observing that two objects can be placed in cell 1 in $\binom{6}{2}$ ways, then two more placed in cell 2 in $\binom{4}{2}$ ways. This gives a total of $\binom{6}{2} \binom{4}{2} = 90$ in agreement with the generating function result. In the same way we see that the distribution specified by $x_1 x_2 x_3^2 x_4^2$ has 180 cases and so forth.

This completes the present introduction to combinatorial concepts. Now we turn to some important, more specialized topics in combinatorics and then to some graph theory of a particularly combinatorial character. For further study of combinatorial theory the reader can refer to Ryser [9], Riordan [13], or Liu [64].

SYSTEMS OF DISTINCT REPRESENTATIVES (Much of the material in this section is based on Ryser [9], and Hall [65]).

Questions of the existence of various designs lead to a number of important theorems. These are basic to theoretical structural combinatorics and are used in studying many problems such as the above Euler conjecture, types of matrices of zeros and ones, etc. Many of these theorems are related, imply one another and so forth and are also related to important aspects of linear programming, particularly the duality concept and through this to parts of the Theory of Finite Games. Thus theorems from the latter subjects prove the combinatorial theorems and conversely. In this area the three subjects are closely connected. It should be remarked that in some sense mathematical programming is a part of combinatorial analysis as it deals with the explicit construction of some mathematical structure, that is, "the program" subject to structural definition (the constants and form of the objective function).

We shall speak about only one, the most fundamental, of the distinct representation theorems; that of Philip Hall as expressed and established by Mann and Ryser. Various other forms exist and several different proofs (for example, see Hall, M., [65]). The expression given here includes an enumerative aspect that is not part of the original form of Philip Hall's theorem.

Let M be a set and denote the set of all subsets of M by $S(M)$. Let H be an n-sample of $S(M)$, that is, a collection of subsets of M which may contain the same subset several times. Denote the elements of H by $S_1, \ldots, S_n$. Let V be an n sample of M and denote its elements by $v_1, \ldots, v_n$ (note the same elements of M may occur several times in V). Then V is called a *system of distinct representatives* for H or, alternatively, for $S_1, \ldots, S_n$ provided the v_i are distinct, (i.e., the sample is a subset of M) and that $v_i \in S_i$ for $i = 1, \ldots, n$. We shall use the designation S.D.R. for system of distinct representatives.

Philip Hall's Theorem (extended to enumeration result). Let $S_1, \ldots, S_n$ denote n subsets of a set M. For each $k = 1, \ldots, n$ suppose that every k of these sets contain between them at least k distinct elements of M. Then there exists a system of distinct representatives (S.D.R.) for those subsets. Moreover, let r be a fixed integer less than or equal to the minimal number of elements in each S_i. Then if $n \geq r$, there are at least $r!$ S.D.R.'s. If $n < r$, then there are at least $r!/(r - n)!$ S.D.R.'s.

Proof. The theorem is established by an inductive argument.

The theorem is true for $n = 1$. Let $n > 1$ and suppose that for each k, $1 \leq k < n$, (note all cases are needed for the induction) every set of k, subsets S_i, contain at least $k + 1$ distinct elements. Take the set S_1 and select from it any representative a_1. There are at least r choices for a_1. Now form the sets

$$\bar{S}_2 = S_2 - \{a_1\}, \ldots, \bar{S}_n = S_n - \{a_1\}.$$

Any k of the sets $\bar{S}_i$ contain between them at least k elements. By the induction hypothesis, if $r \leq n$, then $r - 1 \leq n - 1$ and there are at least $(r - 1)!$ S.D.R. for the $\bar{S}_i$ (i.e., we are assuming the theorem true for $n - 1$). If $n < r$, then $n - 1 < r - 1$, and then there are at least $(r - 1)!/(r - n + 1)!$ S.D.R.'s. Hence there are at least $r!$ or $r!/(r-n)!$ S.D.R.'s.

This shows that on the assumption of truth for $n - 1$ and the fact of each set of k-subsets S_i containing at least $k + 1$ distinct elements, the result for n follows. It remains to show the result if some set of k-subsets S_i contains exactly k distinct elements.

Suppose for some k, $1 \leq k < n$, there are k-subsets, say $S_1, \ldots, S_k$ which contain exactly k distinct elements. Here we must have $k \geq r$, and by the induction hypothesis, these sets have at least $r!$ S.D.R.'s. Let $T = \{a_1, \ldots, a_k\}$ be any such S.D.R. Now form the sets

$$\bar{S}_{k+1} = S_{k+1} - T, \ldots, \bar{S}_n = S_n - T$$

Any h of the $\bar{S}$ sets, $1 \leq h \leq n - k$, cannot contain fewer than h elements. For otherwise the S sets corresponding to these $\bar{S}$ sets along with $S_1, \ldots, S_k$ would contain fewer than $h + k$ elements this cannot be by the induction hypothesis. Apply the induction hypothesis to the $n - k$ sets $\bar{S}$ to complete the proof. (i.e., to say the sets $\bar{S}$ have S.D.R. and combine the two S.D.R.'s gives the full S.D.R. for the n sets.)

This establishes sufficiency. It is obvious that the condition is necessary. ▲

Consider the basic form of Hall's theorem which states the necessary and sufficient condition without giving bounds on the number of S.D.R.'s. There are a number of proofs, discussions, and extensions of that theorem. In particular, the theorem may be proven by use of linear programming methods. That proof illustrates some of the connections between combinatorics and mathematical programming. Additional material is given in Hoffman, [66], and Hoffman and Kruskal, [8]. We shall give only the basic theorem form from those studies.

It is necessary for our further consideration of S.D.R.'s to define the incidence matrix between the elements of M and the subsets S_i. We assume that M has m elements denoted $e_1, e_2, \ldots, e_m$. The incidence matrix $A = (a_{ij})$ is defined as follows:

$$a_{ij} = 1 \text{ if } e_i \in S_j,$$
$$= 0 \text{ otherwise.}$$

Philip Hall's Theorem (Basic Form). Let M be a set of elements $\{e_1, \ldots, e_m\}$, H the n-sample $\{S_1, \ldots, S_n\}$ from $S(M)$ of subsets of M and V a subset of M containing n elements $e_{h_j}, j = 1, \ldots, n$. We say that V is an S.D.R. of H if $e_{h_j} \in S_j, j = 1, \ldots, n$.

H has a S.D.R. if and only if for each $k = 1, \ldots, n$ every k members of H have at least k distinct elements between them.

Proof. (Hoffman, [66]).

Consider the following linear programming problem:

$$\text{maximize} \quad \sum_{i,j} a_{ij} x_{ij} = z$$

$$\text{subject to} \quad \sum_i x_{ij} \le 1, \ \sum_j x_{ij} \le 1, \ x_{ij} \ge 0 \text{ for all } i, j.$$

The system of constraint equations here have the unimodular property so that in this case the solutions will all be in terms of $x_{ij} = 0$ or 1. The x_{ij} indicate the possibility of assigning the ith element to set S_j and the constraints require any assignment to be distinct. Thus the maximum value of z will be n if and only if M contains a set V such $e_{h_j} \in S_j$ for $j = 1, \ldots, n$. This shows that Hall's theorem is equivalent to solving the above problem and obtaining n as the maximum.

The condition of Hall's theorem is clearly necessary. We wish to show it is sufficient to insure maximum $z = n$ in the above linear programming problem,

The dual problem is as follows:

$$\text{minimize} \quad \sum_i u_i + \sum_j v_j = w$$

$$\text{subject to} \quad u_i + v_j \ge a_{ij}$$

$$u_i \ge 0, \ v_j \ge 0 \qquad \text{for all } i \text{ and } j.$$

It follows from the duality theory of linear programming that minimum $w = n$ will establish the result for the original problem.

In the dual structure all solutions of interest have u_i and v_j equal to 0 or 1. A u_i equal to 1 means the selection of the ith row of A and a v_j of 1 means selection of the jth column of A. The minimum will be n if each constraint can be satisfied by setting a total of n u's and v's equal to 1. If the condition of Hall's theorem is satisfied all the dual constraints can be satisfied in that way. Thus the sufficiency of the condition is shown. ▲

It should be remarked that extensions of the above methods lead to considerations in network flows (a topic we shall discuss in Chapter 8) so we are not as far removed from graph theory as the reader might begin to think!

When S.D.R.'s exist we may ask how many there are for a given sample H of subsets. We may also wish to form (some of) them. For these purposes it is useful to introduce the permanent of a rectangular array (or matrix). The *permanent* of an $m \times n$ array $A = (a_{ij})$, written per (A) is defined as the following combinatorial sum:

$$\text{per } (A) = \sum a_{1\,i_1}, \ldots a_{m\,i_m},$$

where the sum is over all m permutations of the integers $1, 2, \ldots, n$. Though the permanent may seem similar to the idea of the determinant the concepts are really very different. Of course, a determinant is only defined for a square array but in addition the combinatorial sum that defines a determinant has a very different form, that is, alternating signs, that allow special results. Thus the permanent is in general very hard to evaluate. One of the few general properties of permanents is that for a square matrix A per $(A) = $ per (A^t) where A^t is the transpose of A. Permanents can be evaluated by means of the results in the following theorem; however, the work involved can be considerable, even for small matrices.

Theorem (Ryser). Consider an $m \times n$ array A in which (without loss of generality) it is assumed that $m \le n$. Replace all elements in r columns of A by zeros and call the resulting array A_r. Let $P_s(A_r)$ denote the product of all the row sums of A_r. If $\sum P_s(A_r)$ stands for the sum of all the $P_s(A_r)$ over all arrays of the form A_r then:

$$\text{per } (A) = \sum P_s(A_{n-m}) - \binom{n-m+1}{1} \sum P_s(A_{n-m+1})$$

$$+ \binom{n-m+2}{2} \sum P_s(A_{n-m+2}) - \cdots +$$

$$+ (-1)^{m-1} \binom{n-1}{m-1} \sum P_s(A_{n-1}).$$

A proof of the theorem may be found in Ryser, [9].
Let

$$A = \begin{pmatrix} 1 & 1 & 0 & 0 & 0 & 0 & 1 & 1 \\ 1 & 1 & 1 & 1 & 0 & 0 & 0 & 0 \\ 0 & 0 & 1 & 1 & 1 & 1 & 0 & 0 \\ 0 & 0 & 0 & 0 & 1 & 1 & 1 & 1 \end{pmatrix},$$

then we may apply Ryser's method for per (A):

$$\text{per } (A) = \sum S(A_4) - \binom{5}{1} \sum S(A_5) + \binom{6}{2} \sum S(A_6) - \binom{7}{3} \sum S(A_7) = 136.$$

A useful corollary to the theorem is the following:

Corollary. When A is an $n \times n$ square array:

$$\text{per } (A) = P_s(A) - \sum P_s(A_1) + \sum P_s(A_2) - \cdots + (-1)^{n-1} \sum P_s(A_{n-1}).$$

Figure 103

If we let J_n denote an $n \times n$ matrix with every element equal to one and I_n denote the identity matrix of order n then: per $(J) = n!$ from the corollary. Then one gets an interesting expression for the derangements or perfect permutations of n elements which we have denoted by S_n: per $(J_n - I_n) = S_n$.

Now let us return to a consideration of S.D.R.'s. By definition of the incidence matrix A of an S.D.R. problem every entry in the combinatorial sum expansion of per (A) corresponds to an S.D.R. for that problem. The reader should recall that an S.D.R. problem involves a set M and a given n-sample H of subsets S_j and asks for S.D.R.'s of H, if any. Thus we have the important results:

Theorem. If A is the incidence matrix for an S.D.R. problem then per (A) equals the number of S.D.R.'s for that problem. Two S.D.R.'s are distinct if their elements are assigned in different ways to the subsets S_j even though the elements themselves may be the same.

By working out per (A) term by term one can generate specific S.D.R.'s. When per $(A) = 0$ there are no S.D.R.'s so that in such a case the conditions of Hall's theorem are not satisfied.

Systems of distinct representatives may be used to enumerate the number of complete edge assignments in a graph G. Such an assignment associates exactly one edge with each vertex, assigning each edge to at most one vertex. Let E denote the set of edges of G and E_i denote the subset of E containing all edges incident on vertex v_i. Assume G has n vertices and m edges. An S.D.R. for the collection $\{E_i\}$ is an edge assignment for G. We should note that such an S.D.R. is not the same as an edge covering of G. In an edge covering each vertex is incident on some edge of the covering but the order of selection does not give distinct coverings. Figure 103 shows two distinct edge coverings at A and two distinct edge assignments at B where assignment is indicated by an arrow on the edge directed toward the vertex to which it is assigned. Clearly for any graph G (except a vertex graph) there are more edge assignments than there are edge coverings.

The enumeration of edge assignments is seen to be given by per (I_g) where I_g is the $n \times m$ incidence matrix of G. Since I_g is also the incidence matrix for the collection $\{E_i\}$ per (I_g) gives the number of S.D.R.'s for the sets E_i which

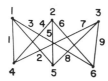

Figure 104

are exactly the edge assignments for G. As an example, let us consider the complete bipartite graph $K_{3,3}$ on six vertices. The numbering of vertices and edges is as shown in Figure 104. Then for the incidence matrix we obtain (5).

$$I_g = \begin{pmatrix} 1 & 1 & 1 & 0 & 0 & 0 & 0 & 0 & 0 \\ 0 & 0 & 0 & 1 & 1 & 1 & 0 & 0 & 0 \\ 0 & 0 & 0 & 0 & 0 & 0 & 1 & 1 & 1 \\ 1 & 0 & 0 & 1 & 0 & 0 & 1 & 0 & 0 \\ 0 & 1 & 0 & 0 & 1 & 0 & 0 & 1 & 0 \\ 0 & 0 & 1 & 0 & 0 & 1 & 0 & 0 & 1 \end{pmatrix} \tag{5}$$

Applying Ryser's theorem on permanent evaluation yields:

$$\sum P_s(A_3) = 3408, \ \sum P_s(A_4) = 1233, \ \sum P_s(A_5) = 180, \text{ and } \sum P_s(A_6) = 6.$$

From which we obtain per $(I_g) = 3408 - \binom{4}{1} \ 1233 + \binom{5}{2} \ 180 - \binom{6}{3} \ 6 = 156$. Thus $K_{3,3}$ has 156 distinct edge assignments and the collection $\{E_i\}$ has that number of S.D.R.'s.

There is a fairly close relation between S.D.R's and bipartite graphs which can be described with the help of the concept of *deficiency*. Let the bipartite graph G have as its vertex set $V = V_1 + V_2$ and the edge set E such that an element of E is of the form (v^1, v^2) where $v^1 \in V_1$ and $v^2 \in V_2$. Consider a subset $S \subseteq V_1$ of vertices. Let $V_2(S)$ be the subset of V_2 consisting of all vertices in V_2 which are edge connected to at least one vertex in S. As usual we let $|A|$ denote the number of elements in the set A. For any finite subset S of V_1 the deficiency is defined as: $d_f(S) = |S| - |V_2(S)|$. For a finite bipartite graph there is an upper bound for the deficiencies and hence a maximal deficiency d_f^*. A set $S \subseteq V_1$ is called critical (or of maximal deficiency) if $d_f(S) = d_f^*$. For the empty set we define $d_f(\Phi) = 0$ so that $d_f^* \geq 0$. Ore gives several results for these concepts in [15] and in [67]. We shall only state some of those results with particular interest directed toward the S.D.R. concept. For the graph $K_{3,3}$ shown in Figure 104 the sets $\{v_i\}$, $i = 1, 2, 3$ have deficiency -2, the sets $\{v_i, v_j\}$ $i \neq j$, $i, j = 1, 2, 3$ have deficiency -1, and the set $\{v_1, v_2, v_3\}$ has maximal deficiency zero. A complete bipartite graph will always have maximal deficiency zero and the graph itself will be the only critical set. Such vertex sets are said to be without deficiency.

A subset S_1 of V_1 is matched to a subset S_2 of V_2 if there exists a one to one correspondence between S_1 and S_2 such that corresponding vertices are edge connected in G, a bipartite graph with vertex set $V = V_1 + V_2$. A basic theorem is the following.

Theorem. Let V_1 be locally finite (each vertex of finite degree) and have maximal deficiency d_f^*. Then there exists a subset S of V_1 with d_f^* vertices such that $V_1 - S$ is matched to a subset of V_2. The quantity d_f^* is the smallest number of vertices with this property.

A necessary and sufficient condition that V_1 can be matched to a subset of V_2 is that V_1 have zero deficiency. Alternatively, the condition may be stated as $d_f^*(V_1) = 0$.

Let us compare the last part of this theorem to P. Hall's theorem for a family of subsets $\{E_i\}$ of an m-set H with elements $\{e_j\}$. Let E_i $i = 1, \ldots, n$ be the vertices of V_1 and if e_j is an element in any of the sets E_i let e_j be a vertex in V_2. Let $r = |\cup_{i=1}^{n} E_i|$ then $|V_1| = n$ and $|V_2| = r$. Construct a bipartite graph G with $V = V_1 + V_2$ and let (v_i^1, v_j^2) be an edge of G if and only if $e_j \in E_i$ for all i and j. A matching of V_1 to a subset of V_2 constitutes an S.D.R. for $\{E_i\}$. By the above theorem such a matching (S.D.R.) will exist if and only if $d_f^* = 0$ for V_1 in the graph G. Consider a subset S of V_1 having a total of k elements. If the k subsets E_i comprising S have at least k elements e_j of H between them then $d_f(S) \leq 0$. If this is true for all such subsets S then $d_f^* = 0$ (since $\Phi \subseteq V_1$ and $d_f(\Phi) = 0$) so that V_1 is without deficiency. In that case V_1 has a matching and $\{E_i\}$ has an S.D.R. We see that the requirements are the same as expressed in the standard form of P. Hall's theorem. Some further results relating permanents, determinants, and bipartite graphs are given by Harary [68].

RAMSEY'S THEOREM

In this section we discuss a result that arose from abstract considerations in (mathematical) logic by F. P. Ramsey. The result is fundamental for many combinatorial studies and in particular leads to various considerations in graph theory.

Let S be a set of exactly n elements and $P_r(S)$ be the set of all r element subsets of S (so that it contains $\binom{n}{r}$ elements) where r is a positive integer less than or equal to n. We will use the notation m-set to denote a set having exactly m elements.

Consider a partition of $P_r(S)$ into t parts (where t is a positive integer) $A_1, A_2, \ldots, A_t$. Thus

$$P_r(S) = A_1 \cup A_2 \cup \cdots \cup A_t$$

and $A_i \cap A_j = \Phi$ for all i, j. The theory allows $A_i = \Phi$ but such cases are so trivial that we shall assume $A_i \neq \Phi$ for any i. Further, consider t integers q_i, $i = 1, \ldots, t$ such that $1 \leq r \leq q_i$ for each i.

If there exists a q_i-subset of S such that all of its r-subsets ($\binom{q_i}{r}$ in number) lie in A_i then we call that q_i-subset a (q_i, A_i)-subset of S.

In terms of the concepts introduced above we may now state the basic theorem.

Theorem (Ramsey as given by Ryser, [9]). Let $q_1, \ldots, q_t$ and r be given integers satisfying the conditions above. Then there exists a minimal positive integer $N(q_1, \ldots, q_t, r)$ such that the following is valid for all integers $n \geq N(q_1, \ldots, q_t, r)$.

Let S be an n-set and let $A_1 \cup \cdots \cup A_t$ be an arbitrary ordered partition of $P_r(S)$ into t-components. Then S contains a (q_i, A_i)-subset for some $i = 1, 2, \ldots, t$.

The quantities $N(q_1, \ldots, q_t, r)$ are called Ramsey numbers.

We shall discuss some special cases and applications of this theorem; a proof may be found in Ryser, [9].

When $r = 1$ Ramsey's theorem yields a careful expression of what is sometimes called the "pigeon-hole principle" which in its rough form states: if a set of sufficiently many elements is partitioned into not too many subsets, that at least one of those subsets must contain many of the elements. Since $r = 1$, $P_r(S)$ is S (except for subtle logical distinctions that need not concern us). For a partition of P_r into t disjoint parts, given any particular numbers $q_1, \ldots, q_t$ we wish to have some subset of S with q_i elements which will be in one of the partition parts of P_r.

When $t = 1$ there is only one partition, it is all of $P_r \equiv P_1$ and hence is all of S. If q_1 is given we can certainly find a q_1-subset of S which is in P_1 since $q_1 \leq n$ is specified. This illustrates the result in the simplest case.

Let $t = 2$ and suppose q_1 and q_2 are given. P_1 is still S but it now has a partition into two parts. Any partition into two parts must be such that some q_i-subset of S ($i = 1$ or 2) will be in a corresponding partition of P_1. Here we see that the critical partition that just fails has $q_1 - 1$ in one part and $q_2 - 1$ in the other for a total of $q_1 + q_2 - 2$ elements for which the condition fails [i.e., there are no (q_i, A_i) subsets satisfying the theorem result]. One sees that there will be no such critical partition with one more element hence we require $q_1 + q_2 - (t - 1)$ elements in this case.

For general t in the $r = 1$ case the critical partition has $q_1 - 1, q_2 - 1, \ldots, q_t - 1$ elements in each partition. By adding an element we get no such critical partition and hence the lower bound on elements. Thus we obtain

$$N(q_1, \ldots, q_t, 1) = q_1 + q_2 + \cdots + q_t - t + 1.$$

When $t = 2$ and r is general one has an important special case of Ramsey's theorem. The theorem has been stated and proven in various places for this special case and more results exist for this case than for any other. The reason for interest in this case is that it is one of the simplest non-trivial cases and in addition is closely related to situations involving binary division of elements. Such situations arise in applications. For this case we have the special results:

$$N(q_1, r, r) = q_1 \quad \text{and} \quad N(r, q_2, r) = q_2$$

Another special case is when $q_i = q \ge r \ge 1$ for $i = 1, \ldots, t$. In that case we consider the r-subsets of S to be partioned into t components in an arbitrary way. Then (if n is large enough) there exists a q-subset of S with all of its r-subsets in one of the t components. For example suppose $r = 2$, $q = 3$, and $t = 3$. Form all $\binom{n}{2}$, 2-subsets of the elements of S, call this set $P_2(S)$. Partition $P_2(S)$ into 3 parts. Then there exists a 3-subset of S with all its 2-subsets ($\binom{3}{2}$ in number) in one of the components of $P_2(S)$. Call the elements of S, $a_1, \ldots, a_n$. There are $n(n-1)/2$ elements in $P_2(S)$.

Divide them into 3 parts such that if their indices are both even they are in A_1, both odd they are in A_2, and one even and one odd they are in A_3. Then we say that for some 3-subset of S all 3 of its 2 sets are in A_1 or in A_2 or in A_3. We need $n \ge 5$ for this to be true. For example, consider $n = 6$. Then $A_1 = \{(a_2, a_4)(a_2, a_6)(a_4, a_6)\}$, $A_2 = \{(a_1, a_3), (a_1, a_5), (a_3, a_5)\}$ and A_3 contains the rest of the 2-subsets of S. Clearly the 3-subset $\{a_2, a_4, a_6\}$ has all of its 2-subsets in A_1.

Ramsey's theorem may be used to establish an interesting result in structural combinatorial theory.

Theorem. Let p be an integer greater than or equal to three. There exists a smallest positive integer N_p such that for all $n \ge N_p$ the following is true: if n points are in general position in the plane (i.e., no three points are colinear) then p of the points form vertices of a convex p-gon.

This theorem is established by using two lemmas.

Lemma A. If five points are in general position in the plane, then there are some four points which form vertices of a convex quadrilateral.

Lemma B. If p points are in general position in the plane and all quadrilaterals formed from those p points are convex, then the p points form the vertices of a convex p-gon.

Now we establish the theorem by using Ramsey's theorem. Let $p \ge 4$ and let $n \ge N(5, p, 4)$. Partition the 4-subsets of the set of n points into the concave and the convex quadrilaterals. By Ramsey's theorem there exists either a 5-gon with all quadrilaterals concave or else a p-gon with all quadrilaterals convex. By Lemma A the first situation is impossible and by Lemma B the p-gon is convex as required. This shows that N_p exists and that $N_p \le N(5, p, 4)$.

It is known that $N_3 = 2 + 1$, $N_4 = 2^2 + 1$, and $N_5 = 2^3 + 1$.

In general, not many Ramsey numbers are known. Some of the known values are $N(3, 3, 2) = 6$, $N(3, 4, 2) = 9$, $N(3, 5, 2) = 14$, and $N(3, 3, 3, 2) = 17$

Now we give a graph theory result that uses Ramsey's theorem. For a related discussion see Turán, [69].

Theorem. There exists a minimal positive integer $f(k, h)$ such that if a simple graph G has at least $f(k, h)$ vertices, then either G has a complete k-graph as a subgraph or $\bar{G}$ has a complete h-graph as a subgraph.

In this theorem $\bar{G}$ denotes a graph with the same vertex set as G such that (v_i, v_j) is an edge of $\bar{G}$ if and only if (v_i, v_j) is not an edge of G. Thus G is an n-graph (without loops or multiple edges), $G \cup \bar{G} = K_n$.

Proof. Form all pairs of vertices. This collection (of edges) may be divided into two parts, one containing all edges of G and the other containing all edges of $\bar{G}$. This forms a partition into two parts of the edges of K_n.

For all pairs (2-subsets) of a set of k vertices and all pairs of a set of h vertices, Ramsey's theorem insures a smallest integer $N(k, h, 2)$ such that the result holds. ▲

This theorem leads us to consider how many edges a graph may have without having various complete subgraphs. The study of that question centers around a basic theorem of Turán.

The Theorem of Turán concerns the maximum number of edges a graph may have without having a complete k-graph as subgraph, for various integers k. It considers graphs without self loops or multiple edges (which have no relation to the problem under study). The discussion is centered on a particular type of graph, denoted $D(n, k)$, which will now be defined (see Turán, [69]).

Let $D(n, k)$ be a simple graph on n vertices and k an integer such that $3 \leq k \leq n$. Define two integers r and t by the equation:

$$n = (k - 1) t + r,$$
$$1 \leq r \leq k - 1,$$

so that t acts like an integral divisor of n with r the integral remainder. Thus n and k determine t and r uniquely. Divide the vertices of $D(n, k)$ into $k-1$ classes. Divide the classes into two sets putting r classes into one set and $k - 1 - r$ classes into the other set. Make the assignment so that each class in the r-set has $t+1$ vertices and each class in the $(k - 1 - r)$-set has t vertices (thus the total number of vertices is $r(t + 1) + (k - 1 - r) t = n$). Any two vertices of different classes are connected by an edge. Any two vertices of the same class are not connected by an edge. This completes the definition of the graph of type $D(n, k)$.

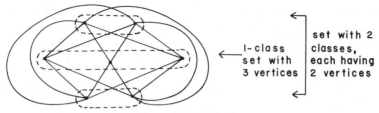

Figure 105. $D(7, 4)$.

Figure 105 shows the graph $D(7, 4)$ for which $t = 2$, $r = 1$, and there are 3 classes. One set of classes has one element (class) with $t + 1 = 3$ vertices. The other set has two classes, each having $t = 2$ vertices. Though $D(7, 4)$ has a number of complete 3-graphs as subgraphs it has no complete 4-graph as a subgraph. However, the addition of a single edge will result in a complete 4-graph subgraph. Thus $D(7, 4)$ is a graph on 7 vertices having a maximum number of edges, (i.e., 16) without having a complete 4-graph as subgraph. This illustrates a property of the $D(n, k)$ graphs that we will establish for the general case.

The graph $D(n, k)$ contains no complete k-graph as a subgraph. To prove this we observe that there are $k - 1$ classes so that in the formation of a k graph we must select two vertices (at least) from one class. Such a pair of vertices is not edge connected by definition. Hence, we cannot construct a complete k-graph from the edges of $D(n, k)$.

Denote the number of edges in $D(n, k)$ by $d_k(n)$. We wish to evaluate this number as a function of n and k. Let us call the two sets of classes A and B. Set A has r classes, each class having $t + 1$ vertices. Set B has $k - 1 - r$ classes, each class having t vertices. The edges of $D(n, k)$ are of three types. Some edges connect vertices lying only in set A, some connect vertices lying only in set B, and some connect vertices such that one vertex is in A and the other vertex is in B.

Edges lying within set A are enumerated as follows. Select a class in set A. Select a vertex in that class. That vertex is incident on $(r - 1)(t+1)$ edges lying within set A. Thus the class contains vertices incident on a total of $(t + 1)(r - 1)(t + 1)$ edges. There are r classes so remembering that each edge is counted twice in the above procedure we find a total of $r(r - 1)(t + 1)^2/2$ edges in set A.

By similar reasoning there are $(k - 1 - r)(k - 2 - r)t^2/2$ edges in set B.

In set B any vertex is incident on $r(t + 1)$ edges that are also incident to vertices in set A. This gives $t\,r\,(t + 1)$ edges per class in B. Thus there is a total of $(k - 1 - r)\,r\,t\,(t + 1)$ edges between vertices of sets A and B.

Addition of these results and some algebraic simplification using the relation $n = (k - 1)t + r$ yields:

$$d_k(n) = \frac{r(r - 1)}{2} + \frac{(k - 2)(n^2 - r^2)}{2\,(k - 1)}.$$

Let $A_k(n)$ denote the class of simple graphs of order n (i.e., on n vertices) which have no complete subgraph of order k where $3 \le k \le n$. We have seen above that $D(n, k)$ belongs to $A_k(n)$. Let $M_k(n)$ denote the maximum number of edges that any simple graph can have when it is a member of the class $A_k(n)$. Then we have the following lemmas for $M_k(n)$.

Lemma A. $M_k(n) > M_{k-1}(n)$ ($k = 4, 5, \ldots, n$).

Proof. Consider a graph E in $A_{k-1}(n)$ having $M_{k-1}(n)$ edges. Add an edge to that graph (which is possible since $k \le n$) and consider the resulting graph E' having $M_{k-1}(n) + 1$ edges. The lemma will be established if we show that E' is in $A_k(n)$. Suppose E' is not in $A_k(n)$. Then E' contains a complete subgraph of order k. Denote that graph by E'' and its vertices by $v_1, \ldots, v_k$. Either the new edge belongs to E'' or it does not. If the new edge does not belong to E'' then E'' is a complete k graph as a subgraph of E, which is impossible since E is in class $A_{k-1}(n)$. Thus the new edge is in E''. We may suppose it connects v_1 and v_2. Then $v_2, \ldots, v_k$ would form a complete $k - 1$ subgraph of E which is impossible. ▲

Lemma B. $M_k(n) \le \binom{k-1}{2} + M_k(n - k + 1) + (k - 2)(n - k + 1)$

Proof. Consider an extremal graph B_1 in class $A_k(n)$. Since B_1 is extremal it has $M_k(n)$ edges. By Lemma A $M_{k-1}(n)$ is strictly less than $M_k(n)$ hence it is not possible to have $M_k(n)$ edges without having a complete subgraph of order $k - 1$. Thus B_1 has a complete subgraph B_2 of order $k - 1$. Denote the vertices of B_2 by $v_1, v_2, \ldots, v_{k-1}$ and denote the remaining vertices of B_1 by the set $C = \{v_k, \ldots, v_n\}$.

At most $(k - 2)$ edges can be incident at each vertex of C and have a vertex of B_2 as their second incident vertex. For example, if $k - 1$ edges were incident on v_k and also to vertices of B_2 then $v_1, v_2, \ldots, v_{k-1}, v_k$ would form a complete subgraph of B_1 of order k, contrary to the definition of B_1. Thus the number of edges connecting vertices of C with those of B_2 is at most $(k - 2)(n - k + 1)$.

Since B_2 is a complete $k - 1$ graph it has $\binom{k-1}{2}$ edges.

The number of edges connecting vertices of C is at most $M_k(n - k + 1)$ since we must have no complete k graph within C (and C has $n - k + 1$ vertices). ▲

We shall now state the major result of the section as the following theorem.

Theorem (Turán). In the class $A_k(n)$, the graph $D(n, k)$ and only that graph has the maximum number of edges.

The theorem states that $d_k(n) = M_k(n)$ and there is a unique graph with the property.

Proof. The proof is by induction. We fix $k \geq 3$ and proceed from n to $n + k - 1$. The possible cases form a $k - 1$ by infinite table as possible values for n (6).

$$
\begin{array}{cccc}
k & 2k-1 & 3k-2 & \ldots \\
k+1 & 2k & 3k-1 & \ldots \\
\cdot & \cdot & \cdot & \\
\cdot & \cdot & \cdot & \\
\cdot & \cdot & \cdot & \\
2k-2 & 3k-3 & 4k-4 & \ldots
\end{array}
\tag{6}
$$

The theorem is established by showing it holds for the cases in the first column and then for each fixed row. These are the two standard steps in the induction.

We shall assume the theorem has been proven for the 1st column and deduce its validity for each row. After this we shall establish the theorem for the 1st column. These steps shall be called Part I and Part II of the proof.

PART I

Consider the theorem for n defined by $n = (k - 1) t + r_1 \ (1 \leq r_1 \leq k - 1)$ with r_1 and k fixed, $t = 1, 2, \ldots,$ We suppose the theorem is true for $t \leq T(T \geq 1)$ (this is the induction assumption) and investigate the case of $t = T + 1$. This is the case $n_1 = (k - 1)(T + 1) + r_1$.

By Lemma B and the induction hypothesis for $M_k(n_1 - k + 1)$ we obtain:

$$
M_k(n_1) \leq \binom{k-1}{2} + d_k[(k-1)T + r_1] + (k-2)(n_1 - k + 1)
$$

Denote the three terms on the right by F_1, F_2, and F_3, respectively. Now we must see when equality can occur in this expression. From the proof of Lemma B an extremal graph B_1 has the following properties:

1. From each vertex of C exactly $(k - 2)$ edges go to vertices of B_2. Term F_3.

2. After a suitable numbering of the vertices of C the edges connecting the vertices belonging to C form the graph $D((k - 1) T + r_1, k)$. Term F_2.

3. The complete graph B_2 belongs to B_1. Term F_1. Each of these properties is associated with one and only one of the terms indicated by F_1, F_2, or F_3 above. Thus any extremal graph B_1 should have the properties and if we show the only graph of order n_1 having (1), (2), and (3) and no complete k-subgraph is $D((k - 1)(T + 1) + r_1, k)$ then the equality will be established and Part I of the proof completed.

Thus we consider an extremal graph B_1 on n_1 vertices satisfying (1), (2), and (3).

Consider the part C of B_1 (set of vertices not in a $k-1$ subgraph B_2). By (2) and definition of $D((k-1)T+r_1,k)$ its vertices can be split into $k-1$ classes, the classes into two sets, one set having r_1 classes with $T+1$ vertices each, the other set having $k-1-r_1$ classes with T vertices each. By (1), to each vertex v_ν of C there corresponds a uniquely determined vertex v_μ of B_2 which is not connected to v_ν (since v_ν is connected to the "other" $k-2$ vertices in B_2). Call v_μ the associate of v_ν.

Observation 1.   The associates of the vertices of C belonging to different classes are distinct from one another, that is, if v_1 and v_2 are in different classes of C then they cannot have the same associate (in B_2).

If the statement was false there would be two vertices of C say v_k and v_{k+1} belonging to different classes and having the same associate v_1. Then by definition of associate (element of B_2 not connected to v_ν) both v_k and v_{k+1} are connected with $v_2, v_3, \ldots, v_{k-1}$.

Since v_k and v_{k+1} belong to different classes they are connected by definition of $D((k-1)T+r_1, k)$. All pairs of vertices of B_2 are connected by (3) (B_2 complete). Hence $v_2, v_3, \ldots, v_{k-1}, v_k, v_{k+1}$ is a complete k-subgraph of B_1 which is impossible. This establishes *observation 1*.

Observation 2.   The associates of the vertices of C belonging to the same class are identical, (i.e., if v_1 and v_2 belong to the same class in C, they have the same associate.)

If the statement is false v_k and v_{k+1} are a pair of vertices in the same class of C with different associates v_1 and v_2, respectively.

Let $v_{k+2}, \ldots, v_{2k-1}$ be a representative system of the remaining $(k-2)$ classes of C (one from each class). By observation 1 their associates are distinct from each other as well as from v_1 and v_2. Thus we have k different associates which is impossible since B_2 has only $k-1$ different vertices (and associates are vertices of B_2).

By the two observations above, it follows that a uniquely determined associate belongs to each class of C.

Adjoin to each class of C the common associate of its vertices. The new classes now contain every vertex in B_1, there are $k-1$ classes, r_1 of these have $T+2$ vertices per class and the remaining $k-1-r_1$ classes have $T+1$ vertices each. Two vertices of the same class are not connected (by definition of associate). It remains to show that vertices of different classes are always connected for this will show that B_1 with properties (1), (2) and (3) is identical with $D((k-1)(T+1)+r_1, k)$.

Consider two vertices belonging to different classes. If both originally belonged to C then they are always connected by (2) and the definition of $D((k-1)T+r_1, k)$. If both belonged to B_2 then they are connected since B_2 is complete (by (3)). The remaining case is where one vertex, say v_k, belonged to C and the other, say v_1, belonged to B_2. By the construction of the new classes v_1 is not the associate of v_k, hence they are connected as required.

PART II

We must now complete the proof by establishing the 1st column results. This corresponds to cases where $T = 1$, that is,

$$n = k - 1 + r_1, \ (1 \leq r_1 \leq k - 1).$$

For these cases Lemma B yields (by direct substitution of $n = k - 1 + r_1$ in the formula)

$$M_k(k - 1 + r_1) \leq \binom{k - 1}{2} + M_k(r_1) + (k - 2)r_1$$

since $r_1 < k$ we can never get a complete k graph on r_1 vertices hence $M_k(r_1)$ can equal the maximum number of edges which can be placed on r_1 vertices. This corresponds to a complete r_1-graph with $\binom{r_1}{2}$ edges, hence:

$$M_k(k - 1 + r_1) \leq \binom{k - 1}{2} + \binom{r_1}{2} + (k - 2)r_1.$$

Once again we must see when the equality will hold. This occurs if and only if:

1. C is a complete subgraph of order r_1 which corresponds to the center term.

2. Each vertex of C is connected with exactly $(k - 2)$ vertices of B_2 which corresponds to the last term.

Thus we define a maximal graph B_1 with properties (1) and (2) and wish to show the only such graph with no complete k-subgraph is $D(k - 1 + r_1, k)$. The procedure (and concept) is similar to the Part I proof. Note however that in this case we have utilized a direct evaluation of $M_k(r_1) = \binom{r_1}{2}$ (simple but very important as it "gets us started" so to speak).

Define the associate of a vertex as in Part I. As before associates of different classes of C are distinct. Now since there are less vertices not all vertices of B_2 are associates (in general).

Let us label the vertices so that v_1 is the associate of v_k, v_2 is the associate of $v_{k+1}, \ldots$ and v_{r_1} is the associate of v_{k-1+r_1}; if $r_1 < k - 1$ then $v_{r_1+1}, \ldots, v_{k-1}$ are not associates of any vertices.

Form $k - 1$ classes, the 1st consisting of v_k and v_1, the 2nd of v_{k+1} and $v_2, \ldots$, the r_1-th of v_{k+r_1-1} and v_{r_1}, and if $r_1 < k - 1$ each of the remaining $(k - 1 - r_1)$ classes consisting of the single vertices $v_{r_1+1}, \ldots, v_{k-1}$ respectively. To identify this graph with $D(k - 1 + r_1, k)$ we must show that two vertices of different classes are always connected, all else follows by construction. All pairs of vertices are connected by an edge except the ones in C with their associates by construction ((1) and (2) and the definition of B_2.)   ▲

Figure 106. $D(5, 3)$.

The graph $D(n, 3)$ is bipartite. There are 2 classes of vertices and r is 1 or 0 when n is odd or even respectively. When n is odd there is one class with $t + 1$ vertices and one class with t vertices. When n is even both classes are of the type having t vertices ($r = 0$). For example, Figure 106 shows $D(5, 3)$ for which $r = 1$ and $t = 2$. The graph $D(6, 3) = K_{3,3}$ is the utilities graph that forms the basic representation of Kuratowski Type II (nonplanar) graphs.

We conclude this section with brief discussions of two topics that are closely related to Turan's theorem.

Dirichlet's principle states that if a set of $n + 1$ elements is such that each element has exactly one of n given properties then there exists at least two different elements having a property in common.

If we think of the $n + 1$ elements as vertices of a graph and form an edge between two vertices if and only if the corresponding elements have a property in common the principle states there will be at least one edge. Now suppose there are $kn + 1$ elements for an integer k, then an immediate extension of the principle states that at least $k + 1$ elements have a property in common. In that case the graph on $kn + 1$ vertices would contain a complete subgraph of order $k + 1$.

Now let us consider the following very general result.

Theorem (Marczewski). An arbitrary simple graph (finite or infinite) always corresponds to a family of sets, one and only one set to each vertex, so that two vertices are connected by an edge if and only if the corresponding sets have no elements in common.

For example a complete k-graph corresponds to a collection of k disjoint sets.

By using Turan's theorem and Marczewski's theorem we obtain the following result for sets. If there is a family of n sets such that at most $d_k(n)$ pairs of sets are disjoint then there exists no collection of k sets from the family in which the sets are all disjoint.

As an example consider the family of sets: (a, b, c), (b, e, f), (d, s, m), (a, e, g), (m, u, h) with $n = 5$ and let $k = 3$ so that $d_3(5) = 6$. Then $t = 2$, $r = 1$ and the graph associated with this family of sets has one class with 3 sets (vertices) and one class with 2 vertices. There is no complete 3 graph as a subgraph. However, if one more edge is added (one more disjoint pair present in the family of sets) a complete 3 graph will occur. This will correspond to a collection of 3 sets all disjoint.

ENUMERATION PROBLEMS FOR LINEAR GRAPHS

There are two major types of enumeration problems concerned with graphs. One type is counting the number of graphs of some particular kind such as the total number of graphs on n vertices (excluding loops and multiple edges). Even this problem must be treated with care since we must identify like graphs (by some definition) and count them only once. As we have discussed in Chapter 1 most of graph theory identifies like graphs by isomorphism.

The second type of enumeration is to count some quantities associated with a particular given graph. For example, one may wish to know the number of paths of length k between two particular vertices in a graph.

Enumeration of graph types finds application in statistical mechanics and enumerations of quantities within a given graph find application in network theory. Of course, many applications exist for both types of enumeration. We treat the enumeration of graphs first and state some general concepts of combinatorics as well. The enumeration within a graph may really be classified as part of graph theory itself since the matrix representation for the graph are used in the enumeration; in addition to this section we shall treat some of these problems in other parts of the text for example in Chapter 9.

Trees are the simplest general class of linear graphs to enumerate (Riordan treats them in [13] Chapter 6). If a particular vertex is identified while the other vertices are unidentified we call the result a rooted tree and the identified vertex is the root. The other vertices may also be labelled (i.e., identified). Thus we can enumerate the following classes of trees: trees with unlabelled vertices, trees with labelled vertices, rooted and labelled trees, rooted and unlabelled trees. The simplest is the last.

Following Riordan [13] we let r_n = number of rooted (unlabelled) trees with n vertices and $r_n(m)$ = the number of such trees with m edges meeting at the root. Clearly $r_n = \sum_{m=1}^{n-1} r_n(m)$. It is an immediate result that $r_n(1) = r_{(n-1)}$. The case of two lines at the root is divided into two different relations, one for n even and one for n odd. The root is one vertex so when n is even the remaining vertices can never divide into two parts each with the same number of vertices. Thus, for n even ($n = 2p$) we get at once:

$$r_{2p}(2) = r_1 r_{2p-2} + r_2 r_{2p-3} + \cdots + r_{p-1} r_p.$$

In the case of n odd ($n = 2p + 1$) however, we may have two parts with p vertices in each. These must be combined according to the formula for selecting with repetition r_p things two at a time. Thus we have:

$$r_{2p+1}(2) = r_1 r_{2p-1} + r_2 r_{2p-2} + \cdots + r_{p-1} r_{p+1} + \binom{r_p + 1}{2}.$$

For $m > 2$ the difference relations become complicated and it is necessary to utilize generating functions in the formulation. A particular case of m is treated by considering the m part partitions of $n - 1$ since the $n - 1$ vertices are to be divided up into m sets, each set will make a (rooted) tree and be connected by an edge from its root to the root vertex of the full (n-vertex) tree. Such a partition is represented by $[1^{k_1} 2^{k_2} \cdots (n-1)^{k_{n-1}}]$ where $k_1 + k_2 + \cdots + k_{n-1} = m$ and $k_1 + 2k_2 + \cdots + (n-1)k_{n-1} = n - 1$. The number of possibilities corresponding to this partition is:

$$\binom{r_1 + k_1 - 1}{k_1} \binom{r_2 + k_2 - 1}{k_2} \cdots \binom{r_{n-1} + k_{n-1} - 1}{k_{n-1}}$$

therefore we find for r_n the combinatorial formula:

$$r_n = \sum_P \prod_{j=1}^{n-1} \binom{r_j + k_j - 1}{k_j}$$

where P is the set of all partitions of $n - 1$, that is, $k_1 + 2k_2 + \cdots + (n-1)k_{n-1} = n - 1$. Note we do not have any restriction on parts since we are evaluating r_n.

This result is typical of many combinatorial results. In a sense it is the solution to the enumeration of rooted, unlabelled trees yielding a direct formula for r_n. Many people do in fact declare this to be a (or the) solution. However, it is clearly very difficult to say, on the basis of this result that $r_{25} = 2{,}067{,}174{,}645$ for example! The difficulty in using the above formula is in the evaluation of the combinatorial sum. We are summing over partitions rather than just letting some indices run through certain values as takes place in simple algebraic sums. There is also substantial computational difficulty in evaluating expressions of the form:

$$\prod_{j=1}^{n-1} \binom{r_j + k_j - 1}{k_j}$$

even when the r_j and k_j are known. In practice one most often builds up numerical tables of the quantity being enumerated in step by step fashion utilizing the concepts leading to the "abstract" solution. Sometimes generating functions can be used as we now indicate. Let $r(x)$ be the generating function for the r_n then $r(x) = r_1 x + r_2 x^2 + \ldots$. From our theoretical formulation above the function $r(x)$ can be expressed as follows (using the basic combination rules for generating functions):

$$r(x) = x(1 - x)^{-r_1}(1 - x^2)^{-r_2} \ldots$$

This solution also is hard to use in practice since to find the coefficient of powers of x greater than 5 is very tedious to say the least (the reader may wish to show by expansion of $r(x)$ that $r_6 = 20$).

Now the case of labelled rooted trees will be considered ([13], page 133). R_n stands for the number of such trees and $R_n(m)$ for the number with m edges at the root. The label for the root can be any one of the n labels available and then the other points can be labeled using the remaining $n - 1$ labels. This case is somewhat simpler than the unlabelled case (it has more structure and in combinatorial analysis this often simplifies things; many problems arise from lack of structure). The $m = 2$ situation in particular is simpler since the two parts are always different (even when they have the same number of vertices) since the vertices are all different (this is what calling them labelled means.) One has the following results, including the generating function $R(x) = R_1 x + R_2 x^2/2! + R_3 x^3/3! + \cdots$

$$R_n(1) = n \, R_{n-1}$$

$$R_n(2) = \frac{n}{2!} \sum_{j=1}^{n-2} \binom{n-1}{j} R_j R_{n-1-j}$$

$$R(x) = x \exp R(x).$$

Note that $R(x)$ is a generating function of the so-called exponential type, where the coefficient of x^k is not simply R_k but is modified by $1/k!$ Such generating functions sometimes lead to simpler relations, as in this case.

To enumerate R_n we can expand $R(x)$, that is, solve the functional equation given above for $R(x)$. The result is:

$$R(x) = x + 2\frac{x^2}{2!} + 3^2\frac{x^3}{3!} + \cdots + n^{n-1}\frac{x^n}{n!} + \cdots$$

from which we see that R_n can be simply expressed by the formula

$$R_n = n^{n-1}.$$

Note the extreme difference in complexity between R_n and r_n due to the presence of added structure introduced by labelling (i.e., distinguishing) all the vertices. The solution to the functional equation is the big step here and is far from simple. One way to proceed is as shown in (7).

$$
\begin{aligned}
R(x) = xe^{R(x)} &= x\left[1 + R(x) + \frac{R^2(x)}{2!} + \frac{R^3(x)}{3!} + \cdots\right] \\
&= x\left[1 + x\,e^{R(x)} + x^2\frac{e^{2R(x)}}{2!} + x_3\frac{e^{3R(x)}}{3!} + \cdots\right] \\
&= x\left[1 + x\left(1 + R + \frac{R^2}{2!} + \cdots\right) + x^2\left(1 + 2R + 2^2\frac{R^2}{2!} + \cdots\right) + \cdots\right] \\
&= x + x^2 + \left(\frac{1}{2!} + 1\right)x^3 + \cdots \\
&= x + 2\frac{x^2}{2!} + 3^2\frac{x^3}{3!} + \cdots
\end{aligned}
\tag{7}
$$

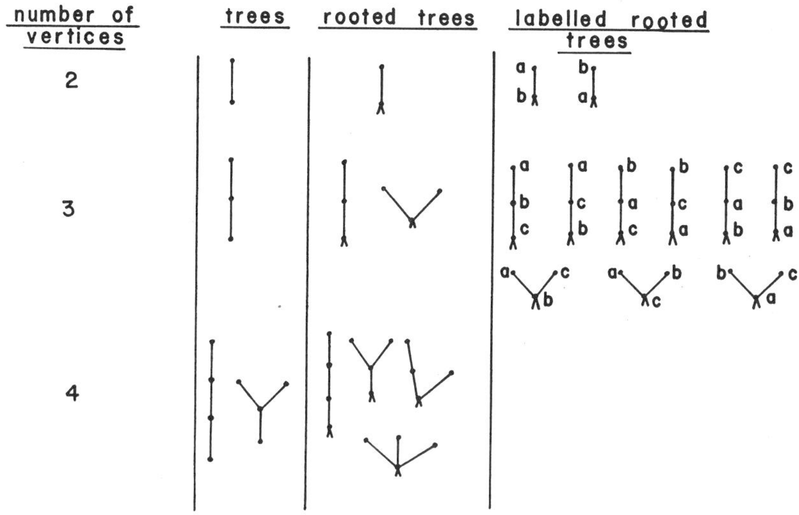

number of vertices	trees	rooted trees	labelled rooted trees

Figure 107

One may induce the general term getting the series stated previously then substitute the series into the equation to verify it.

The number of trees with distinct vertices, T_n, is found at once from R_n since any vertex can be taken as the root (all points having labels) so that one obtains the Cayley formula;

$$n\,T_n = R_n \qquad \text{or} \qquad T_n = n^{n-2}.$$

Further tree enumerations require more advanced combinatorial methods. These and many other linear graph enumerations are based on Polya's Theorem. This powerful theorem utilizes concepts of group theory to specify identification among the objects being enumerated. The statement of the theorem is beyond the scope of this book and we shall omit it. For treatment of Polya's theorem see Riordan [13], DeBruijn [70], or Liu [64]. Some further results in linear graph enumeration will be given, they are obtained with the help of Polya's theorem or equally advanced techniques. Figure 107 and Table 6 show some results for trees (as given by Riordan [13], p. 138.)

Table 6.   Number of Trees t_n and Rooted Trees r_n

n	1	2	3	4	5	6	7	8	9	10	11	12	13	14
t_n	1	1	1	2	3	6	11	23	47	106	235	551	1301	3159
r_n	1	1	2	4	9	20	48	115	286	719	1842	4766	12486	32973

Enumeration of linear graphs with no self loops or multiple lines is indicated by Table 7 (the method is by generating functions derived from Polya's theorem as obtained in Riordan ([13], p. 146).

Table 7.   Linear Graphs with n Vertices and k Edges

k/n	2	3	4	5	6	7	8	9
0	1	1	1	1	1	1	1	1
1	1	1	1	1	1	1	1	1
2		1	2	2	2	2	2	2
3		1	3	4	5	5	5	5
4			2	6	9	10	11	11
5			1	6	15	21	24	25
6			1	6	21	41	56	63

Special graphs of particular interest are generalizations of trees. In fact, what we have called trees are more properly called Cayley trees. A general or Husimi tree is formed by using some fixed finite graph as we used lines in forming a Cayley tree. For example, Figure 108 shows a Husimi tree formed from rectangles. When triangles are used as the generating configuration the resulting trees are called cacti. If C_n = number of cacti with n triangles, we have the generating function (Riordan [13], p. 159)

$$c(x) = 1 + C_1 x + C_2 x^2 + \ldots.$$

This generating function satisfies the functional equation: (Harary and Uhlenbeck 1953)

$$c(x) = \exp\left[h(x) + \frac{h(x^2)}{2} + \cdots \frac{h(x^k)}{k} + \cdots \right]$$

where $h(x) = x[c^2(x) + c(x^2)]/2$. Two cacti with three triangles are shown in Figure 109 the reader may wish to look for others that are not isomorphic to these (i.e., distinct in the basic graph theory definition of distinct).

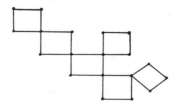

Figure 108.   Husimi tree.

Figure 109. Cacti.

As a final example of enumeration of graphs, we consider connected graphs with a single cycle. In such a graph the single cycle serves as a foundation and trees (we always mean Cayley trees unless we state otherwise) are attached to its vertices. Two examples are given in Figure 110. The enumeration of such objects is based on the length p of the single cycle. In Table 8 we list some results given by Riordan ([13], p. 150):

Table 8. **Number of Connected Graphs on n Vertices with One Cycle of Length p**

p/n	3	4	5	6	7	8	9
3	1	1	3	7	18	44	117
4		1	1	4	9	28	71
5			1	1	4	10	32
6				1	1	5	13
7					1	1	5

Enumeration within a given graph presents rather different problems and are utilized in different ways than the enumerations of linear graphs illustrated above. Some such enumerations utilize methods similar to those mentioned (generating function, Polya's theorem, etc.) but we shall speak of only some simple fixed graph enumerations. These are based on the matrix representation of the graph. Further results of this type are given in Chapter 9.

For a given graph G we may want to know the number of paths of length k from one vertex to another or the number of cycles of length k which contain a given vertex or similar quantities. These can be enumerated as follows.

Figure 110

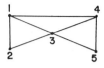

Figure 111

If A is the adjacency matrix of a graph G (directed or undirected) then p_{ij} represents the paths of length k from vertex i to vertex j when p_{ij} is the (i,j) element of $P(k) = A^k$. Note that the diagonal elements p_{ii} represent cycles of length k containing the vertex i. Also note that the paths may be directed or undirected depending on G.

This is a simple result and it is easy to use. The result is well known and was probably originated by Cayley. The reader can see the reason for the result by following one or two matrix multiplications of A by itself. Unfortunately the p_{ij} do not enumerate simple paths or cycles, thus paths may contain cycles and the cycles may be traversed in multiple ways. This is the price we must pay for having such a simple result.

To illustrate the procedure consider the graph in Figure 111, to count the number of cycles and paths we form the adjacency matrix (8) and then $P(k)$ for whatever length k of paths we are interested in.

$$
A = \begin{array}{c c} & \begin{array}{c c c c c} 1 & 2 & 3 & 4 & 5 \end{array} \\ \begin{array}{c} 1 \\ 2 \\ 3 \\ 4 \\ 5 \end{array} & \begin{pmatrix} 0 & 1 & 1 & 1 & 0 \\ 1 & 0 & 1 & 0 & 0 \\ 1 & 1 & 0 & 1 & 1 \\ 1 & 0 & 1 & 0 & 1 \\ 0 & 0 & 1 & 1 & 0 \end{pmatrix} \end{array}, \quad A^2 = \begin{pmatrix} 3 & 1 & 2 & 1 & 2 \\ 1 & 2 & 1 & 2 & 1 \\ 2 & 1 & 4 & 2 & 1 \\ 1 & 2 & 2 & 3 & 1 \\ 2 & 1 & 1 & 1 & 2 \end{pmatrix} = P(2). \quad (8)
$$

Notice the non-simple character of the enumeration, the off diagonal elements do in this case all represent simple paths of length 2; in fact, here, they stand for the number of edges incident at each vertex (this will always be the case when there are no simple cycles of length 2).

Though the method is straightforward it is hard in practice to multiply a matrix by itself to high powers (e.g., to compute A^5 for the above example is unpleasant). It is possible to express A^k in terms of the characteristic values of A and certain simple matrices derived from A by the so-called Lagrange interpolation formula. This result for unequal characteristic values can be stated rather simply, for equal characteristic values the treatment is more involved and we omit it (note this is a topic in linear algebra). Let λ_r denote the rth characteristic value of A, (i.e., a solution of the polynomial $|A - \lambda I| = 0$) then:

$$
A^k = \sum_{r=1}^{n} \lambda_r^k U_r
$$

where the matrices U_r are defined by the Lagrange interpolation formula:

$$U_r = \prod_{\substack{j=1 \\ j \neq r}}^{n} \frac{A - \lambda_j I}{\lambda_r - \lambda_j}.$$

This result is extremely useful in finding A^k once the U_r have been determined (and of course the λ_r); one need not compute all the lower power matrices and in fact no matrix multiplication (other than the original formation of the U_r) is needed. Since all we get are general paths and cycles the enumeration is not as useful as one would like. Enumeration of simple paths and cycles is more involved. Such enumerations will be discussed in Chapter 9.

Another disadvantage of the above method is that the use of the Lagrange interpolation is complicated by having to solve an nth degree polynomial.

Some enumerations on graphs can be carried out directly to yield useful explicit results. For example, we may study the cycles of length 3 in a complete antisymmetric graph. These are directed graphs such that the adjacency matrix elements a_{ij} satisfy: either $a_{ij} = 1$ and $a_{ji} = 0$ or $a_{ij} = 0$ and $a_{ji} = 1$ for $i \neq j$ and $a_{ii} = 0$ for all i. In such a graph there is a directed edge between every pair of vertices (complete) but only one such (antisymmetric). For example consider Figure 112. Let $C_3(n)$ denote the maximum number of cycles of length 3 possible in a complete antisymmetric graph on n vertices. Then

$$C_3(n) = \frac{n^3 - n}{24} \text{ for } n \text{ odd,}$$

$$= \frac{n^3 - 4n}{24} \text{ for } n \text{ even.}$$

If $h_i = \sum_{j=1}^{n} a_{ij}$ then the number of cycles of length 3 is

$$\frac{n(n-1)(2n-1)}{12} - \frac{1}{2} \sum_{i=1}^{n} (h_i)^2.$$

These results are obtained by direct combinatorial reasoning rather than advanced methods but we omit the proofs here (see Berge [6], Chapter 14).

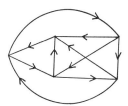

Figure 112

EXERCISES

1. Describe the 4-part partitions of 7. Show that each case leads to a class of sub-graph partitions of K_7 into 4 parts. Enumerate the number of cases in each class.

2. Define a tribracket as an L shaped figure that covers three squares on a chess-board. One square is covered by the apex region and each arm covers one square. Find conditions on m and n such that an $n \times m$ chessboard can be covered by tribrackets. How many coverings are there for the 3×4 board? Give a careful definition of how you distinguish different coverings.

3. In how many different ways can n (regular) paper clips be strung together?

4. Formulate the enumerative generating function for the number of ways of distributing m distinct objects into n distinct cells with at most two objects per cell. Give a detailed illustration of results for the $m = 6$, $n = 4$ case. Study some special cases; for example when objects are divided two per cell among three cells show there are 90 cases.

5. The generating function for the Legendre Polynomials $P_n(x)$ is $(1 - 2xu + u^2)^{-1/2} = \sum_{n=0}^{\infty} P_n(x)u^n$. Show that the derivative $P_n'(1) = \sum_{k=1}^{n} k$, the sum of the positive integer through n.

6. Consider the set of integers 1 through 9. Select five 4-samples of integers by using the last 4 digits of numbers selected at random (e.g., from a phone book). Form the simple incidence matrix for these sets. Does an S.D.R. exist? If S.D.R.'s exist enumerate them using permanents (by Ryser's method). Write five 4-samples for which an S.D.R. does not exist.

7. A simple path is a bipartite graph, find the maximal deficiency for such a graph having n vertices. When n is even a simple cycle is bipartite. Find the maximal deficiency for such a graph having $n = 2k$ vertices.

8. By placing points in a plane, illustrate Lemmas A and B associated with Ramsey's theorem. Study the numbers $f(k, h)$ for small values of k and h; in particular, show that $f(2, 3) = 3$.

9. Illustrate Turán's theorem by drawing the graphs $D(7, k)$ for $k = 2, 3, 5, 6, 7$. Compute $d_k(n)$ for $k \le n$ and $n = 2, 3, 4, 5$. Interpret the graphs $D(n, k)$ as representing families of sets, and thus illustrate Marczewski's theorem.

10. Use the methods of this chapter to enumerate the number of paths of length r in K_n as a function of r and n. Illustrate with some numerical examples, in particular consider $n = 3, 4$, and 5.

11. (Due to the referee). Let n be a positive integer and let the vertices of a graph G consist of all n letter words from a 2 letter alphabet. Two vertices are joined by an edge if the words agree in as many positions as they disagree. Show that G contains the complete 4-graph if and only if n is divisible by 4.

   This exercise is related to the topic of Hadamard matrices. An Hadamard matrix H of order n is an $n \times n$ matrix with elements $+1$ or -1 which satisfies the relation $HH^T = nI$. Such matrices are known to exist for $n = 4t$ when $t = 1, 2, \ldots, 28$. The Hadamard matrix conjecture states that Hadamard matrices of order $4t$ exist for all integral values of t (see M. Hall [71]).

Chapter 7

Random Graphs

RANDOMNESS IN GRAPH THEORY

The concept of randomness can enter into graph theory in a number of ways. It seems to be helpful to categorize these into three major types of considerations. Thus, by the term *random graphs* we shall mean any topic that can reasonably be put into one of the following formulations:

1. Problems of a probabilistic form that utilize the concepts of graph theory in their formulation or solution.
2. Specific graphs for which the vertices or edges are assigned probabilities of occurrence or removal.
3. Graphs that develop their form (structure) as the result of random addition of vertices or edges. Such graphs may be generalized to graphs that develop their form as the result of the random replacement of vertices by new randomly generated parts that become subgraphs. Graphs that develop in these ways may be called evolutionary random graphs.

In this section a problem of type 1 will be discussed to illustrate that class of formulations. Each of the other sections of this chapter will deal with other types of formulations. It is the goal of this chapter to introduce the reader to these major types of random graph considerations. The subject may be seen to be an extensive one and many problems and applications may occur to the reader. The subsequent chapters will also contain some topics that will be recognized as properly belonging within the subject of random graphs.

As an introduction to randomness in graphs we shall discuss the problem of the distribution of crossings (the intersection of edges at points other than vertices) in random complete graphs due to Moon ([72]). It is most convenient to suppose that the graphs are drawn on the surface of a sphere (as we have seen this is equivalent to their being drawn on a plane).

Suppose n vertices, denoted by i, j, etc., are distributed uniformly and independently over the surface of a sphere. Connect i and j by the shortest path between them to form a representation of the complete n-graph (note special arrangements of the points that fail to yield such a representation have probability zero of occurring and are not considered). The number of crossings of edges in such a configuration is denoted by the random variable C_n.

Consider the unit sphere with surface area 4π and angular coordinates α and θ shown in Figure 113. Let

$$F_A(\alpha) = Pr[\text{distance between two points} \leq \alpha],$$

where we shall use Pr for "probability". Since the points are located at random we must find the surface area S on the unit sphere with its boundary lying a distance α from some fixed point. We take the North pole as the fixed point, then:

$$S = \int_0^\alpha 2\pi \sin\theta d\theta = 2\pi[1 - \cos\alpha].$$

Therefore, $F_A(\alpha) = \frac{1}{2}[1 - \cos\alpha]$ and the random variable α has density function $\frac{1}{2}\sin\alpha$. That random variable represents the (random) distance between vertices.

We now define the following set of random variables:

$$x(i, j; k, l) = 1 \text{ if edges } (i, j) \text{ and } (k, l) \text{ intersect}$$
$$= 0 \text{ otherwise.}$$

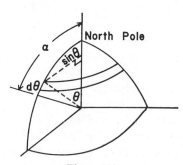

Figure 113

Now $Pr[x(i, j; k, l) = 1] = \frac{1}{2} \int_0^\pi \alpha/2\pi \frac{1}{2} \sin \alpha d\alpha$, since the probability that k and l are on opposite sides of the great circle determined by edge (i, j) is $\frac{1}{2}$, and edge (k, l) intersects that great circle in the segment between i and j with probability $\alpha/2\pi$ (by uniformity), and edge (i, j) has length between α and $d\alpha$ with probability $\frac{1}{2} \sin \alpha d\alpha$. The integral includes all possible cases for α length.

Now we observe that $C_n = \sum x(i, j; k, l)$ where the sum is taken over all pairs of edges. Thus the mean is given by:

$$\mu = E[C_n] = \sum E[x(i, j; k, l)].$$

Now $E[x(i, j; k, l)] = Pr[x(i, j; k, l) = 1] = 1/8\pi \int_0^\pi \alpha \sin \alpha d\alpha = 1/8$ and the number of pairs of edges is $\frac{1}{2}\binom{n}{2}\binom{n-2}{2}$ so that

$$\mu = \frac{1}{16}\binom{n}{2}\binom{n-2}{2}.$$

The variance of C_n is defined to be $\sigma^2 = E[(C_n - \mu)^2]$. To compute the variance we introduce the modified random variables $y(i, j; k, l) = x(i, j; k, l) - \frac{1}{8}$ so that:

$$\sigma^2 = E\{[\sum y(i, j; k, l)]^2\}.$$

The terms in the expansion will yield zero expected value except in cases where they are dependent. To be dependent a pair must have at least two vertex indices in common and may have as many as four. Thus there are four types of dependency situations. To compute σ^2 we must find the expected value of each type and how many cases there are of each type.

Type 1 is of the form $y(i, j; k, l)y(i, j: k, l)$. In this case

$$E\{[x(i, j; k, l) - \tfrac{1}{8}][x(i, j; k, l) - \tfrac{1}{8}]\} = E\{[x(i, j; k, l)]^2\} - \tfrac{1}{64} = \tfrac{7}{64}.$$

There is a term of this type for every possible pair of edges, hence there are $\frac{1}{2}\binom{n}{2}\binom{n-2}{2}$ terms.

Type 2 is of the form $y(i, j; k, l)y(i(j; u, v)$. In this case

$$E\{[x(i, j; k, l) - \tfrac{1}{8}][x(i, j; u, v) - \tfrac{1}{8}]\} = \{P_r[x(i, j; k, l)x(i, j; u, v)] = 1\} - \tfrac{1}{64}.$$

For the required event it is necessary that both (k, l) and (u, v) cross (i, j) and that (i, j) have length in an interval α to $\alpha + d\alpha$. Thus

$$\{P_r[x(i, j; k, l)x(i, j; u, v)] = 1\} = \int_0^\pi \left(\frac{1}{2}\frac{\alpha}{2\pi}\right)^2 \frac{1}{2} \sin \alpha d\alpha = \frac{\pi^2 - 4}{32\pi^2}.$$

Thus $E[\text{Type 2 case}] = (\pi^2 - 8)/64\pi^2$. The number of such cases is the number of pairs of edges times the ways one can select the third edge. There is a factor of two to take care of the similar cases with $y(k, l; i, j)y(u, v; i, j)$ (involving the same selection of edges). Thus there are $\frac{1}{2}\binom{n}{2}\binom{n-2}{2} 2\binom{n-4}{2}$ cases of Type 2.

Type 3 is of the form $y(i, j; k, l)\, y(i, j; k, t)$ In this case

$$E\{[x(i, j; k, l) - \tfrac{1}{8}][x(i, j; k, t) - \tfrac{1}{8}]\} = \frac{\pi^2 - 8}{64\pi^2} \quad \text{as in Type 2.}$$

The number of cases is the number of pairs of edges, $\frac{1}{2}\binom{n}{2}\binom{n-2}{2}$, times the ways of selecting another vertex, $(n-4)$, times the number of ways of forming type 3 situations. The last factor depends on where the twice used vertex falls. It is vertex k in the typical case shown above but it could be i, j, or l just as well so we need a factor of 4. Thus the total number of type 3 cases is $\frac{1}{2}\binom{n}{2}\binom{n-2}{2}(n-4)4$.

Type 4 is of the form $y(i, j; k, l)\, y(i, k; j, l)$. In this case

$$E[\text{type 4 case}] = E[x(i, j; k, l)\, x(i, k; j, l)] - \tfrac{1}{64} = -\tfrac{1}{64}$$

since the two random variables involved can never both be one. The number of Type 4 cases is the number of pairs of edges times 2 since either pair of edges can be the uncrossed pair.

Combining all the results obtained above we find:

$$\sigma^2 = \frac{1}{2^7\pi^2}\binom{n}{2}\binom{n-2}{2}[5\pi^2 + (n-1)(n-4)(\pi^2 - 8)].$$

Thus we have the expected value and the variance of C_n the number of crossings in a uniformly placed complete n-graph on the unit sphere. By studying the higher moments one can show that $(C_n - \mu)/\sigma$ has a distribution function which tends to the distribution of the standard normal random variable $N(0, 1)$.

For $n = 4$, $\mu = \frac{3}{8}$ and one expects very few crossings for this planar graph. Higher values of n yield non-planar graphs; for $n = 5$, μ is close to 2 and continues to increase with n.

It is interesting to speculate on what the effect of changing the distribution of the vertices would be. However, the uniform distribution requires a fairly detailed analysis as indicated above (the involved study of the higher moments has been omitted). It may be expected that detailed analysis for other distributions would be extremely involved; but of course one might proceed by simulation to generate statistical estimates on the distribution of C_n in such cases.

DOUBLE CORRESPONDENCE GRAPHS

In this section we illustrate the kind of random graph problem in which vertices or edges of a particular graph are removed by a random event. The example that is discussed is a special case from the general topic of multiple correspondences.

A number of situations arising in operations research and elsewhere can be represented as correspondences between two sets of objects; one a set of service facilities and the other a set of installations requiring service. In case the service facilities are subject to removal from the correspondence according to some probabilistic law, the study of such arrangements has been called statistics of multiple correspondences by B. O. Koopman [73].

The expected number of installations able to receive service after the random removal of stations can be used as a selection criterion for the original correspondence. As the number of installations and facilities increase this measure also may be expected to increase. It is therefore useful to define an index that is more closely reflective of the basic structural properties of the arrangement. For that purpose an index I is defined as the expected numbers given above divided by the total number of installations.

Using the same probability law of station removal described above, one may define the probability that every installation will be able to receive service after the random removal of service facilities has occurred. This full service probability offers an alternative criterion for the selection of correspondences and gives a useful statistical description of a given correspondence. The full service probability has been discussed for some simple arrangements but seems difficult to deal with in general (Marshall [74]). In this section we will discuss the expected service index I for a simple type of correspondence.

When each service facility is restricted to serve exactly two installations, the correspondence can be represented as a linear graph whose points represent installations and whose lines represent service facilities. Such a graph can conveniently be specified by a general adjacency matrix A with elements a_{ij} equal to the number of lines (service facilities) connecting vertices (installations) i and j. We shall call these double correspondence graphs. Even within this special type of correspondence there are many possibilities. Let n denote the number of vertices (installations) and m denote the number of edges (stations). We restrict our attention to those cases in which $m = 2n$. Even within this highly restricted class there are a number of cases possible and the graph having the best properties is difficult to select (as a function of n). Of course, for any fixed n (not too large) one can study the class of graphs by monte carlo simulation. However we are interested in general results as functions of n. In addition to the theoretical interest of such results they tell us more about the structure of the graphs than simulations do (in general).

The random nature of the problems discussed here can be introduced as follows. Suppose an arrangement is to be made at time t_0 such that, assuming maximum use of service facilities, a "best" result for installation service can be obtained at time t_2 if a random event occurs at t_1 that removes some number of stations, where $t_0 < t_1 < t_2$. A station can service one and only one of its associated installations at t_2. Hence at t_2 it will be necessary to assign the remaining stations in such a way as to give maximum service to installations subject to the restriction that the association between station and installation has previously been established. The probability of any individual station being removed by the random event is a constant p independent of every other station. It is assumed that maximum use is always made of the remaining stations.

Because of the difficulty of obtaining general results for this problem, even for the restricted class of double correspondence graphs with $m = 2n$, bounds are developed for the index I(A). Here I(A) = $E(A)/n$ is the expected number of vertices to which edges can be assigned (each edge being assigned to exactly one vertex) in a graph A after random removal of edges, divided by the number of vertices. Lower bounds are obtained by evaluating $I(A)$ for special graphs A. Upper bounds depend on general considerations that do not rely on specific graph properties. Therefore the upper bound problem is not of direct interest to us here and we shall only discuss some lower bounds. Upper bounds are given by Finch [75].

First the simplest possible type of arrangement will be discussed followed by a somewhat more interesting case and finally a more complicated arrangement will be discussed. All of these examples are double correspondences for which $m = 2n$. We shall denote these as the class D^*. They have been studied by Koopman [73], Finch [75], and Marshall [76], respectively. Koopman has also studied the second case by implication though it is not discussed directly in his paper.

Suppose n is even, say $n = 2r$, then a simple arrangement in D^* is a disconnected double correspondence graph with r components where each component has the form shown as A in Figure 114. It is clear that double correspondence graphs are not simple (in general); they have no self loops but

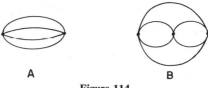

A   B

Figure 114

many have multiple edges. Let us denote this r component graph by G_1. Since each component of G_1 is independent the value $I(G_1)$ will not depend on n but it will give a (crude) lower bound on $I(A)$ for A in D^* since we can achieve $I(G_1)$ by using arrangement G_1.

Let x denote the number of vertices that cannot be assigned an edge.

$$p(x = 2) = p^4 \quad \text{all edges lost}$$

$$p(x = 1) = \binom{4}{3} p^3(1 - p) \quad \text{only one edge left.}$$

$$p(x = 0) = \binom{4}{2} p^2(1 - p)^2 + \binom{4}{1} p(1 - p)^3 + (1 - p)^4$$

where the terms on the right correspond to 2 edges left, 3 edges left, and 4 edges left. Thus: $E(x) = 2 p^4 + \binom{4}{3}p^3(1 - p) = 2 p^3(2 - p)$, and $E(G_1) = n/2$ $E(x) = n p^3(2 - p)$. The index

$$I(G_1) = \frac{\text{expected number of vertices to receive edges}}{n} = \frac{n - E(G_1)}{n}.$$

$$I(G_1) = 1 - p^3(2 - p)$$

For $p = \frac{1}{2}$, $I(G_1) = 0.813$ and for 100 vertices the expected number to which edges can be assigned is 81.3.

A similar analysis for a graph G_2 based on components of the kind shown as B in Figure 114 yields:

$$I(G_2) = 1 - p^4(2p^2 - 6p + 5)$$

which is also independent of n (as it should be). For $p = \frac{1}{2}$ this graph yields $I(G_2) = 0.844$. It is not possible to construct a graph like G_2 with 100 vertices since each component requires 3 vertices. However a G_2 on 99 vertices is very close so far as these crude bound graphs are concerned.

Improved results can be obtained by considering connected graphs belonging to D^*. The simplest such graph is called a regular arrangement. An example for $n=6$ vertices is shown as A in Figure 115. Number the vertices

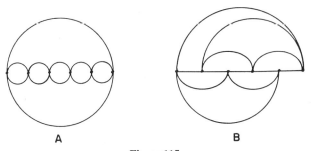

A   B

Figure 115

of a double correspondence graph with values 1 to n. A regular arrangement may be defined in terms of its adjacency matrix $E = (e_{ij})$ where

$$e_{ij} = 2 \quad \text{where } j = i - 1 \text{ or } i + 1 \text{ (mod } n)$$
$$= 0 \quad \text{otherwise,} \quad \text{for } i = 1, \ldots, n.$$

We denote a regular arrangement by G_3. The index for regular arrangement of size n (i.e., n vertices representing—installation) is:

$$I(G_3) = 1 - p^4 - \frac{p^4 q(1 - q^{n-2})}{1 - q} + p^2 q^{n-1} \quad \text{where} \quad q = 2p(1 - p)$$

For $n = 100$ and $p = \frac{1}{2}$ this gives the lower bound of $I(G_3) = 0.875$ which is a considerable improvement over the crude results obtained using G_1 or G_2. This result is established as follows:

Let B_i denote the (random) event that vertex i loses all its incident edges. Let A_1 denote the number of vertices that lose all their edges. Let $S_i = 1$ if B_i occurs and $S_i = 0$ otherwise. Then

$$A_1 = \sum_{i=1}^{n} S_i \quad \text{and} \quad E(A_1) = n \, P_r(B_i) = np^4.$$

In general, if A_i is the number of sets of i vertices that share $i - 1$ edges (number of trees on i vertices), then $E(A_i) = np^4 q^{i-1}$ for $i = 1, 2, \ldots, n - 1$ where $q = 2p(1 - p)$. In each such case one vertex is without an assigned edge when the (maximal) assignment of edges is made.

The case $i = n$ is special since only 2 edges need to be removed in addition to the occurrence of $n - 1$ special pairs. Therefore $E(A_n) = np^2 q^{n-1}$ since the n vertex tree can begin at any vertex.

Thus

$$E(G_3) = \sum_{i=1}^{n} E(A_i) = n \sum_{i=1}^{n-1} p^4 q^{i-1} + np^2 q^{n-1}.$$

Therefore

$$I(G_3) = 1 - \frac{E(G_3)}{n} = 1 - p^4 \sum_{i=1}^{n-1} q^{i-1} - p^2 q^{n-1}$$

which gives the result for $I(G_3)$.

The graph G_3 was fairly simple to study because each set of vertices is the same as any other set. One can split the graph into pieces as was done above to obtain the total result. Once edges spread beyond the neighboring vertices it becomes much more difficult to obtain results. The simplest case of that kind is a one-spread arrangement such as the one illustrated on 6 vertices at B in Figure 115. Such a graph will be denoted G_4.

We shall describe a method for studying G_4 and give some results. Further details may be found in Marshall [76].

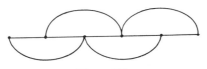

Figure 116

A one-spread arrangement for n installations is a special double correspondence defined by the adjacency matrix with elements:

$$e_{ij} = 1 \quad \text{when } j = i - 2, i - 1, i + 2 \ (\text{mod } n),$$
$$e_{ij} = 0 \quad \text{otherwise} \quad i = 1, \ldots, n.$$

An incomplete one-spread arrangement on n vertices is formed by removing the stations shared by installations n and 1, installations $n - 1$ and 1 and installations n and 2. The incomplete one-spread arrangement of size six is represented by the double correspondence graph shown in Figure 116.

Let D_n denote the number of installations (vertices) unable to have edges assigned after a random removal of stations (edges) where the arrangement is of the one spread type G_4. Let $E(D_n)$ and $I(D_n)$ express the expected value and index for the incomplete one spread case. These are used to get $I(G_4)$ and are also useful as bounding values themselves even though they are 3 edges short of a fully defined arrangement for each n. Certainly for large n the absence of the 3 edges does not have a significant effect.

Loss of Service Probability. Let P_i denote the probability that vertex i is unable to receive service after the random event, subject to the assignment rule (discussed below). The index i takes integral values 1, $\ldots$, n for an arrangement of size n.

In addition to P_i the following quantities are used, where in each case the assignment rule is assumed to operate:

Q_i = the probability that vertex i is unable to receive service from its left edges.

Q_{i0} = the probability that vertex i is unable to receive service from its left edges if the edge jumping vertex i (ith jump edge) is not present.

Q_{i1} is similar to Q_{i0} with the condition that ith jump edge is present.

R_i = the probability that vertex i is unable to receive service from its left edges and the ith jump edge (given as present) is needed for service at the left.

The assignment rule required to give unambiguous formulation to the quantities defined above is as follows: In any chain where one vertex must lose service the vertex which will lose service shall be at the extreme right of the chain.

Let p denote the probability of removal of an edge (station) in the random event. The state of each edge is assumed to be independent of the state of all other edges.

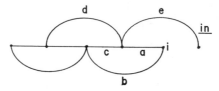

Figure 117

General recurrences for the quantities of interest may now be developed. Special initial values will be collected at the end.

Clearly

$$P_i = p^2 \, Q_i \qquad i = 2, \ldots, n-2,$$

$$P_{n-1} = pQ_{n-1}, \tag{1}$$

$$P_n = Q_n = Q_{n0},$$
$$Q_i = pQ_{i0} + (1-p)Q_{i1} \qquad i = 2, \ldots, n-1, \tag{2}$$

In case the ith jump edge (edge e) is present the situation is as shown in Figure 117 which, recalling the assignment rule, leads to relations (3), and (4).

$$\underset{ab}{Q_{i1}} = p^2 + \underset{a}{p}(1-\underset{b}{p})\,\underset{c}{pQ_{i-2}} \qquad i = 3, \ldots, n-1, \tag{3}$$

$$R_i = \underset{ab}{p^2} \, Q_{i-1,0} + \underset{a}{p}(1-\underset{b}{p})\,R_{i-1} \; (i=3, \ldots, n-1). \tag{4}$$

The cases leading to various terms of the relations are indicated by the placement of letters, representing edges of the diagram, below a p when that edge is removed and a $(1-p)$ when that edge is not removed by the random event.

When the jump edge e is not present the situation is as shown in Figure 118 from which relation (5) is obtained.

$$Q_{i0} = \underset{ab}{p^2} + \underset{b}{p}(1-\underset{a}{p})Q_{i-1,0} + \underset{a}{p}(1-\underset{b}{P})\left\{ \underset{d}{p}Q_{i-2,0} \right.$$

$$\left. + (1-p)\left[\underset{c}{p}Q_{i-2,0} + (1-p) \underset{c}{R_{i-2}} \right] \right\} \tag{5}$$

$$+ (1-p)^2 \underset{ab}{p} \left[\underset{c}{p}Q_{i-2,0} + (1-p)R_{i-2} \right] \qquad (i = 3, \ldots, n-1).$$

The term indicated by a vertical arrow in relation (5) and corresponding to edges a and c out and edges b and d in, arises as follows. Edge b is to be needed at vertex $i-2$ in every case to be considered. A factor representing the most general situation leading to the assignment of edge b to vertex $i-2$

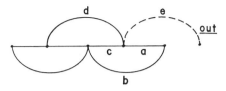

Figure 118

is therefore required. Because of the special conditions imposed on some edges the quantity Q_{i-2} cannot be used as the required factor.

There are two types of events that must be considered. In one type the assignment of edge d to vertex $i - 3$ will allow an arrangement to be made (subject to the assignment rule) in which service is supplied to vertex $i - 2$ from the left, whereas no such arrangement would be possible without this assignment of edge d. In the other type such an assignment of edge d will not produce any such arrangement giving service to vertex $i - 2$ from the left.

For the first type of events, since vertex $i - 1$ is isolated except for edge d, it is consistent with the general assignment rule to assign edge d to vertex $i - 1$ and overcome the resulting loss of service at vertex $i-2$ by assigning edge b to vertex $i - 2$. For the second type of events the presence of edge d does not affect the situation at vertex $i - 2$. Therefore in either case the events contributing to loss of service from the left at vertex $i - 2$ are contained in the factor $Q_{i-2,0}$.

Direct evaluation leads to the following initial values for the quantities introduced above. It is consistent with the recurrence relations to define R_1, Q_{11} and Q_{10} to be unity. Then,

$$P_1 = p^2, Q_1 = 1, Q_{11} = Q_{10} = 1, R_1 = 1 \qquad (i=1).$$

$$P_2 = p^2[p + p(1 - p)], Q_2 = p + p(1 - p),$$
$$Q_{20} = 1, Q_{21} = p, R_2 = p \qquad (i=2).$$

$$Q_{30} = p^2 + 2p(1 - p) + p(1 - p)^2,$$
$$Q_{31} = p^2 + p^2(1 - p), R_3 = p^2 + p^2(1 - p) \qquad (i=3).$$

The equations and values developed above are combined to form appropriate difference equations which yield Q_{i0} and Q_{i1}.

The resulting solutions are of the form

$$Q_{i0} = c_1 m_1{}^i + c_2 m_2{}^i + c_3 m_3{}^i + c, \qquad (6)$$

$$Q_{i1} = k_1 r_1{}^i + k_2 r_2{}^i + b_0 + b_1 m_1{}^i + b_2 m_2{}^i + b_3 m_3{}^i \qquad (7)$$

where c is a particular solution to the difference equation so that

$$[1 - 2p(1 - p) - p^2(1 - p)(2 - p) + p^3(1 - p)^2]c = p^2 - p^3(1 - p). \qquad (8)$$

The value b_0 is found to satisfy the condition

$$[1 - p^2(1 - p)^2]b_0 = p^2 + p^3(1 - p)c. \qquad (9)$$

The constants c_1, c_2, c_3, k_1 and k_2 are determined from initial values of Q_{i0} and Q_{i1}; $m_1, m_2,$ and m_3 are functions of p and b_1, b_2, b_3 are determined as part of the solution procedure for Q_{i1}.

For our purposes the values c and b_0 and the general form of the solution given above are all that is required. The detailed solution is straightforward.

From the above solutions for Q_{i0} and Q_{i1} the quantity Q_i is obtained, which in turn gives P_i at once. The general solutions lead to P_i except in the special case $i = n$; for this value we have

$$Q_n = Q_{n0}$$
$$P_n = Q_n.$$

Let e_i be a random variable associated with the event "vertex i does not receive service" (subject to the assignment rule); assign the value unity to e_i if this event occurs and the value zero otherwise. The value assumed by the random variable e, defined by

$$e = \sum_{i=1}^{n} e_i$$

is the number of vertices that fail to receive edge assignment. Then

$$E(D_n) = E(e) = \sum_{i=1}^{n} E(e_i),$$

where $E(e_i) = P_i + 0(1 - P_i) = P_i$ so that the required value of $E(D_n)$ is given by

$$E(D_n) = \sum_{i=1}^{n} P_i.$$

This may be written in terms of Q_{i0} and Q_{i1} as follows:

$$E(D_n) = S + p^2 Q_{n-1,0} + p(1 - p)Q_{n-1,1} + Q_{n0}$$

where $S = p^3 \sum_{i=1}^{n-2} Q_{i0} + p^2(1 - p) \sum_{i=1}^{n-2} Q_{i1}$.

Insertion of the known values of Q_{i0} and Q_{i1} as functions of i and summing the resulting geometric series gives

$$S = (n - 2)[p^3 c + p^2(1 - p)b_0] + f(p),$$

where $f(p)$ depends on the various parameters occurring in the detailed solutions for Q_{i0} and Q_{i1} and is a function of the probability p. The function $f(p)$ also contains n as an exponent of quantities which are positive and less than unity; it does not depend on n in any other way. Thus

$$\lim_{n \to \infty} \frac{f(p)}{n} = 0.$$

The index of an incomplete one-spread arrangement of size n is given by

$$I(D_n) = 1 - \frac{E(D_n)}{n}.$$

Limiting values of the index. The limiting value of the index, denoted by $I(D_\infty)$, defined by $I(D_\infty) \equiv \lim_{n \to \infty} I(D_n)$ may be obtained as follows:

$$I(D_\infty) = 1 - [p^3 c + p^2(1 - p)b_0].$$

Substituting the value of b_0 into this expression and simplifying, results in

$$I(D_\infty) = 1 - \frac{p^4(1 - p) + p^3 c}{1 - p^2(1 - p)^2},$$

where c is given by (8).

This expression gives a lower bound on the index of arrangements from class D^*. For $p = \frac{1}{2}$ we find $I(D_\infty) = 0.894$ which is a significant improvement over the previous result for $n = 100$ (i.e., large) in $I(G_3) = 0.875$ for the regular arrangement. Finch has shown that for $p = \frac{1}{2}$ an upper bound in the class D^* is 0.916 so there is reason to suppose that a further spreading of edges would yield an arrangement with a higher index. Of course $I(G_4)$ will be a little larger than the incomplete value $I(D_\infty)$ but the difference is not expected to be significant. Detailed analysis such as the above is extremely involved for more complex arrangements. Therefore one may turn to simulation studies. The arrangement's random development is simulated by a computer program and many runs are used to estimate the various quantities of interest such as the index or full service probability for a given arrangement.

A major part of such a simulation is forming the optimal arrangement after the random removal of edges. We must assign edges to vertices in such a way that as many vertices as possible have an edge assigned to them. The graph remaining after the random removal of edges is represented by its incidence matrix. An algorithm is utilized to operate on this incidence matrix and form an optimal assignment, it yields the maximum number of vertices to which edges can be assigned. Such an algorithm could be made to yield a specific assignment yielding that optimum but we are not interested in such specific assignments. A flow chart for such a program is given in Figure 119.

As an example of optimal assignment the arrangement with incidence matrix (10) as final matrix gives a count of 1 service loss which is correct.

$$\begin{pmatrix} 1 & 0 & 0 & 0 & 0 \\ 1 & 1 & 0 & 0 & 1 \\ 0 & 1 & 1 & 0 & 0 \\ 0 & 0 & 0 & 1 & 0 \\ 0 & 0 & 0 & 1 & 0 \\ 0 & 0 & 1 & 0 & 1 \end{pmatrix} \xrightarrow{\text{yields}} \begin{pmatrix} 1 & 0 & 1 \\ 1 & 1 & 0 \\ 0 & 1 & 1 \end{pmatrix} \qquad (10)$$

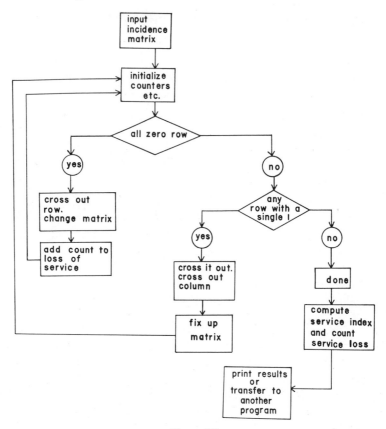

Figure 119

A program of this type can be put together with a random column selector to yield a monte carlo simulation scheme for studying multiple correspondences. The random selector operates on the original incidence matrix, removing columns with probability p (formed from a random number generator). This part of the program yields a final incidence matrix which is introduced into the optimal assignment program described above. The service index and number of possible service assignments are obtained from this program. The complete operation is carried out many times to produce estimates of service index and various probabilities of service.

We shall terminate our discussion of this type of random graph with the above example. It should be observed that the improvement in index value depends on a balance between the value of having many edges packed into a small set of vertices so that the set is less likely to lose service and the value of having edges spread out so that a local excess of edges can be used to supply service to more distant vertices where a great loss of edges has (at random) occurred.

EXAMPLES OF RANDOM GRAPHS

One formulation of the random graph process is that the edges are formed by selecting a pair of vertices from the n vertices (we take the view of forming a graph but one could similarly remove edges from the complete n-graph); this can be done at each step in $\binom{n}{2}$ ways and hence at each step a given edge has probability $1/\binom{n}{2}$ of being introduced into the graph structure. Clearly self loops are not allowed but multiple edges are. Reference [77] gives some results for this model which we now summarize.

We start with n vertices and after m steps have m edges, the resulting graph may have various components. The number of components is denoted by r. For convenience we set $N = \binom{n}{2}$. At any trial one of N edges is selected so the total number of possible graphs that can be formed in m trials is N^m. Let T_{nmr} be the number of graphs with m edges and r components and $C_{nm} = T_{nm1}$ be the number of connected graphs. The respective probabilities are T_{nmr}/N^m and C_{nm}/N^m. Enumerating generating functions are defined as follows

$$T(x, y, z) = \sum_n \sum_m \sum_r T_{nmr} \frac{x^n y^m}{n!\,m!} z^r = \sum_n \sum_m T_{nm}(z) \frac{x^n y^m}{n!\,m!}$$

$$= \sum_n T_n(y, z) \frac{x^n}{n!}, \quad n, m = 0, 1, \ldots; r = 1, 2, \ldots, n.$$

$$C(x, y) = \sum_n \sum_m C_{nm} \frac{x^n y^m}{n!\,m!} = \sum_n C_n(y) \frac{x^n}{n!}.$$

These enumerators are related by $T(x, y, z) = \exp\left[zC(x, y)\right]$. $T_{nm}(1)$ is the sum over r of all terms T_{nmr} and hence equals all possible graphs; therefore $T_{nm}(1) = N^m$. Thus $T(x, y, 1)$ is a known function of x and y. From it $C(x, y)$ is given by the relation $T(x, y, 1) = \exp C(x, y)$. From $C(x, y)$ one can determine $T(x, y, z)$ and from it obtain formulas for T_{nmr} and C_{nm}. These formulas are typical combinatorial formulas and are hard to evaluate numerically; nevertheless they represent complete mathematical solutions to the problem of determining T_{nmr} and C_{nm} from which the probabilities for various component graphs can be given.

Let

$$\tau_{nk}(m) = \sum_{(k)} \frac{n!\left[k_1\binom{1}{2} + k_2\binom{2}{2} + \cdots + k_n\binom{n}{2}\right]^m}{k_1! \cdots k_n!\, 1!^{k_1}\, 2!^{k_2} \cdots n!^{k_n}}$$

with summation, indicated by (k) over all k-part partitions of n, so that $k_1 + 2k_2 + \cdots + nk_n = n$ and $k_1 + k_2 + \cdots + k_n = k$. Also denote by $s(k, r)$ the Stirling number of the first kind defined by the generating function:

$$z(z - 1) \cdots (z - k + 1) = \sum_r s(k, r) z^r.$$

Then the desired quantities are given by the following expressions:

$$T_{nmr} = \sum_{k=r}^{n} \tau_{nk}(m)s(k, r)$$

$$C_{nm} = \sum_{k=1}^{n} (-1)^{k-1}(k-1)!\, \tau_{nk}(m)$$

Special cases are:

$$T_{n,n-1,1} = (n-1)!\, n^{n-2}$$

A connected graph on n vertices with $n-1$ edges is a tree (spanning the vertices) there are n^{n-2} such trees. In the random construction process each edge is distinct hence any given tree can be formed in $(n-1)!$ ways; this being the different orders of edge selection. This gives the above result.

$$T_{n,n,1} = \tfrac{1}{2} n!\,(n-1)! \left[1 + n + \frac{n^2}{2!} + \cdots + \frac{n^{n-3}}{(n-3)!} + \frac{n^{n-2}}{(n-2)!} \right].$$

This result is related to the enumeration of connected graphs with exactly one cycle of specified length. For length 2 the number of such is $n(n-1)n^{n-3}$, for length of cycle $k > 2$ the number of such graphs is $n(n-1)\ldots(n-k+1)n^{n-k-2}/2$.

The expected number of components can also be computed. Exact results are involved but certain approximations can be given that are quite good. For example if we let M_{nm} denote the expected number of components for n vertices and m edge selections an approximation for M_{nn} can be given for various ranges of n as follows:

$$M_{nn} \sim 1 + \frac{n}{e^2}, \qquad\qquad n < e^4$$

$$M_{nn} \sim 1 + \frac{n}{e^2} + \frac{n}{e^4}, \qquad\qquad e^4 < n < e^6$$

$$M_{nn} \sim 1 + \frac{n}{e^2} + \frac{n}{e^4} + \frac{2n}{e^6}, \qquad e^6 < n < e^8$$

and so forth.

Considerations of random graphs often fall within the subject matter of discrete probability as demonstrated by the above example. Therefore enumerative combinatorial results play an important role in the study of random graphs. In [78], Gilbert has developed several combinatorial formulas of considerable intrinsic interest and he has utilized some of his results in considering a class of random graphs in [79]. We shall present some of Gilbert's results from those references as further examples of random graphs and to provide the reader with useful combinatorial theorems which may be employed in a variety of additional investigations.

Gilbert's combinatorial results are expressed in terms of generating functions which we have previously observed represent one type of enumerative expression. Two main results from [78] are the following:

Theorem A. The number of connected graphs with n distinct (labeled) vertices and r indistinguishable (unlabelled) edges is $n!$ times the coefficient of $x^n y^r$ in the series expansion of the generating function

$$F(x, y) = \log \left(1 + \sum_{i=1}^{\infty} \frac{(1 + \alpha y)^{\alpha \beta(i)} x^i}{i!} \right),$$

where

$$\alpha = \begin{cases} -1 \text{ if multiple edges are allowed,} \\ 1 \text{ otherwise,} \end{cases}$$

and

$$\beta(i) = \begin{cases} \binom{i}{2} & \text{for undirected graphs without self-loops,} \\ \binom{i+1}{2} & \text{for undirected graphs with self-loops,} \\ i(i-1) & \text{for digraphs without self-loops,} \\ i^2 & \text{for diagraphs with self-loops.} \end{cases}$$

A more specialized result deals with graphs in which every cycle is of even length. This includes trees where the cycles are of zero (and hence even) length. We have seen that such graphs may be called bichromatic by a characterization theorem of König given in Chapter 5. Thus Gilbert's result may be stated as follows:

Theorem B. The number of bichromatic graphs with r edges labeled $1, 2, \ldots,$ r and n unlabeled vertices is $r!$ times the coefficient of $x^r y^n$ in the series expansion of the generating function:

$$G(x, y) = \frac{1}{2} \left\{ y^2 (e^x - 1) + \log \left(1 + \sum_{i=1}^{\infty} T_i(y) \frac{x^i}{i!} \right) \right\}$$

where $T_i(y) = (\sum_k S(r, k) y^k)^2$ with $S(r, k)$ a Stirling number of the second kind (discussed in Chapter 6).

In Theorem A the enumeration distinguishes graphs that are isomorphic but have the vertices labeled differently. By setting $y = 1$ in $F(x, y)$ we obtain the total number of graphs allowing any number of edges. Gilbert gives the values:

n	1	2	3	4	5
$F(n, 1)$	1	1	4	38	728

For $n = 4$ the 38 graphs correspond to 6 classes of (unlabelled) isomorphic graphs; within each class the number of isomorphic graphs is

Class	C_1	C_2	C_3	C_4	C_5	C_6
Number of graphs	4	12	3	12	6	1

the graphs of C_1 are stars, those of C_2 are paths of length 4, C_3 are cycles of length 4, C_4 a cycle of length 3 with one pendant edge, C_5 has a cycle of length 4 and 2 cycles of length 3, and C_6 is the complete 4 graph. The reader may wish to sketch these graphs and demonstrate the above values.

The type of random graph treated by Gilbert in [79] is based on n vertices which are labelled $1, \ldots, n$ and hence are distinct. Multiple edges are not considered so that one selects or does not select each of the $\binom{n}{2}$ possible edges. Thus the sample space under consideration consists of $2^{\binom{n}{2}}$ possible graphs. Each edge (uniquely specified by its labelled terminal vertices) is selected with probability p independently of all other edge selections. Gilbert studies the probability P_n that the random graph generated by this process is connected and the probability R_n that two designated vertices (v_r and v_s) are connected. In addition to the rather detailed specific results that are obtained for these probabilities some asymptotic values are also given in [79].

Let $C_{n,r}$ denote the number of connected graphs with n labelled vertices and r edges (unlabelled). In the probability model described above any such graph will occur with probability equal to $p^r\, q^{\binom{n}{2}-r}$ where $q = 1 - p$. Thus the probability that a connected graph with r edges will occur is $C_{n,r}\, p^r\, q^{\binom{n}{2}-r}$ where $n - 1 \le r \le \binom{n}{2}$ so that

$$P_n = \sum_r C_{n,r}\, p^r\, q^{\binom{n}{2}-r}.$$

Since multiple edges and self loops are not allowed in the class of graphs under consideration the generating function $F(x, y)$ for the numbers $C_{n,r}$ is given by setting $\alpha = 1$ and $\beta(i) = \binom{i}{2}$ in Theorem A above. Thus one has:

$$\sum_{n,r} C_{n,r} \frac{x^n y^r}{n!} = \log\left(1 + \sum_{i=1}^{\infty} \frac{x^i (1 + y)^{i(i-1)/2}}{i!}\right).$$

A generating function for P_n is developed as follows:

$$P(x) = \sum_{n=1}^{\infty} P_n \frac{x^n q^{-\binom{n}{2}}}{n!} = \sum_n \sum_r C_{n,r} \frac{x^n}{n!}\left(\frac{p}{q}\right)^r$$

so that by letting $y = \dfrac{p}{q}$ in $F(x, y)$ we obtain $P(x)$ directly since $p + q = 1$.

$$P(x) = \log\left(1 + \sum_{i=1}^{\infty} \frac{x^i\, q^{-\binom{i}{2}}}{i!}\right).$$

This is a formal generating series whose terms yield the values P_n, it does not converge as a function of x in the usual sense. In addition it is difficult to use the above expression to compute the P_n. Gilbert gives the following recurrence relation,

$$1 + P_n = \sum_{i=1}^{n-1} \binom{n-1}{i-1} P_i q^{i(n-i)}.$$

Some initial values of P_n are

$$P_1 = 1$$
$$P_2 = 1 - q$$
$$P_3 = 1 - 3q^2 + 2q^3$$
$$P_4 = 1 - 4q^3 - 3q^4 + 12q^5 - 6q^6.$$

It is interesting to develop the result for P_4 directly. The reasoning required is contained in the following table.

Number of edges selected	6	5	4	3
Probability of connected graph	p^6	$6p^5q$	$15p^4q^2$	$16p^3q^3$

Thus $P_4 = p^6 + 6p^5q + 15p^4q^2 + 16p^3q^3$. If one sets $p = 1 - q$ in this expression one obtains by direct algebraic calculation the expression for P_4 given above (as a function of q only).

The quantity R_n can be expressed in terms of P_n and [79] gives the formula:

$$R_n = 1 - \sum_{i=1}^{n-1} \binom{n-2}{i-1} P_i q^{i(n-i)}$$

where the ith term in the sum gives the probability that designated vertex v_r is connected to exactly i of the non-specified vertices (not v_r or v_s). The sum gives the probability that v_r and v_s are not connected so that R_n results as the probability of the complement of that event.

Comparison of the expressions for P_n and R_n yields

$$R_n - P_n = \sum_{i=1}^{n-1} \left[\binom{n-1}{i-1} - \binom{n-2}{i-1} \right] P_i q^{i(n-1)}$$

$$R_n - P_n = \sum_{i=1}^{n-1} \frac{(n-2)!}{(i-2)!\,(n-i)!} P_i q^{i(n-1)}$$

which shows that $R_n > P_n$ for all $n > 2$ and that $R_2 = P_2$ both results agreeing very well with intuitive reasoning. As a numerical example for $q = 0.5$ Gilbert gives $P_5 = 0.71094$ and $R_5 = 0.85353$. He includes a table of illustrative values which are also useful as checks against his asymptotic results. The asymptotic formulas for large n given below were found to be very good (within 3 percent) for $q \leq 0.3$ when $n \geq 6$. As q becomes larger the formulas apply only for larger values of n as one would expect. Gilbert's asymptotic results are:

$$P_n = 1 - n\,q^{n-1} + 0\,(n^2\,q^{3n/2}),$$
$$R_n = 1 - 2\,q^{n-1} + 0\,(n\,q^{3n/2}).$$

The asymptotic expressions given above are extremely simple and provide some idea of the values that are assumed by P_n and R_n for large n. For small values of n those quantities may be computed from the recurrence relations given so that one can obtain a wide range of values by utilizing both types of expressions. When the recurrence expressions yield values that are close to the asymptotic values one is assured of the accuracy of the latter expressions. This fact can serve as a useful guide for numerical calculations.

We have discussed two types of random graphs. In [77] Austin, Fagen, Penney, and Riordan utilized a random structure that selected one of the $\binom{n}{2}$ possible vertex pairs in an n-graph at each of m trials. This allowed multiple edges in the resulting random graph. In [79] Gilbert selected each edge at most once with a given probability, independently of all other edge selections. These two types of random graphs are most directly contrasted by observing that in the first type edges are selected with replacement whereas in the second type the selection is without replacement but with fixed probability of selection.

Erdös and Renyi have studied a rather different type of random graph in [80]. To formulate their graph structure we let $G_{n,N}$ be a random graph on n vertices with N edges. Thus the number of edges is fixed and the random character derives from the possibility of selecting one of the $\binom{\binom{n}{2}}{N}$ sets of N edges. Each such selection is independent and equally likely. The term isolated vertex is used to denote a vertex that has no incident edges in the random graph. A random graph without isolated vertices is said to be completely connected if it is connected.

The random graphs studied by Erdös and Renyi are related to Gilbert's combinatorial work on $C_{n,N}$ the number of connected graphs with n vertices and N edges which has been discussed above. In terms of those combinatorial numbers the probability that $G_{n,N}$ is completely connected is equal to $C_{n,N} / \binom{\binom{n}{2}}{N}$. As we have seen the quantities $C_{n,N}$ are given in terms of generating functions and recursion relations rather than simple formulas so that the study of the random graphs $G_{n,N}$ can not utilize them.

In [80] several questions are discussed for the type of random graph discussed with particular attention given to asymptotic results. The material presented utilizes the quantity N_c which is defined as follows. For any arbitrary real number c

$$N_c = [\tfrac{1}{2}n \log n + cn]$$

where the square brackets denote the greatest integer function (integral part of the quantity enclosed).

The theorems depend on a key lemma which is of considerable interest in itself. A random graph G_{n,N_c} is said to be of type A if it has a connected subgraph on $n - k$ vertices and k isolated vertices, for $k = 0, 1, \ldots$. Any other form of random graph is said to be of type $\bar{A}$.

Lemma. Let $P_0\,(\bar{A}, n, N_c)$ be the probability that G_{n,N_c} is of type $\bar{A}$. Then $\lim_{n \to \infty} P_0(\bar{A}, n, N_c) = 0$.

This lemma states that for large n most random graphs G_{n,N_c} are of type A. That is they have exactly one connected component.

Let $P_k(n, N_c)$ denote the probability that the maximum set of connected vertices of G_{n,N_c} contains $n - k$ vertices so that $k = 0$ corresponds to a completely connected graph as in the lemma. Several results on the random graphs G_{n,N_c} are contained in the following theorems.

Theorem.

$$\operatorname*{Lim}_{n \to \infty} P_0(n, N_c) = \exp\left(- e^{-2c}\right).$$

Proof. Let $N_0(n, N_c)$ denote the number of completely connected graphs of type G_{n,N_c}. Let $\bar{N}_0(n, N_c)$ denote the number of type G_{n,N_c} graphs with no isolated vertices. We may now apply the principle of inclusion and exclusion.

$\bar{N}_0(n, N_c) =$ Total number of type G_{n,N_c} graphs $-$ all graphs with (at least) one isolated vertex $+$ all graphs with (at least) two isolated vertices— and so forth. This may be written

$$\bar{N}_0(n, N_c) = \sum_{i=0}^{n} (-1)^i \binom{n}{i} \binom{\binom{n-i}{2}}{N_c}.$$

Consider a typical term of this sum. For any i we have

$$A_i = \frac{\binom{\binom{n-i}{2}}{N_c} \binom{n}{i}}{\binom{\binom{n}{2}}{N_c}}$$

so that after some simplification one obtains

$$i!\, A_i =$$

$$\frac{n(n-1)\cdots(n-i+1)[(n-i)(n-i-1)][(n-i)(n-i-1)-2\cdot 1]\cdots[(n-i)(n-i-1) - 2N_c + 2]}{[n(n-1)][n(n-1) - 2 \cdot 1]\cdots[n(n-1) - 2N_c + 2]}$$

For large values of n this can be expressed in a much simpler form by representing all of the first i factors by n and all of the remaining N_c factors by $(n-i)(n-i-1)/n(n-1)$. Then we have

$$i! \, A_i \approx n^i \left[\frac{(n-i)(n-i-1)}{n(n-1)} \right]^{n/2 \, \log n + nc}$$

It is simple to show that

$$\lim_{n \to \infty} \left[\frac{(n-i)(n-i-1)}{n(n-1)} \right]^{nc} = e^{-2ic}.$$

Consider the remaining factor and call it V, then

$$\lim_{n \to \infty} \log V = \lim_{n \to \infty} \frac{i + n/2 \, \log \, (n-i)(n-i-1)/n(n-1)}{1/\log n}$$

which is indeterminant of the form $0/0$. For large values of n we can just as well use the simpler expression

$$\lim_{n \to \infty} \log V = \lim_{n \to \infty} \frac{i + n \, \log \, (1 - i/n),}{1/\log n}$$

applying L'Hospital's rule and doing some simplification yields:

$$\lim_{n \to \infty} \, \log V = - i^2 \lim_{n \to \infty} \frac{(\log n)^2}{n - i} = 0 \qquad \text{so that} \qquad \lim_{n \to \infty} V = 1.$$

Thus we obtain $\lim_{n \to \infty} A_i = e^{-2ic}/i!$ for each value of i. Utilizing this result in the sum for $\overline{N}_0(n, N_c)$ we obtain:

$$\lim_{n \to \infty} \frac{\overline{N}_0(n, N_c)}{\binom{\binom{n}{2}}{N_c}} = \sum_{i=0}^{\infty} (-1)^i \frac{e^{-2ic}}{i!} = e^{-e^{-2c}}$$

By the definition of $N_0(n, N_c)$, $\overline{N}_0(n, N_c)$, and $P_0(\overline{A}, n, N_c)$,

$$0 \le \frac{\overline{N}_0(n, N_c) - N_0(n, N_c)}{\binom{\binom{n}{2}}{N_c}} \le P_0(\overline{A}, n, N_c)$$

for any n. The lemma states that $P_0(\overline{A}, n, N_c)$ tends to zero as n becomes large. Thus

$$\lim_{n \to \infty} P_0(n, N_c) = \lim_{n \to \infty} \frac{N_0(n, N_c)}{\binom{\binom{n}{2}}{N_c}} = \lim_{n \to \infty} \frac{\overline{N}_0(n, N_c)}{\binom{\binom{n}{2}}{N_c}} = e^{-e^{-2c}} \qquad \blacktriangle$$

We now have a statement of the general result for the limit distribution of the quantities $P_k(n, N_c)$. The previous theorem is seen to be a special case of this result for $k = 0$.

Theorem.

$$\lim_{n \to \infty} P_k(n, N_c) = (e^{-2c})^k \frac{e^{-e^{-2c}}}{k!}.$$

These results show that the type of random graphs under consideration here behave in a very specific way for large n. The number of vertices that fail to belong to the maximum connected component of G_{n,N_c} satisfy a Poisson distribution with mean equal to e^{-2c}.

Erdös and Renyi give several additional results for random graphs of type G_{n,N_c} of which the following is illustrative.

Theorem. Let $S_k(n, N_c)$ denote the probability that G_{n,N_c} has exactly $k + 1$ disjoint (connected) components. Then

$$\lim_{n \to \infty} S_k(n, N_c) = (e^{-2c})^k \frac{e^{-e^{-2c}}}{k!}.$$

Thus $P_k(n, N_c)$ and $S_k(n, N_c)$ have the same Poisson limiting distribution.

These results illustrate one method of studying random graphs and of providing insight into graph theory concepts, such as connectedness, by means of probability theory (a major technique utilized by Erdös and others in graph theory studies). The reader may wonder how the quantity N_c is thought of in such studies. There is no doubt that the above results depend strongly upon the definition of N_c which must seem to come "out of the blue" to many readers. Erdös and Renyi provide us with some insight into the selection of N_c. In a footnote of [80] they define $R_k(n, N)$ as the probability that exactly k boxes remain empty upon the distribution of N balls among n boxes. Then they state that

$$\lim_{n \to \infty} R_k(n, n \log n + xn) = \frac{(e^{-x})^k e^{-e^{-x}}}{k!}$$

for $k = 0, 1, \ldots$. This relates the present random graph study to problems in distributions, a standard topic in probability. The quantity $n \log n + xn$ suggests N_c and our proof of the first theorem above will indicate the need for the factor $\frac{1}{2}$ if the reader fills in some omitted detailed calculation.

In [81] and [82] Erdös employs probabilistic (combinatorial) reasoning to obtain some results in graph theory that relate to Ramsey's theorem. This work does not deal with random graphs in the sense we have been using that term. Never the less it is much in the same spirit of development. As an indication of what sort of questions are studied in these papers we shall give some definitions and one of the major results, the extremely detailed proofs are beyond the scope of our presentation here.

Erdös points out that one consequence of Ramsey's theorem (see Chapter 6) is that to every n there exists a smallest integer $g(n)$ such that every $g(n)$ graph contains either a set of n independent vertices (no two vertices of the set connected) or contains a complete n-graph. There exists a $(g(n) - 1)$-graph which does not contain a complete n-graph or a set of n independent vertices. Reference [81] states one of the few results known about the Ramsey numbers $g(n)$

$$2^{n/2} < g(n) \leq \binom{2n - 2}{n - 1}$$

which was obtained much earlier by Erdös (in 1947).

Erdös and Szekeres have defined $f(k, r)$ as the smallest integer such that every $f(k, r)$-graph contains either a complete k-graph or a set of r independent vertices. We note that $f(k, k) = g(k)$ by definition. Szekeres has shown an upper bound expression

$$f(k, r) \leq \binom{k + r - 2}{k - 1}$$

and Erdös remarks that this is close to being a best possible upper bound.

Reference [81] defines $h(k, r)$ as the smallest integer such that every $h(k, r)$-graph contains either a cycle of at most k edges or contains r independent vertices. Thus $h(3, r) = f(3, r) \leq r(r + 1)/2$ by the above inequality.

In [81] Erdös arrives at a major result for this kind of study which is stated in the following theorem.

Theorem. Let α be a fixed, sufficiently large number. Then there exists a positive integer N such that for all $n > N$ there is an n-graph containing no cycle of length three (triangle) which also does not contain a set of $[\alpha n^{1/2} \log n]$ independent vertices.

If we take the critical case of r independent vertices in this theorem as approximately (dropping the greatest integer stipulation for simplicity) $\alpha n^{1/2} \log n = r$ or

$$n = \frac{r^2}{\alpha^2 (\log n)^2}.$$

Taking logs of the critical equality yields:

$$\log \alpha + \tfrac{1}{2} \log n + \log \log n = \log r$$

which gives the inequality $\tfrac{1}{2} \log n < \log r$. Squaring this expression, substituting into the critical expression in place of $(\log n)^2$, and identifying the critical *n* value as $f(3, r)$ we obtain Erdös' result:

$$f(3, r) > \frac{r^2}{4\alpha^2(\log r)^2}$$

which gives a lower bound on the Ramsey like numbers $f(3, r)$. The bound is of limited utility because of its explicit dependence on α but it does indicate the order of dependence on r, the size of the independent vertex set specified in $f(3, r)$. In particular since r increases far more rapidly than does log r the quantity $f(3, r)$ will be forced to increase rapidly with increasing r. This agrees with our intuitive understanding since as we require a larger independent set of vertices as the alternative to the presence of a triangle we will expect to need more vertices. However the inequality quantifies our intuition to a considerable extent indicating a rather strong rate of increase in the number of vertices required.

This result concludes our discussion of random graphs and probability concepts in graph theory. In this chapter we have only touched on an extremely interesting and important aspect of graph theory. The topics described and their extensions and generalizations may be expected to play an increasingly important role in both the theory and application of linear graphs.

EVOLUTIONARY RANDOM GRAPHS

When the number of vertices or edges are not fixed but rather form by random addition (or removal) we have the concept of an *evolutionary random graph*. The simplest types are those formed by the random addition of vertices and their edge connections to the existing graph. A more complicated type of evolutionary random graph is formed when entire subgraphs are added at random. This can be done in two ways. One way is to introduce such a graph as a component then add connecting edges (at random or according to some rule). The other way is to replace an existing vertex by a subgraph (of fixed or random structure) which must then be connected in some appropriate way to the original structure. This latter type of evolutionary random graph may be called a *hierarchy graph* since it provides a useful model for various levels of detail. One can start with a simple model of a situation dealing only with major aspects and then expand the detail as information increases or as need for more intensive analyses develop. One example of such an application will be given at the end of this section. Though very useful in providing model structure the hierarchy graph is in general far too complex to allow much in the way of direct analysis. One must usually resort to simulation to carry out studies with such structures.

We shall now give an elementary illustration of the random graph concept. Since analyses of these types of random graphs are so involved even the elementary examples soon take us beyond the scope of this book. Thus we must be content to introduce the ideas and indicate the areas of application.

A branching process may be described as a random graph developing in time. The times at which development events occur is not important in our model; they only serve to provide us with a basic foundation on which to describe the process. At such a development instant the event which occurs is the generation by a vertex of other vertices to which it becomes connected at once. The random graph of a branching process is a random tree; changes in structure taking place at each development instant at only the terminal vertices of the tree. It is therefore a very special random graph. The number of terminal vertices at the nth instant is denoted by X_n. Then X_0 is the number of initial vertices. As the graph develops each initial vertex grows into a tree so that the graph is in fact a set of trees. The number of such trees depends on the value of X_0 which is in general taken to be a random variable such that $Pr[X_0 = k] = p_k$ where the range of k is determined by an index set I and $\sum_{k \in I} p_k = 1$. Thus a branching process may be called a random forest, that is, a random number of random trees.

At the development instant n vertex j may give rise to $Y_j(n)$ vertices. The random variables $Y_j(n)$ have distributions of the form $Pr[Y_j(n) = k] = r_{jn}(k)$ where $\sum_k r_{jn}(k) = 1$. In many applications the $Y_j(n)$ distributions are all the same but they need not be for the general model. We do require that the mean and variance of each $Y_j(n)$ be finite.

Branching processes can be studied in great detail and form a large area of both theoretical investigation and application. We simply define them here in terms of random graphs and illustrate their use as a model. A branching process at three stages of development is shown in Figure 120. It should be remarked that once a vertex fails to split at a development instant for which it is in the set of terminal vertices it may never split thereafter. It then becomes an end vertex and is no longer included in any terminal vertex sets.

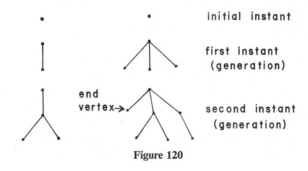

Figure 120

An example (from Genetics) which makes use of a branching process is the generation of offspring in a population. We consider the offspring from a single individual so that the branching model has $X_0 = 1$ and all vertex splitting distributions equal. Thus we can write $Pr[Y_j(n) = k] = r_k$. This is a relatively simple case of a branching process and leads to a single random tree. The size of the nth generation is the number of elements in the terminal vertex set at the nth instant.

Let the individual offspring mean and variance be defined by:

$$\mu = \sum_{k=0}^{\infty} k\, r_k \quad \text{and} \quad \sigma^2 = \sum_{k=0}^{\infty} (k - \mu)^2\, r_k$$

both of which we assume converge to finite values. Then we wish to find the mean and variance of the nth generation. We use these two quantities to characterize the nth generation distribution. The distribution itself is very involved.

By using conditional expectations defined as random variables in probability theory one can obtain difference equations for the nth generation mean m_n and variance σ_n^2. Note these are conditional values dependent on $X_0 = 1$. The equations, though simple to derive use rather advanced probabilistic methods so we shall merely state them here. By reasoning in two different ways two different equations are obtained for σ_n^2 both are stated below:

$$m_n = \mu\, m_{n-1} \qquad n \geq 2 \tag{11}$$

$$\sigma_{n+1}^2 = \mu\, \sigma_n^2 + \mu^{2n}\, \sigma^2 \qquad n \geq 1 \tag{12}$$

$$\sigma_{n+1}^2 = \mu^2\, \sigma_n^2 + \mu^n\, \sigma^2 \qquad n \geq 1 \tag{13}$$

By solving such equations one can study the expected population sizes at various instants and also estimate the extent to which those expectations may represent the actual (random) population size.

We shall conclude this section with an illustration of how hierarchy graphs form useful models. The example is from the field of Management Science and relates to the important and difficult problem of technological management.

Technological management necessarily operates over a wide spectrum of application areas. The relative location on this spectrum of a particular management area is largely determined by the degree of exactness with which the associated problems, goals and objectives can be formulated. To a considerably lesser extent the directness of problem resolution by analytic or other means contributes to the location of a management area within what may be called the finer structure of the technological management spectrum.

Any particular technological management area can be considered within a single extremely abstract format. On one hand are the objectives and associated goals involved. These are directly related to the responsibility or mission of the management area. On the other hand there are the resources available to management which it may utilize to carry out its mission, achieve its goals and proceed successfully toward its objectives. Between these aspects of the management situation is the management process itself, the possibilities of utilization of analytic procedures, the role of executive judgement and decision and so forth.

In this context the various technological management areas can be fairly well located in relative positions along the spectrum. At one end the responsibility is very restricted. Such restriction results in well formulated specific goals relatively few in number (often one in number) with mostly fixed objectives. Likewise the available resources can easily be assigned relative value or priority and the management process can be based on direct analytic procedures or utilize effectively experienced executive judgement. One then proceeds along the spectrum away from this region of extremely well specified problems. The further one goes the less definite each aspect of the management process becomes. Areas are reached in which the mission is extremely broad and the goals largely undefined. In such cases the objectives are evolutionary in nature and broad in description. Corresponding difficulties occur with regard to the resources available to deal with such poorly specified situations. It is not clear what categories of resource should be utilized or with what relative degrees of utility those selected should be valued. With this general lack of bounding specification to an area of technology management the opportunity to employ any techniques, particularly analytic methods toward the formulation and solution of appropriate problems becomes an extremely important and challenging problem. Though analytic or semi-analytic methods seem very difficult to formulate and perhaps even hard to implement, the gain from such methods in these fluid areas is sufficient to justify every effort to formulate appropriate methods.

The general formulation of an area of technological management as discussed above can be described and studied analytically by means of the theory of graphs. Not only is such description possible, but it affords what is probably the most complete and powerful method of representation and analysis. Representing the various aspects of a technological area by the vertices of a graph and interactions or relations between such aspects by connecting edges (directed or not) results in a very general method of representation. Complex interactions are clearly indicated and the absence of important aspects or relations often manifest themselves almost at sight. Such a graph representation presents the structure of a technology area. A clear formulation of this structure is by itself extremely useful. For example it may greatly assist in the formation of qualitative judgements.

Analytic investigations can be formulated by assigning values (which may be denoted by the general name of capacities) to the various edges of the structural graph. The resulting mathematical model is a network which represents some area of the technology which is to be investigated. The values or capacities may be of any type, representing definite physical quantities, degrees of information of some type, monetary values, time delays or whatever may be appropriate to the particular investigation. For the study of some technological area a number of such network formulations may be required. Each such network is based on the same linear graph which represents the structure of the technology itself. Analyses are then carried out on the networks to provide quantitative measures directed toward answering the various questions of interest regarding the technology. In this way such analyses are generated by and addressed to the management problem for the particular area of technology under investigation.

A methodology utilizing the classical graph theory concepts can be made appropriate to the study of technological management problems by the addition of the concepts of evolutionary random graphs and hierarchy graphs. Together these provide for the structural formulation and analysis of dynamic and evolutionary aspects of a technology at the several levels of detail that may be required for a particular study. Hierarchy graphs are primarily a structural concept and allow for increases in detail of the structural formulation in a graph representation of a technological study area.

One may speak of various levels of detail for the graphical representation of the structure of the technology. The possibility of expanding vertices in this way to produce more and more detailed representations allows great flexibility in analysis. At a particular level some vertex may be expanded into a higher level to enable more detailed study of analyses initiated at the former level of presentation (e.g., more clearly formulate the composition of network values) or such expansion may be effected to allow for analytic studies of the higher level aspects which are absent altogether at the former level. In complex situations of the kind that occur in technological management it is advantageous to begin analyses with a fairly low level graph, keeping the number of vertices as small as possible. Studies with such representations are relatively simple and serve the dual purpose of yielding useful information and forming a basis for expansion into more detailed studies. Thus the representations of such situations clearly call for the conceptual framework afforded by hierarchy graphs. The levels of hierarchy investigated are determined in any investigation by the type of problems to be studied, the complexity of the technology, the degree of exactness of specification possible and by the amount of funds available for carrying out the analyses. Each added level increases the complexity and difficulty of analysis and most often will increase the number of individual analyses required as well.

It should also be observed that in the graphs under consideration here it is often useful and in our formulation entirely possible, to have several lines between one pair of nodes.

Now let us consider the introduction of random graph concepts. We wish to include cases in which edges enter the graph according to a probability law and cases in which edges drop out of some fixed linear graph. In technology analysis both these situations are necessary to reflect the dynamic nature of the technology. Dynamic properties of the technology are reflected in the changes which take place in the relations and correspondences between vertices; they are therefore well described by means of random graphs. In addition to their value in representing dynamics of the technology, random graphs can be useful in certain analytic investigations. For example they can assist in studying questions concerning the importance of certain connecting edges or the value to the technology of establishing edges which may not be present.

Random graphs can be formulated in two ways or as a combination of these ways if appropriate. One way to treat such a graph is to consider it to be changing in time in a manner prescribed by the probability laws. In this view a random graph is treated as a stochastic process of rather simple theoretical nature but great complexity of detail. Such a formulation well represents the dynamics of technology which also are dependent on temporal variation. Alternatively, the element of time need not be considered. A random graph is then a graph with probability values assigned to its edges. Any possible graph within the given structure of vertices and edges has a certain probability of being the proper representative of the technology. Questions about the technology are then discussed in terms of the distribution of possible representations. For example in this formulation one can speak of the most probable representation or an expected number of paths between a particular pair of vertices.

As described above random graphs are not sufficient to represent all properties of a technology. Evolutionary behavior is in many ways one of the most important and characteristic properties of technology. Such behavior is well represented by the introduction of new vertices to represent new aspects of the technology. Less often but occasionally a vertex may also drop out of the representation due to lack of feasibility or obsolescence. The drop out of vertices is more usual at higher levels of a hierarchy representation since at low levels the vertices represent such basic, far reaching aspects of the technology that they seldom cease to be important unless of course the entire technology itself dies out. The changes in vertices can be governed by probability laws as is the case for edges in simple random graphs. Definite changes can be subsumed under this conceptual formulation by assigning unit probabilities where appropriate or simply considering new graphs with the desired vertex structure.

Evolutionary behavior is far more complex than dynamic behavior and has a much greater influence on the technology. This is well reflected in the concepts of a hierarchy representation where the introduction of a vertex carries with it the possibility of essentially infinite expansion into higher and higher levels of detail.

Hierarchy considerations in the context of evolutionary random graphs are seen to yield representations of the structure of technologies reflecting all major aspects of the technology. Analyses on such structures, based on the formation of networks by addition of numerical values to the edges (and if appropriate to the vertices as well) are closely associated with the particular technology under consideration.

The location of a technology within the technology spectrum previously discussed will determine the graph's structure to a large extent and will affect the appropriate network formulations and analyses to an equivalent or even greater extent. At what may be called the low end of the spectrum where problems are straightforward to formulate, the appropriate analyses can be described with corresponding simplicity. As one moves along the spectrum of technologies it becomes unclear how analyses should be carried out, structures formulated, and needs of technology management reflected. In this region it is uncertain that the methodology can be formulated so as to carry out the requisite analytic investigations required by these levels of technology management.

In order to represent evolutionary random graphs the type of subgraphs must be fully specified including the way in which such graphs fit into the original graph. Thus specific vertices are designated as points of attachment in the formation of new graphs. It is not sufficient to give the subgraphs alone. An illustration is shown in Figure 121 where the original technology graph is shown as T and vertex v is to be subject to the random replacement by a new subgraph. Three such subgraph denoted A_1, A_2, and A_3 are shown with

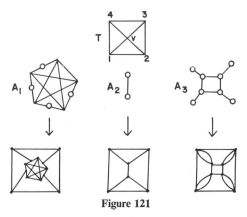

Figure 121

designated vertices shown as open circles and these occur with probabilities p_1, p_2, and p_3 respectively. The results are also shown in Figure 121 where we see that the new graph will be non-planar with probability p_1, will be cubic with probability p_2, or will be an euler graph with probability p_3.

To interpret the graph of Figure 121 we may consider the vertices as representing several aspects related to the state of technology and technological management. Vertex v denotes processing; it is directly related to materials (vertex 3), testing (vertex 2), applications (vertex 1), and design (vertex 4). Moreover, we see that design is directly related to materials and applications but not to testing. Testing is also directly related to applications and materials. In the situation being represented by the graph, applications are not directly related to materials.

Evolutions of vertex v represent expansions of the processing vertex. These may be due to a desire to look into greater detail or to the occurrence of new features requiring introduction into the model. For example an evolution of type A_1 may represent an expansion of detail in processing while A_2 would indicate a new process technique in which part of the process relates to design and materials while the other part relates only to applications and testing.

EXERCISES

1. Consider Moon's problem of expected crossings in a complete graph with vertices placed at random on a sphere. Place the vertices in such a way that four vertices are in fixed position and select the location of the others by means of a uniform distribution.
2. Let K_5 represent a double correspondence graph. Find the full service probability and expected service index for K_5. Do the same for $K_{3,3}$ and for K_6. Which of these two 6-graphs is better according to the two criteria?
3. Compute T_{nmr} and c_{nm} for $n \leq 6$ and make a table of the graphs being counted. Get an exact result for $M_{4,4}$ and compare this with the approximate formula.
4. Verify Gilbert's result that $F(5, 1) = 728$ and consider the detailed composition of the graphs enumerated (by classes). Evaluate $c_{n,r}$ for $n \leq 5$ and make a table of the graphs being counted. Work out p_5 directly and compare with the result given by Gilbert's generating function.
5. For the type of random graph studied by Erdös and Renyi compute $P_k(n, N_c)$ for small values of the parameters. Compare results with the limiting values given in the text.
6. Find the probability that a graph selected at random from all graphs (multiple edges allowed but loops not allowed) having 7 vertices and 15 edges is connected. Similarly find the probabilities for other number of edges.
7. For any arrangement A (multiple correspondence) in which exactly K installations are associated with each service station show that an upper bound on the service index is $I(A) \leq 1 - p^{2K}$ where p is the probability of removal of a station (Finch's result).

Chapter 8

Applications in Operations Research

INTRODUCTORY EXAMPLES

Operations Research may be defined as the application of scientific principles and analytic techniques to problems of system operation. The system may be a production plant, economic model, military operation, biological entity or any number of other things. The usual situation is that the system is complex, being large in size, (i.e., involving many components) or intricate in structure or both. A major step in the study of such systems is the formation of a clear picture of interrelationships between the system components. Here a component is taken to be a logical distinct unit which can be specified as contributing some function to the system operation. The division of a system into interconnected components can often be carried out in several different ways by defining the components differently. For example, two possible divisions of a cargo ship are shown in Figure 122.

Sometimes the different formulations are made to facilitate different analyses but they may also result from different research teams treating the same problem in alternative ways. Certain representations may be superior to others. When this is so and in what way it is so may be of considerable importance.

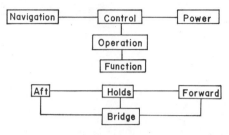

Figure 122

The formulation of a system in terms of its components can often be carried out by forming a linear graph whose nodes represent the components and whose lines represent relationships between the components. These relations can be information or command flow, dependence of operation, material transformation and so forth. In many cases digraphs are called for to specify direction.

In 1939 H. Robbins [83] preceded by at least two decades the subsequent extensive application of mathematical techniques to problems in traffic flow. His result is an interesting theorem in graph theory that finds possible application in a wide variety of situations. The origin of the theorem is in the following question that Robbins attributes to S. Ulam: when may the edges of a connected graph be assigned directions in such a way that at least one directed path exists between any pair of vertices, (i.e., the resulting digraph is strongly connected)?

Robbins casts the question into one associated with flow of traffic on the streets of a city. In that form one asks when it is possible to make every street (between two intersections) one way and be able to move from any intersection to any other, following prescribed direction, on the street network. The answer is that it is possible when and only when any undirected street (two-way) on the undirected street network can be blocked off without preventing passage between any two intersections (points of the network). This result is stated in graph theory terminology in the following theorem of Robbins: A graph is said to be orientable if it can be converted into a strongly connected digraph by some assignment of directions to its edges.

Theorem. A graph is orientable if and only if it remains connected after the removal of any edge.

It is interesting to note that it is not sufficient for a graph to be isthmus free to be orientable. For example a simple cycle graph with one pendant edge is isthmus free but is not orientable. This is because an isthmus disconnects the graph into (at least) two non-vertex graph components. Of course the presence of an isthmus is sufficient for non-orientability. The reader may wish to attempt a proof of the theorem as an exercise. Necessity is simple to prove and an effective proof of sufficiency is given in [83].

As the initial topic in this chapter we consider the calculation of distances in a graph by particularly effective methods. Need for such calculations arises in many applications of graphs to operations research and systems analysis.

MINIMAL DISTANCE AND CASCADE ALGORITHMS

One may consider the concept of distance between vertices in a graph when non-negative, numerical measures of length have been assigned to the edges. The resulting distance network can be based on directed or undirected graphs. In the former case one only deals with directed paths between vertices. A standard problem for such networks is to determine the shortest distance between two vertices. That is the sum of edge lengths in a connecting path such that no other path has a smaller total length. In case the length assignments are all unity one obtains the basic idea of path length and vertex separation distance in a graph which has been discussed elsewhere in the book from other points of view. Here we present an efficient method for the computation of shortest distances. The method is given in terms of digraphs but applies to graphs as well by making obvious identifications. When considering non-directed graphs in this context we exclude multiple edges and (for convenience) self loops.

Distance aspects of the network (edge length assigned graph) introduced above are specified by the distance matrix $D = (d_{ij})$ in which d_{ij} is the length assigned to the edge from vertex i to vertex j. In the context of a graph, D becomes the adjacency matrix. When there is no edge between vertices i and j one assigns infinity to d_{ij} (in practice a large positive number is employed).

A number of authors have made contributions to the method discussed in this section. Probably the most significant presentation of the topic is to be found in [84] which includes an historical discussion and references. Our material is based primarily on [85] which gives particularly clear formulations and proofs.

The distance calculation method presented here is based on the concepts of mini-addition of two matrices and cascading of matrix elements. The mini-addition operation has been referred to as a special kind of matrix multiplication by Hu [85] and also by others. This follows common mathematical usage of the term multiplication to denote any abstract operation and should not be thought of as referring to the usual concepts of matrix multiplication in any sense. We shall call the operation mini-addition and indicate it by the symbol $\oplus$ so that:

$$A \oplus B = C$$

where A and B are square of order n and $c_{ij} = \min_k (a_{ik} + b_{kj})$.

Computation of shortest distances may be based on the mini-addition operation $\oplus$ alone and in fact such a method has been widely discussed to the point of becoming a classical technique. To understand how that classical method operates the reader will observe that $D \oplus D \equiv D^2$ has vertices $d_{ij}(2)$ which represent shortest distances between vertices i and j employing at most two edges. Similarly $D^r \equiv (d_{ij}(r))$ has entries which represent shortest distances between vertices i and j employing at most r edges. Since no minimal distance path in an n-graph can use more than $n - 1$ edges there is a limit to the number of mini-additions required on D to obtain all shortest distances. In fact only even power operations need to be computed since $D^2 \oplus D^2 = D^4$ and so forth. Recall that D^4 has all shortest path distances for paths of length at most four (i.e., it includes those of length three). Thus if we form $(D^2)^k$ we shall surely obtain all shortest paths provided $2^k = n - 1$. In case $n - 1$ is odd we let $2^k = n$. Then the number of mini-additions required is at most $k = [\log_2(n - 1)]$ (employing the greatest integer notation).

As an illustration of the mini-addition operation consider a 5-graph with distance matrix (1).

$$D = \begin{bmatrix} 0 & 3 & 4 & \infty & 1 \\ 3 & 0 & 2 & \infty & \infty \\ 4 & 2 & 0 & 1 & \infty \\ \infty & \infty & 1 & 0 & 1 \\ 1 & \infty & \infty & 1 & 0 \end{bmatrix} \tag{1}$$

The mini-addition yields (2).

$$D^2 = \begin{bmatrix} 0 & 3 & 4 & 2 & 1 \\ 3 & 0 & 2 & 3 & 4 \\ 4 & 2 & 0 & 1 & 2 \\ 2 & 3 & 1 & 0 & 1 \\ 1 & 4 & 2 & 1 & 0 \end{bmatrix}, \quad D^4 = \begin{bmatrix} 0 & 3 & 3 & 2 & 1 \\ 3 & 0 & 2 & 3 & 4 \\ 3 & 2 & 0 & 1 & 2 \\ 2 & 3 & 1 & 0 & 1 \\ 1 & 4 & 2 & 1 & 0 \end{bmatrix} \tag{2}$$

The cascade concept allows the development of an algorithm requiring only two mini-addition operations in contrast with the k such operations that may be required by the classical technique described above. In a cascade process one replaces the entries in the original matrix by the values obtained by mini-addition as they are computed. Thus d_{12} is replaced by $d_{12}(2)$ as soon as the latter is computed. The matrices being operated on change with each element calculation. In such a process the order of calculation will, in general, result in different answers. Two types of cascade process orderings are of particular interest and are known as the forward process and the backward process respectively.

In the forward process mini-addition with cascade replacement of computed elements takes place from left to right and top to bottom. In the backward process the operations take place from right to left and bottom to top.

As an example of the forward cascade process one may form one forward process matrix F from the distance matrix D used above. The reader may verify that $F = D^2$ in this case.

For a time it was felt that two applications of the forward cascade process would produce the shortest distance matrix. This conjecture is discussed by Hu in [85] and was shown to be false by a counterexample given in [84]. The counterexample due to Farbey, Land, and Murchland employs a simple digraph with distance (3).

$$D = \begin{pmatrix} 0 & \infty & \infty & \infty & 1 & \infty & \infty \\ \infty & 0 & \infty & \infty & \infty & \infty & 1 \\ \infty & \infty & 0 & 1 & \infty & 1 & \infty \\ \infty & \infty & 1 & 0 & 1 & \infty & \infty \\ 1 & \infty & \infty & 1 & 0 & \infty & \infty \\ \infty & \infty & 1 & \infty & \infty & 0 & 1 \\ \infty & 1 & \infty & \infty & \infty & 1 & 0 \end{pmatrix} \tag{3}$$

Two forward cascade processes yield the array (4).

$$\begin{pmatrix} 0 & \infty & 3 & 2 & 1 & 4 & 5 \\ \infty & 0 & 3 & 4 & 5 & 2 & 1 \\ 3 & 3 & 0 & 1 & 2 & 1 & 2 \\ 2 & 4 & 1 & 0 & 1 & 2 & 3 \\ 1 & 5 & 2 & 1 & 0 & 3 & 4 \\ 4 & 2 & 1 & 2 & 3 & 0 & 1 \\ 5 & 1 & 2 & 3 & 4 & 1 & 0 \end{pmatrix} \tag{4}$$

Array (4) fails to give the shortest distance between vertices 1 and 2 thus disproving the conjecture that two forward processes always suffice to obtain shortest distances.

One forward process followed by a backward process yields the same values as in the above matrix except for the (1, 2) and (2, 1) entries which have the correct minimum distance of 6. Except for this counterexample and others very similar to it two forward cascade processes seem to yield the distance array for most graphs. The reader may wish to attempt to construct other types of counter examples. A detailed description of what types of graphs do not yield correct distance arrays under two forward processes seems to be missing from the literature.

It has been established by Farbey, Land, and Murchland in [84], by Hu in [85], and possibly by others that one forward process followed by one backward process will always yield the correct distance array. The following proof of this result utilizes Hu's very systematic development from [85].

Let the minimum distance between vertices v_r and v_s be denoted by L_{rs}. Denote the results of one forward cascade process by $d_{rs}(F)$, of one forward followed by one backward process by $d_{rs}(F, B)$, and so forth.

The sequence of vertices lying on a path connecting two vertices may be ordered by the natural ordering of their indices. Thus we say v_i precedes v_j when $i < j$. (Hu uses the term "less than" where we use precede.) Any sequence of vertices $v_{j_1}, v_{j_2}, \ldots, v_{j_k}$, in a connecting path is said to form an ascending sequence if the indices form a monotone increasing sequence (i.e., $j_1 < j_2 \cdots < j_k$). They form a descending sequence if $j_1 > j_2 > \cdots > j_k$, and they form a valley if $\min(j_1, j_k) > \max_i j_i$ for $i = 2, \ldots, k - 1$. In a path, a vertex is minimal if it precedes its two neighbors in the path; it is maximal if it is preceded by them. Cascade process results are developed by employing these concepts to the study of minimal distance paths.

Any sequence S of vertices lying on a connecting path can generate a sequence in reduced form, denoted by S_r. To obtain S_r consider any valley subsequence of S and remove all intermediate vertices from the valley sequence. Carry out such a reduction for every valley sequence in S. When no valley sequences remain the result is a unique reduced sequence S_r. Moreover S_r is an ascending sequence, or it is a descending sequence, or it consists of an ascending sequence followed by a descending sequence. These results follow directly from the definitions given above.

Development of the cascade process results depend on three lemmas which we now state without proof ([85]).

Lemma 1. If the vertices in a shortest path form an ascending sequence, or a descending sequence, then the shortest distance between the terminal vertices of the path will be obtained by one forward cascade process.

Lemma 2. If the vertices in a shortest path form a valley sequence then the shortest distance between the terminal vertices of the path is obtained by one forward cascade process.

Lemma 3. If the vertices in a shortest path form an ascending sequence, a descending sequence, or an ascending sequence followed by a descending sequence, then the shortest distance between the terminal vertices of the path is obtained by one forward process followed by one backward process.

Theorem. For any shortest path $d_{rs}(F, B) = L_{rs}$.

Proof. For any shortest path S consider the reduced sequence S_r. In the reduced sequence either two neighboring vertices are connected by an edge of minimal length or they are the end vertices of a valley sequence. In either of these cases the forward process correctly calculates the minimal distance between those vertices by Lemma 2. The minimal distances for the vertices of S_r itself are given by the backward process (i.e., (F, B)) as stated by Lemma 3. Thus the total minimal distance for S will be given as the combination of the S_r distance and its parts. ▲

Hu has established that three forward cascade processes will suffice to always yield correct shortest distances. Thus one can avoid the backward process if one wishes. For computer application there is little to be gained from using one method over the other. In using the backward process it has been found convenient to reorder the matrix and apply the forward process on the modified matrix.

As another example of the use of the cascade process in computation of minimum distances consider the graph with distance matrix (5).

$$\begin{pmatrix}
0 & 6 & \infty & 8 & \infty & 1 & 2 & \infty & 10 & \infty \\
6 & 0 & 1 & \infty & 1 & \infty & \infty & \infty & \infty & \infty \\
\infty & 1 & 0 & \infty & \infty & \infty & 1 & \infty & \infty & \infty \\
8 & \infty & \infty & 0 & 2 & \infty & \infty & \infty & \infty & 1 \\
\infty & 1 & \infty & 2 & 0 & \infty & \infty & \infty & \infty & \infty \\
1 & \infty & \infty & \infty & \infty & 0 & \infty & 10 & \infty & 10 \\
2 & \infty & 1 & \infty & \infty & \infty & 0 & \infty & \infty & \infty \\
\infty & \infty & \infty & \infty & \infty & 10 & \infty & 0 & 10 & \infty \\
10 & \infty & \infty & \infty & \infty & \infty & \infty & 10 & 0 & \infty \\
\infty & \infty & \infty & 1 & \infty & 10 & \infty & \infty & \infty & 0
\end{pmatrix} \tag{5}$$

For this example the minimal distance matrix is (6).

$$\begin{pmatrix}
0 & 4 & 3 & 7 & 5 & 1 & 2 & 11 & 10 & 8 \\
4 & 0 & 1 & 3 & 1 & 5 & 2 & 15 & 14 & 4 \\
3 & 1 & 0 & 4 & 2 & 4 & 1 & 14 & 13 & 5 \\
7 & 3 & 4 & 0 & 2 & 8 & 5 & 18 & 17 & 1 \\
5 & 1 & 2 & 2 & 0 & 6 & 3 & 16 & 15 & 3 \\
1 & 5 & 4 & 8 & 6 & 0 & 3 & 10 & 11 & 9 \\
2 & 2 & 1 & 5 & 3 & 3 & 0 & 13 & 12 & 6 \\
11 & 15 & 14 & 18 & 16 & 10 & 13 & 0 & 10 & 19 \\
10 & 14 & 13 & 17 & 15 & 11 & 12 & 10 & 0 & 18 \\
8 & 4 & 5 & 1 & 3 & 9 & 6 & 19 & 18 & 0
\end{pmatrix} \tag{6}$$

Certain problems in operations research and reliability utilize the concept of a random graph type network where the probability that an edge connects vertices v_i and v_j is p_{ij}. For example one may consider a graph representation of a communications network in which the vertices indicate stations and edges specify links between stations. A link is present between stations v_i and v_j with probability p_{ij} and one wants to know the probability of a most likely path between two stations v_r and v_s, each such path will be present with a probability value dependent on the edges comprising that path. Some of the paths may in fact have the same probability of occurring. The path or paths having maximum probability of occurrence constitute most likely paths between v_r and v_s. The probability value associated with such paths gives an upper bound on the probability of communication between v_r and v_s. It may be observed that calculation of the probability of some communication link between v_r and v_s is in general an extremely difficult (usually impossible) task. This accounts for some of the necessity to use monte carlo simulation in dealing with random network problems.

Hu has remarked that the most likely probability of connection values can be obtained from a modification of the cascade and special matrix operations discussed above. Let P denote the matrix of edge probabilities in which p_{ij} = probability that edge (i, j) is present for $i \neq j$ and $p_{ii} = 1$ for all i. Then if the matrix operation denoted by $C = A \oplus B$ is specified by the rule:

$$c_{ij} = \max_k (a_{ik} \cdot b_{kj})$$

the cascade process (FB) will yield the most likely probability matrix P^*.

As an example in which $p_{ij} = p$ for each edge that is present with non-zero probability, consider the graph with edge probability matrix (7).

$$\begin{pmatrix} 1 & p & 0 & 0 & 0 & p \\ p & 1 & p & 0 & p & 0 \\ 0 & p & 1 & p & 0 & 0 \\ 0 & 0 & p & 1 & p & 0 \\ 0 & p & 0 & p & 1 & p \\ p & 0 & 0 & 0 & p & 1 \end{pmatrix} \tag{7}$$

One forward cascade process yields the most likely probability matrix P^* (8)

$$\begin{matrix} 1 & p & p^2 & p^3 & p^2 & p \\ p & 1 & p & p^2 & p & p^2 \\ p^2 & p & 1 & p & p^2 & p^3 \\ p^3 & p^2 & p & 1 & p & p^2 \\ p^2 & p & p^2 & p & 1 & p \\ p & p^2 & p^3 & p^2 & p & 1 \end{matrix} \tag{8}$$

The graph with probability matrix (9) yields (10) after one forward cascade.

$$\begin{pmatrix} 1 & 0 & 0 & 0 & p \\ 0 & 1 & 0 & p & 0 \\ 0 & 0 & 1 & p & p \\ 0 & p & p & 1 & 0 \\ p & 0 & p & 0 & 1 \end{pmatrix} \tag{9}$$

$$
\begin{matrix}
1 & 0 & p^2 & p^3 & p \\
0 & 1 & p^2 & p & p^3 \\
p^2 & p^2 & 1 & p & p \\
p^3 & p & p & 1 & p^2 \\
p & p^3 & p & p^2 & 1
\end{matrix}
\tag{10}
$$

After two forward processes (or after the process (F, B)) one obtains the most likely probability matrix P^* (11).

$$
\begin{pmatrix}
1 & p^4 & p^2 & p^3 & p \\
p^4 & 1 & p^2 & p & p^3 \\
p^2 & p^2 & 1 & p & p \\
p^3 & p & p & 1 & p^2 \\
p & p^3 & p & p^2 & 1
\end{pmatrix}.
\tag{11}
$$

As effective as the cascade process is, particularly in comparison with other methods employed for similar calculation, it can become involved for large graphs. Hu has mentioned that other orders of cascading work just as well as do the forward or backward processes. In particular one can cascade in partitions of the matrix (corresponding to partitions of the graph vertices). Land and Stairs have developed the partition cascade concept for large graphs in [86] which should be consulted by anyone interested in distance or similar calculations on large graphs.

PERT

Because of the wide range of possible applications we can only indicate the utilization of linear graphs in Operations Research. An example that is characteristic of such application is the PERT network concept widely used in management Operations Research.

Though linear graphs are useful in describing the structure of complex systems their use can be considerably extended by forming *networks* based on the linear graph. A network is a linear graph together with an assignment of weights, (i.e., real numbers) to its edges or vertices or both. These weights may represent various physical quantities depending on what the edges (or vertices) themselves represent.

A Program Evaluation and Review Technique (PERT) network is used to describe the development in time of a programmed activity. The activity may be the development of a new product line (such as a color T.V. set) or weapons system (such as a Polaris submarine for which the PERT method was, in fact, developed) or it may be the construction of a complex structure such as a large building or highway system. Though certain aspects of PERT can be utilized for repetitive operations the major idea of PERT is as a development schematic of one-shot systems.

The program for which PERT is to be used is divided into events (these correspond to what are called milestones in the more classical approach to development planning) and activities leading from one event to another. An event represents some important accomplishment in the development, there is no duration of time associated with an event, only times at which it is, or is expected to be, completed. On the other hand the activities require time duration for carrying them out (this represents moving from one event to another). Other quantities such as cost (or expenditure of energy, etc.) can also be associated with each activity. Thus we see that the arrangement of events with their connecting activities can be well represented by a digraph. The assignment of times and/or costs to the activities results in a network. This is called the PERT network, its vertices represent events and its directed edges represent activities. Some vertex is the initiation of the program and another is the completion event. Only outward edges are incident to the initial vertex and only inward edges are incident to the completion vertex.

PERT is used by following a standard application procedure (though of course there are many individual innovations):

1. Formation of the PERT graph (requires definition of the events and activities.) This may not be unique, and different people might formulate different graphs, some being more appropriate (and therefore more useful) than others.

2. Times for completion of activities must be determined, as well as costs for activities for advanced applications (usually as functions of completion time).

3. Analysis of the PERT network is carried out to determine completion dates, scheduling requirements, and so on.

Item 1 depends on the application. Item 2 is statistical in nature. Each activity time is a random variable and must be estimated. The standard procedure is to make three estimates (based on past data for similar activities, the considered opinion of experts and so forth) and use them to compute an expected time. These three times are defined as follows:

Optimistic time. The shortest time in which the activity can be completed, denoted by a.

Most likely time. If only one estimate were given this would be it. For repeated runs of an activity this is the average time, denoted by m.

Pessimistic time. Largest time the activity could take, denoted by b.

These are used to compute t_e and σ_t^2 as estimates of the expected completion time and the variance of the completion time respectively. The estimates are computed from the formulas:

$$t_e = \frac{a + 4m + b}{6} \; ,$$

$$\sigma_t^2 = \left(\frac{b - a}{6}\right)^2 \; .$$

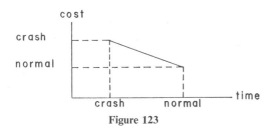

Figure 123

In the statistical theory of PERT the activity times are assumed to follow a Beta distribution for which estimates of expected value and variance are given by the above formulas.

When cost is to be considered each activity is studied to obtain a cost vs. completion time curve. This is most often a straight line based on two points, a normal time − normal cost point and a crash time − crash cost point as shown in Figure 123.

Item 3, the analysis requires the computation of two time values for each event (vertex). These are as follows.

(a) *Calculated expected time* (denoted by T_E). Every path leading from the *Earliest start* initial event to the event for which T_E is being computed is identified and all t_e for activities on such a path are added together. These sums are computed for every such path. The greatest of these sums is T_E for the given event. The value of T_E is computed for each vertex.

(b) *Calculated latest allowable time* (denoted by T_L). For each event the *Latest start* appropriate T_L is computed by subtracting the t_e, along every path from the event to the terminal event, from the T_E value for the terminal event (variations of this definition are used as well). This is done for every such path and the smallest such number is T_L for the event. The T_L indicates a time of arrival at a vertex which will allow the remaining activities sufficient time for completion without altering T_E for the final event.

At each vertex one can compute the slack $T_L − T_E$ which measures how critical the activity times (t_e) are on paths leading to that vertex from the start.

A critical path is one requiring greatest time to go from the initial vertex to the terminal vertex. This path (or paths) together with knowledge of slack values can be used to study how well a program is progressing, methods of improved scheduling, etc. When cost is to be considered the individual cost − time diagrams for each activity are used to obtain a series of points on a project cost vs. project time curve, these points are connected by linear segments yielding a project curve as shown in Figure 124.

One may determine a satisfactory point on this curve for which cost and time requirements are both met. From this a schedule may be computed which will result in a given time (together with its cost) for each activity. The

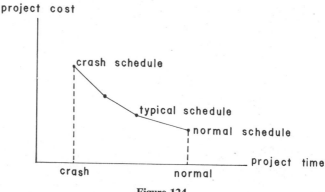

Figure 124

determination of such a schedule utilizes computational algorithmic procedures for which various computer programs have been developed.

The above represents the major aspects of PERT as utilized in industry, management and defense planning. It should be observed however, that since the PERT network is based on a linear graph many aspects of graph theory may be applied to it. This can lead to additional useful applications beyond the basic results. For example, the study of subgraphs of particular types such as trees can give useful representations of parts of the program which may be studied by themselves in detail. Thus though the original formulation of PERT is simple, it contains the nucleus of rather deep investigations into programming developments. Moreover, the network can be given a random graph structure leading to applications in which events or activities change due to unexpected developments as the work progresses.

Consider a small scale example to illustrate the concepts introduced above. The PERT network is shown in Figure 125. In practice PERT networks are considerably more complex.

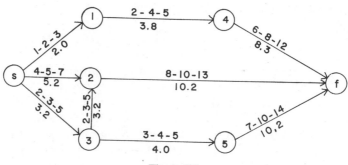

Figure 125

For each activity we show $a - m - b$ on one side and t_e on the other. Each event is labeled; s is the start, f the finish and 1, 2, 3, 4, 5 are intermediate events. For each event we have Table 9.

Table 9

Event	T_E	T_L	Slack $T_L - T_É$
s	0	0	0
1	2.0	5.3	3.3
2	6.4	7.2	0.8
3	3.2	3.2	0
4	5.8	9.1	3.3
5	7.2	7.2	0
f	17.4	17.4	0

In Table 9 we have followed standard PERT practice of rounding off all numbers to tenths. Due to the fact that the data used are (often crude) estimates, there is no point in carrying computation any further. The critical path is the vertex sequence s, 3, 5, f, and requires total time of 17.4 for its completion. We observe that the events lying on the critical path have zero slack. Other events may or may not have zero slack.

To conclude our discussion of PERT networks we consider the beta distribution. It is a logical distribution to use since it is non-zero over only a finite range and has a bell shaped appearance about its mean as we would expect the distribution of activity times to look, it is not symmetric about its mean however, this also reflects the behavior of activity times. A typical beta distribution is shown in Figure 126.

The beta function $B(k, n)$ is defined by

$$B(k, n) = \int_0^1 x^{k-1} (1-x)^{n-1} \, dx$$

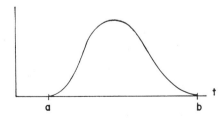

Figure 126

$$= \left[\frac{\kappa}{n+\kappa}\right] \cdot \beta(\kappa, n)$$

The density function, expected value and variance of the beta distribution shown above, for parameters k and n are as follows:

$$f_T\,(t) = \frac{1}{B(k,\,n)}\left(\frac{t-a}{b-a}\right)^{k-1}\left(\frac{b-t}{b-a}\right)^{n-1}\frac{1}{b-a}\quad,$$

using the fact that $B(k+1,\,n) = k/k+n\ B(k,\,n)$ we obtain:

$$E[T] = \frac{k(b-a)}{k+n} + a, \qquad \text{Var}\,[T] = \frac{kn(b-a)^2}{(k+n)^2(k+n+1)}\quad.$$

The estimates of $E[T]$ and Var $[T]$ given by the PERT formulas are statistical estimates based on the range $b-a$ and most likely value m. This latter quantity is not carefully defined in many PERT studies. It may be defined at the t value which yields the maximum of $f_T(t)$. By solving $df_T(t)/dt = 0$ one finds

$$M = \frac{kb+an-b-a}{n+k-2};$$

The estimates assume no detailed knowledge of the distribution hence we assume for them that $k = n$ for PERT applications. For this case, $E[T] = (a+b)/2$. In fact, the formula for t with M defined as above yields this result. PERT statistics assumes the formula continues to give a good estimate for more general beta distributions.

Additional discussion of PERT networks may be found in Miller [87] and Muth and Thompson [88].

FLOWS IN NETWORKS

The standard networks $N(F)$ upon which flow problems are formulated are defined as follows:

1. $N(F)$ is a finite, connected digraph without self loops.
2. A real number $C_{ij} \geq 0$ is assigned to each edge of $N(F)$. It is called the capacity of the edge to which it is assigned; here (i, j) denotes the vertex lables defining the edges.
3. Two special vertices are selected from $N(F)$ and designated as source and sink. Only outward edges are incident on the source vertex and only inward edges are incident on the sink. All other vertices are called intermediate vertices.

A real number $0 \leq x_{ij} \leq C_{ij}$ is called a flow in the edge (i, j). If an edge (p, r) is not in the graph we can set $C_{pr} = 0$ and as a result for (p, r) the only possible flow is $x_{pr} = 0$. The negative of a flow, $- x_{ij}$ represents a reverse flow in the edge (i, j), that is, it proceeds in a direction from vertex j to vertex i against the specified edge direction.

The special properties of inward and outward edge incidence at the sink and source respectively, as stipulated in condition (3) may not be satisfied in a given network but by adding special vertices and appropriate edges they can be satisfied. Therefore, we only consider flows on the standard network $N(F)$. As in any graph we let N stand for the set of vertices which we suppose are labeled (so that N is the set of labels) and vertex 1 is the source and vertex n is the sink, there being a total of n elements in N, (i.e., we formulate our flow network on an n-graph).

A flow F specified by $\{x_{ij}\}$ for $i, j \in N$, on the network $N(F)$ is any set of x_{ij} satisfying the following conditions:

$$\sum_i x_{ij} - \sum_k x_{jk} = 0, \qquad j \neq 1, n \tag{12}$$

$$0 \leq x_{ij} \leq C_{ij} \qquad , \qquad i, j \in N. \tag{13}$$

The network flow is

$$Z = \sum_j x_{1j}, \qquad j \in N.$$

For any flow network $N(F)$ there is always a set of flows satisfying the above conditions. In particular the zero flow $x_{ij} = 0$ all i, j always exists. Thus the first major problem in network flows is to determine the maximum flow and a set $\{x_{ij}\}$ yielding that flow. This is a problem in analysis of flow networks since the network is specified and one is to find a quantity defined by the given specifications. When multiple sources and sinks are allowed and more general capacities, (i.e., may be negative) and/or various specifications of vertex capacities are introduced no flow may exist satisfying all given conditions. These generalizations lead to problems concerning the existence of flows or the construction of networks having specified flows. Such problems are concerned with synthesis in addition to analysis. In treating problems of the latter type considerable notation and detail of concept must be developed hence we restrict our discussion to the problem of maximum flows.

The maximum flow problem can be solved by linear programming and in fact a special algorithm can be utilized. Before treating this solution method the important concept of cuts in a network will be developed so that we can state the major theorem of the subject.

Let x and y be subsets of N then we make the definitions: $(x, y) = \{(i, j) | i \in x, j \in y\}$ so that in particular the set of edges (i, y) is the set of all edges from vertex i to vertices in the set y. We can express the set of all edges into a vertex k by (N, k) and so forth. The sum of all edge capacities assigned to edges of the set (x, y) is denoted by $C(x, y)$ and the sum of flows in these edges is $f(x, y)$. This is a convenient and widely used notation in network flow theory. In this notation the flow problem is:

$$\text{maximize } z = f(1, N)$$

subject to:

$$f(i, N) - f(N, i) = 0, \qquad i \in N, (i \neq 1, n)$$
$$0 \leq f(i, j) \leq C(i, j), \qquad i, j \in N.$$

A cut (in a standard network $N(F)$) that separates the source vertex 1 and the sink vertex n is a set of edges $(x, \bar{x})$ such that $1 \in x$ and $n \in \bar{x}$ where $\bar{x}$ is the (set) complement of x in N. The capacity of a cut $(x, \bar{x})$ is $C(x, \bar{x})$.

If the set of edges $(x, \bar{x})$ of a cut is removed from $N(F)$ no chain from source to sink can be found in the remaining subnetwork. However, in the undirected graph sense, paths from source to sink may remain. Thus a cut separates source and sink but may not disconnect them. These concepts are shown in Figure 127 where $x = \{1, 2\}$, $\bar{x} = \{3, 4, 5, 6\}$ and $(x, \bar{x}) = \{(1, 3), (2, 4), (2, 5)\}$.

After removal of $(x, \bar{x})$ the source and sink are separated (no flow can occur between them except of course the trivial zero flow) but they are still connected in the linear graph sense. Had we defined x as $\{1, 2, 3\}$ the source and sink would become disconnected as well as separated.

For each cut there is an associated capacity value and for each flow there is a total flow value, many important relations hold between these values. We state some major results.

Theorem. If z is a flow from source to sink in a standard network $N(F)$ and $(x, \bar{x})$ is any cut then z is equal to the net flow across the cut and does not exceed the cut capacity; that is,

$$z = f(x, \bar{x}) - f(\bar{x}, x) \leq c(x, \bar{x}).$$

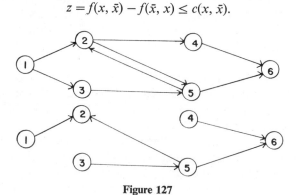

Figure 127

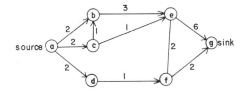

A standard network N. Numbers are edge capacities.

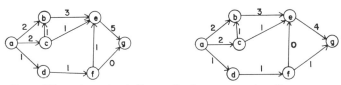

Two different maximal flows. Numbers are edge flows.
In either case, $z^* = 5$.

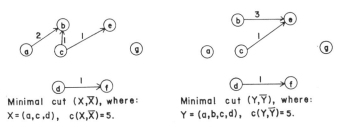

Minimal cut $(X,\overline{X})$, where: Minimal cut $(Y,\overline{Y})$, where:
$X = (a,c,d)$, $c(X,\overline{X}) = 5$. $Y = (a,b,c,d)$, $c(Y,\overline{Y}) = 5$.

Figure 128

Theorem (Max-flow min-cut theorem). The maximum value of a flow from source to sink in a standard network $N(F)$ is equal to the minimum cut capacity of all cuts $(x, \bar{x})$ separating source and sink.

Corollary. A flow f^* is maximal if and only if there is no path from source to sink such that $f^* < c$ on forward arcs of the path and $f^* > 0$ on reverse arcs of the path.

The max-flow min-cut theorem supplies a link between network flow studies and combinatorial analysis concerned with systems of district representatives and related questions. It is also related to linear programming concepts. These connections have been extensively studied by Hoffman and others (e.g., see [66]). The ideas presented above are shown in Figure 128. In each diagram only the edges in the cut are shown; the numbers are the edge capacities.

We now present an algorithmic solution method for finding maximal flows in a standard network. The method is due to Ford and Fulkerson ([89] and [90]).

Let P_1 denote the source, P_n denote the sink and P_i denote intermediate vertices $(i=2,\ldots,n-1)$. The flow is specified by x_{ij} where $i,\ j \in N$ and $i \neq j$ so that x_{ij} is the directed flow from P_i to P_j. Then x_{1j} is a flow from the source to some vertex P_j. The maximal flow problem can then be expressed as follows:

$$\text{maximize total flow from the source } z = \sum_{j=2}^{n} x_{1j}$$

subject to: $\sum_j (x_{ij} - x_{ji}) = 0$, $i = 2,\ldots, n-1$ which is flow balance at each intermediate vertex, and $0 \leq x_{ij} \leq C_{ij}$ which is the edge capacity limitation.

Start with any integral flow (x_{ij} integers), for example one can always start with the zero flow (which exists for every network.)

This initial flow defines a starting matrix $A_1 = (a_{ij})$ where $a_{ij} = C_{ij} - x_{ij} + x_{ji}$, the elements of which specify the excess capacity available on edge (i,j).

Thus if one starts with the zero flow (not efficient if some other initial flow is available) the starting matrix is simply the capacity matrix, $A_1 = C = (C_{ij})$.

For some values of $j = 1,\ldots, n$, labels (indices) v_j, μ_j are defined in a recursive way as follows.

Let $v_1 = \infty$, $\mu_1 = 0$. For all j such that $a_{1j} > 0$, define $\mu_j = 1$ $v_j = \min (v_1, a_{1j}) = a_{1j}$.

From those i which have received lables v_i, μ_i select an i (since this is one step of a general iterative process the i selected is to be one which has not been previously selected). Scan for all j such that $a_{ij} > 0$ and v_j, μ_j have not been defined. For these j, define

$$v_j = \min (v_i, a_{ij}), \ \mu_j = i.$$

(If $v_i = a_{ij}$, select the common value).

Continue this process until v_n, μ_n have been defined, or until no further definition can be made and v_n, μ_n are not defined. In the latter case the process terminates.

In the former case a new matrix (a_{ij}) is defined as follows.

Replace $a_{\mu_n n}$ by $a_{\mu_n n} - v_n$ and $a_{n \mu_n}$ by $a_{n \mu_n} + v_n$. In the general case, replace $a_{\mu_j j}$ by $a_{\mu_j j} - v_n$ and $a_{j \mu_j}$ by $a_{j \mu_j} + v_n$, where each j is the μ_j' of the preceding j' in the backward replacement going down from n. Continue until $\mu_j = 1$ has been achieved. The labels v_j, μ_j are then recomputed using the new matrix and the process repeated. Though this process may seem somewhat involved it is really a detailed specification of a rather simple procedure. One works back along a selected path, trying to increase the flow as much as possible at each step. Thus at each vertex one must move back to a vertex just preceding the current vertex along the directed path under modification.

Since v_n is a positive integer the process terminates. Upon termination, the maximal flow x is given by

$$x_{ij} = \max (C_{ij} - a_{ij}, 0)$$

where A is the final matrix obtained by the above process. Proof that the above procedure gives the maximum flow is contained in two parts.

PART I.   The x Is a Flow

By construction $0 \le x_{ij} \le C_{ij}$. It must be shown that x satisfies $\sum_j (x_{ij} - x_{ji}) = 0$, $(i = 2, \ldots, n-1)$. By construction (of the process) $C_{ij} + C_{ji} = a_{ij} + a_{ji}$. Thus $\sum_j (x_{ij} - x_{ji}) = \sum_j (C_{ij} - a_{ij})$ by the definition of x. Hence it is only necessary to prove that $\sum_j a_{ij} (i = 2, \ldots, n-1)$ is unchanged by the computation process since by definition of the starting flow process one has for it $\sum_j (C_{ij} - a_{ij}) = 0$.

If i (not equal to 1 or n) is the μ_l of some l, then there is a $k = \mu_i$. Thus, for this i, the new a_{ij}' are either equal to the old a_{ij} or are given by

$$a_{ij}' = \begin{cases} a_{ij} - v_n & \text{for } j = l \\ a_{ij} + v_n & \text{for } j = k \\ a_{ij} & \text{otherwise} \end{cases}$$

in any case $\sum_j a_{ij} = \sum_j a_{ij}'$ which establishes this part of the result.

PART II.   The x Is a Maximal Flow

It is sufficient to show that the value of x is equal to the value of a cut. Hence, by the max. flow-min, cut theorem x is the maximal flow. For suppose this is not true, we have a cut value corresponding to the maximal flow which is the minimal cut value, but it is greater than the cut value corresponding to the flow x; this is a contradiction.

At the termination of the process described above there is defined a set S of vertices, consisting of vertices P_i for which v_i, μ_i have been defined. Moreover, P_1 is a vertex of S and P_n is not a vertex of S.

Consider the set Γ of directed edges $P_i P_j$ such that $P_i \in S$ and $P_j \notin S$. The values $a_{ij} = 0$ for such pairs i, j as otherwise values v_j, μ_j would have been defined.

It remains to show that Γ is a cut whose value is equal to $\sum_j x_{1j}$ (the value of the flow.)

Γ is a cut. If there was a chain of the form $P_1 P_{i_1} \cdots P_{i_k} P_n$ with edges $P_1 P_{i_1} \notin \Gamma, \ldots, P_{i_k} P_n \notin \Gamma$ one could deduce (starting with $P_1 \in S$) that

$P_{i_1} \in S, \ldots, P_n \in S$ but this is a contradiction. The value of Γ is equal to the flow value. Observe that

$$\sum_j (C_{ij} - a_{ij}) = \begin{cases} 0 & 1 < i < n \\ \sum_j x_{ij} & i = 1 \end{cases}$$

Sum these equations over those i such that $P_i \in S$ (so it includes the flow value). On the left, if both P_i and P_j are in S then $C_{ij} - a_{ij}$ and $C_{ji} - a_{ji}$ are both in the summation and are negative of each other. All that remains are terms of the form $C_{ij} - a_{ij}$, where $P_i \in S$ and $P_j \notin S$. For these $a_{ij} = 0$ so the sum on the left is the sum of capacities of the elements of Γ, that is, the value of the cut. On the right the sum is the value of the flow $\sum_j x_{1j}$ ▲

An example for network flows will now be given. Consider the network shown in Figure 129.

The capacity values are shown in (14).

$$\begin{array}{llll}
C_{12} = 2 & C_{13} = 3 & C_{14} = 0 & C_{15} = 0 \\
C_{21} = 0 & C_{23} = 1 & C_{24} = 1 & C_{25} = 0 \\
C_{31} = 0 & C_{32} = 2 & C_{34} = 1 & C_{35} = 2 \\
C_{41} = 0 & C_{42} = 0 & C_{43} = 0 & C_{45} = 2 \\
C_{51} = 0 & C_{52} = 0 & C_{53} = 0 & C_{54} = 0
\end{array} \qquad (14)$$

As initial flow we may take (rather than the null flow which is further from the solution) as shown in (15).

$$\begin{array}{llll}
x_{12} = 1 & x_{13} = 1 & x_{14} = 0 & x_{15} = 0 \\
x_{21} = 0 & x_{23} = 0 & x_{24} = 1 & x_{25} = 0 \\
x_{31} = 0 & x_{32} = 0 & x_{34} = 0 & x_{35} = 1 \\
x_{41} = 0 & x_{42} = 0 & x_{43} = 0 & x_{45} = 1 \\
x_{51} = 0 & x_{52} = 0 & x_{53} = 0 & x_{54} = 0
\end{array} \qquad (15)$$

(Here the value of the flow is 2.) The initial A matrix is as in (16).

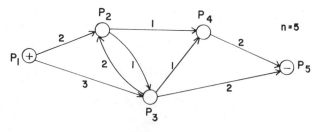

Figure 129

$$A_1 = \begin{pmatrix} 1 & 2 & 0 & 0 \\ 1 & 1 & 0 & 0 \\ 1 & 2 & 1 & 1 \\ 0 & 1 & 0 & 1 \\ 0 & 0 & 1 & 1 \end{pmatrix} = (a_{ij}) \tag{16}$$

Here $j = 1, \ldots, n$ we select labels v_j, μ_j as follows.

Let $v_1 = \infty$, $\mu_1 = 0$. Here $a_{1j} > 0$ for $j = 2$ and $j = 3$ so let $\mu_2 = 1$ and $\mu_3 = 1$. $v_j = a_{1j}$, $v_2 = 1$, $v_3 = 2$. See Table 10.

Table 10

i	defines
1	$v_1 = \infty \; \mu_1 = 0$ gives $j = 2, 3$
2	$v_2 = 1 \; \mu_2 = 1$ came from $i = 1$.
3	$v_3 = 2 \; \mu_3 = 1$

$i = 2$ gives nothing new
$i = 3 \; a_{3j} > 0 \to a_{31} = 1, a_{32} = 2, a_{34} = 1, a_{35} = 1,$

where a_{31} and a_{32} are not new.

$$v_4 = \min(v_3, a_{34}) = 1 \qquad \mu_4 = 3,$$
$$v_5 = \min(v_3, a_{35}) = 1 \qquad \mu_5 = 3.$$

Gives the new matrix (17) ($j = 5$, $\mu_5 = 3$ go to $j = 3$, $\mu_3 = 1$ then the modification is completed.)

$$\begin{pmatrix} a_{12} & a_{13}-1 & a_{14} & a_{15} \\ a_{21} & a_{23} & a_{24} & a_{25} \\ a_{31}+1 & a_{32} & a_{34} & a_{35}-1 \\ a_{41} & a_{42} & a_{43} & a_{45} \\ a_{51} & a_{52} & a_{53}+1 & a_{54} \end{pmatrix}$$

$$A_2 = \begin{pmatrix} 1 & 1 & 0 & 0 \\ 1 & 1 & 0 & 0 \\ 2 & 2 & 1 & 0 \\ 0 & 1 & 0 & 1 \\ 0 & 0 & 2 & 1 \end{pmatrix}$$

Now start second iteration

$$v_1 = \infty, \mu_1 = 0 \qquad a_{12} = 1, \qquad a_{13} = 1 \qquad \text{gives:}$$
$$v_2 = 1, \; \mu_2 = 1 \qquad\qquad\qquad\qquad i = 2 \text{ gives nothing new}$$
$$v_3 = 1, \; \mu_3 = 1 \qquad\qquad\qquad\qquad i = 3, a_{31} = 2, a_{32} = 2, a_{34} = 1$$

gives result for $j = 4$ as

$$v_4 = \min(v_3, a_{34}) = 1, \mu_4 = 3 \qquad a_{45} = 1, \qquad \text{gives for } j = 5$$
$$v_5 = \min(v_4, a_{45}) = 1, \mu_5 = 4$$

Gives new matrix (18) (start at $j = 5$, $\mu_5 = 4$ go to $j = 4$, $\mu_4 = 3$ go to $j = 3$, $\mu_3 = 1$)

$$\begin{pmatrix} a_{12} & a_{13}-1 \ a_{14} & a_{15} \\ a_{21} & a_{23} & a_{24} & a_{25} \\ a_{31}+1 \ a_{32} & & a_{34}-1 \ a_{35} \\ a_{41} & a_{42} & a_{43}+1 \ a_{45}-1 \\ a_{51} & a_{52} & a_{53} & a_{54}+1 \end{pmatrix} \tag{18}$$

$$A_3 = \begin{pmatrix} 1 & 0 & 0 & 0 \\ 1 & 1 & 0 & 0 \\ 3 & 2 & 0 & 0 \\ 0 & 1 & 1 & 0 \\ 0 & 0 & 2 & 2 \end{pmatrix}$$

Now $v_1 = \infty$, $\mu_1 = 0 \ a_{12} = 1 \ v_2 \equiv 1$, $\mu_2 = 1 \ a_{23} = 1 \ v_3 = 1$, $\mu_3 = 2 \ a_{31} = 3$, $a_{32} = 2$ gives nothing new. Hence the process is completed. A_3 will give the maximal flow solution.

$$x_{ij} = \max (C_{ij} - a_{ij}, 0),$$

first form (19).

$$(C_{ij}-a_{ij})= \begin{pmatrix} 1 & 3 & 0 & 0 \\ -1 & 0 & 1 & 0 \\ -3 & 0 & 1 & 2 \\ 0 & -1 & -1 & 2 \\ 0 & 0 & -2 & -2 \end{pmatrix}. \tag{19}$$

Then (x_{ij}) is as shown in (20).

$$\begin{pmatrix} 1^x{}_{12} & 3^x{}_{13} & 0 & 0 \\ 0 & 0 & 1^x{}_{24} & 0 \\ 0 & 0 & 1^x{}_{34} & 2^x{}_{35} \\ 0 & 0 & 0 & 2^x{}_{45} \\ 0 & 0 & 0 & 0 \end{pmatrix}$$

Note the matrix elements are not numbered in the usual way but follow the pattern written in detail above. The variables are shown within the matrix. The "sink row" will always be zero for a flow; it is the last row in the above matrix.

The solution is shown in Figure 130.

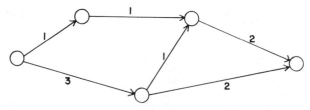

Figure 130

MEASUREMENT OF COMPLEXITY

In many areas of investigation where linear graph models are employed the question of relative complexity for such models arises. One is led to a study of complexity for a variety of reasons. The intrinsic information content of a graph model is one factor associated with complexity. Another factor is the possibility of ranking different graph models and of measuring their relative utility as models by means of information content or complexity. Quantifications of complexity may also form part of the quantified model structure and thus play a direct role in analyses based upon the linear graph model. The role of complexity and information content of graphs has been discussed in chemistry by Karreman [91], in biology by Rashevsky [92], and in various contexts by Trucco [93], Sabidussi [94], and others. In [95] Mowshowitz has remarked on the possibility of such models for the study of friendship relations in populations (the reader is invited to compare this with our group structure model for urban studies later in this chapter) and he has indicated their use as measures of algorithm complexity in [96]. In view of the wide range of interesting interpretations of graph complexity measurement it seems most appropriate to present a discussion of it here under operations research, a field of study which seems better specified by the methods it employs than by specific subject matter. Certainly quantification of complexity for a linear graph is very much in the spirit of operations research methods regardless of specific applications (e.g., to biology).

Though many authors have contributed to the concepts utilized in expressing a quantification of graph complexity the topic is given particularly detailed treatment by Mowshowitz in [95] and [96] (these references were brought to the author's attention by James A. Parsons). In the former reference Mowshowitz deals with undirected, finite graphs, without self loops or multiple edges and we shall base our discussion here on that paper. We shall not discuss his studies of digraphs and non-finite (undirected) graphs which may be found in [96].

The basic idea employed by Rashevsky, Trucco, Mowshowitz, and others is to formulate an entropy like expression utilizing a probability scheme determined by some partition of the vertices of the graph. If G is an n-graph with vertex set V and A_i, $i = 1, \ldots, h$ is a partition of V then the probability scheme is defined by associating the probability $p_i = |A_i|/n$ with the set A_i for $i = 1, \ldots, h$. In terms of these probabilities the information content $I(G)$ is defined to be the entropy like expression

$$I(G) = - \sum_{i=1}^{h} p_i \log_2 p_i$$

in which one should note that the logarithms are taken to the base two in the usual information theoretic way.

Two observations are immediately obvious concerning this definition of information content. The probability scheme employed is extremely simple in form, using direct enumerative terms. For this reason it may not be the most meaningful for some applications and as Mowshowitz points out it cannot be extended to infinite graphs. However, being the simplest type of definition also has many advantages. Often one would not have any reason for using a more involved alternative (complexity of a model should be inverse to one's ignorance about the situation being modeled) and the p_i have interesting and useful properties as several authors have shown. Thus we shall employ these p_i in this introduction to information content and complexity in graphs. The second observation is really much more important for our consideration. It is the fact that complexity and information content are undefined and indeed in each case they become defined by the way in which the partitions A_i are formed. One can form the partitions in many ways. One way to form the partitions is from a chromatic number coloring of a graph in which all vertices of the same color are placed into one set of the partition. Another way is to let the A_i be transitive sets or orbits associated with the automorphism group G_p of the graph G. The first of these methods for forming partitions has been mentioned by Mowshowitz but his main development is in terms of the automorphism groups. These also provide the framework for the work of most other authors on the information content subject, particularly Trucco who utilized the automorphism group in his definition of Rashevsky's information content.

The development of information content may be divided into two parts, the first dealing with basic concepts and their application to individual graphs, the second considering the effect of combining graphs in certain ways. We shall discuss this topic primarily from the first part, giving some indication of results from the second.

We are considering graphs with labelled vertices for which an automorphism of the graph corresponds to a permutation of the vertex labels which preserves adjacency. A graph automorphism is an isomorphism (Chapter 1) of the graph with itself. The set of all automorphisms of an n-graph G form a group which may be considered (is isomorphic to) a subgroup of the symmetric group S_n. The group S_n has $n!$ elements and represents the set of all permutations on n objects. The complete graph K_n on n vertices has S_n as its automorphism group as does the vertix graph with n vertices (and no edges). One may consider the other extreme for which the automorphism group consists of only one element, the identity permutation. Clearly K_1 has the identity as its automorphism group but that is rather a trivial case. The smallest non-trivial graph having the identity as its automorphism group is the 6-graph with adjacency matrix (21) and is attributed to Kagno by Ore in [15]. In [36] Harary calls graphs whose automorphism group is the identity

$$\begin{pmatrix} 0 & 1 & 0 & 0 & 0 & 0 \\ 1 & 0 & 1 & 0 & 1 & 0 \\ 0 & 1 & 0 & 1 & 0 & 1 \\ 0 & 0 & 1 & 0 & 0 & 1 \\ 0 & 1 & 0 & 0 & 0 & 1 \\ 0 & 0 & 1 & 1 & 1 & 0 \end{pmatrix} \tag{21}$$

permutation, identity graphs.

A transitive set or orbit of a group consists of a set of elements which transform into each other under the group elements. If G_p is the automorphism group of the graph G then a subset H of the vertex set V of G is an orbit of G_p if for all $\alpha \in G_p$ and all $i \in H$ $i\alpha \in H$. The orbits associated with an automorphism group form a partition of the vertex set of the graph. Thus the automorphism group and its orbits may be used to define the information content of a graph. It should be understood that when this is done the concept of graph complexity is specified by the orbit concept and may or may not correspond to specific needs in application or to subjective insights as to what complexity "should" mean. In the context of partitions specified by the orbits of automorphism groups the notion of complexity relates to an inability to make substitutions among vertices. All the vertices in an orbit may be thought of as playing a similar role within the structural model. In this sense graph models with a larger number of small size orbits are more complex than those having a few large size orbits. Such models certainly suggest themselves in chemistry and in certain areas of biology and the social sciences. However if we think of complexity as interrelation the automorphism definition seems to be much less useful or even undesirable and something like the coloring of vertices appears more appropriate as we shall illustrate below.

If an n-graph G is an identity graph then its vertices are partitioned into n orbits each of size one so that $I(G) = -\sum_{i=1}^{n} 1/n \log 1/n = \log n$. On the other hand the graph K_n with automorphism group S_n has only one orbit of n elements so that $I(K_n) = \log 1 = 0$ and this is also the information content of any vertex graph (without edges).

Consider the houses and utilities graph $K_{3,3}$ which as we have seen is a nonplanar, bichromatic 6-graph. If we number the vertices of one part (color) of this graph 1, 3, and 5 and those of the other part 2, 4, 6 we can form the automorphism group from the following considerations. The vertices 1, 3, 5 can interchange in 3! ways and, independently of such interchanges, the vertices 2, 4, 6 can interchange in 3! ways. Such permutations of vertices contribute $(3!)^2$ elements to the automorphism group of $K_{3,3}$. In addition the vertices 1, 3, 5 can transform in the vertices 2, 4, 6 in 3! ways and for each such the displaced indices 1, 3, 5 can transform into new positions in 3! ways. Thus the automorphism group of $K_{3,3}$ has 72 elements and is a subgroup of

S_6 which has $6! = 720$ elements. However $K_{3,3}$ only has one orbit of 6 elements under its automorphism group. Hence $I(K_{3,3}) = 0$ which for example is the same as $I(K_6)$. As we have remarked above this may be a satisfactory measure of complexity from the point of view of interchanging vertices but it does not indicate complexity in an interconnectedness context. However the situation changes when the partitions A_i are specified by sets of vertices of the same color and the number of sets is equal to the chromatic number. Let us denote the information content in this case by I_c. Then $I_c(K_6) = \log 6$ and $I_c(K_{3,3}) = \log 2 = 1$ so that K_6 has more complexity than $K_{3,3}$ and both have more than the vertex 6 graph for which the (chromatic) information content is zero. This is considerably more satisfactory as a measure of connectedness complexity but it is not ideal since the 6-cycle also has information content one though we should like to think of it as less (connectedness-wise) complex than $K_{3,3}$. We see from all this that the ideas presented are interesting and useful but in no sense provide a fully satisfactory theory of graph complexity.

The use of automorphism groups to define complexity has an operational disadvantage. One must find the automorphism group for a graph and more important one must find the orbits. At least one must find the number of orbits of each size. This is in fact equivalent to finding a specific partition of n corresponding to the automorphism group of the n-graph under study. There are some facts that are helpful in dealing with these calculations such as the statement that the complement $\bar{G}$ of a graph G has the same automorphism group as G. One may also construct the group from simpler groups when the graph has several relatively simple components. Questions of enlarging the concepts on the basis of simpler results form the second part of information content studies as we have mentioned above and we shall illustrate some results from that body of material. However the problem remains extremely difficult to treat in specific cases of interest once the number of vertices becomes sizeable (i.e., more than six). Computer search employing algorithms may be expected to be of help in solving such problems and of course special structure graphs can be solved by direct reasoning employing the special structure. We illustrate such special results by the following example of Mowshowitz given in [95].

Let G_n be a path of length n with vertices numbered consecutively 1 through n. Vertices 1 and n are of degree one while all other vertices are of degree two. Using the cycle notation for permutations we see that

$$\alpha = \begin{cases} (1, n)(2, n - 1) \cdots \left(\dfrac{n}{2}, \dfrac{n}{2} + 1 \right), & \text{for } n \text{ even,} \\[3mm] (1, n)(2, n - 1) \cdots \left(\dfrac{n + 1}{2} - 1, \dfrac{n + 1}{2} + 1 \right), & \text{for } n \text{ odd,} \end{cases}$$

is an automorphism of G_n. Under the automorphism α the set of elements $(1, n)$ transform into each other and therefore they must form a subset of some orbit of the automorphism group of G_n. Since vertices v_1 and v_n have degree one and no other vertices have degree one the pair $(1, n)$ must in fact be an orbit itself, since adjacency could not be preserved otherwise.

Now consider $G_n - (v_1, v_n)$ as a path with $n - 2$ vertices. By the argument used above the pair $(2, n - 1)$ is an orbit for this path. Since the $n - 2 -$ path is (isomorphic to) a subgraph of G_n the pair $(2, n - 1)$ is an orbit of G_n. Applying this reasoning we find that all the pairs that constitute the cycles of α are orbits. Thus if n is even G_n has $n/2$ orbits and if n is odd G_n has $(n + 1)/2$ orbits. The information content in each case is $I(G_n) = \log n/2$ for n even and $I(G_n) = (n - 1)/n \log n/2 + 1/n \log n$ if n is odd.

As an illustration of how the concepts may be extended by combining graphs consider the cartesian product of two graph H_1 and H_2 written $H = H_1 \times H_2$. If $V(G)$ and $E(G)$ denote respectively the vertex and edge sets of a graph G then $V(H) = V(H_1) \times V(H_2)$ and $E(H) = \{(v_1, v_2), (w_1, w_2)|v_1, w_1 \in V(H_1)$ and $v_2, w_2 \in V(H_2)$ and either $v_1 = w_1$ and $(v_2, w_2) \in E(H_2)$ or $v_2 = w_2$ and $(v_1, w_1) \in E(H_1)\}$. In such combinations one replaces each vertex of H_2 with an isomorphic copy of H_1 and adds edges between those copies whenever there is an edge in H_2 connecting the associated vertices. When one forms the cartesian product of a graph V with itself we have the useful result: the transformation Φ defined by $(i, j) \Phi = (j, i)$ for each $(i, j) \in V(Y \times Y)$, $1 \le i, j \le |V(Y)|$ is an automorphism of $Y \times Y$.

We may apply these definitions and results to the graph $G_n \times G_n$ which is the n by n plane square lattice graph which we denote by $L_{n,n}$. The vertex set of $L_{n,n}$ is $V(G_n \times G_n) = \{(i, j), 1 \le i, j \le n\}$ and the edges are the horizontal and vertical line segments comprising the square lattice. Let O_i be the orbits of G_n which we have found above so that

$$
O_i = \begin{cases}
(i, n - i + 1), & \left(1 \le i \le \dfrac{n}{2}\right) \text{ for } n \text{ even} \\[2mm]
(i, n - i + 1), & \left(1 \le i \le \dfrac{n + 1}{2} - 1\right) \\[2mm]
\left(\dfrac{n + 1}{2}\right), & \left(i = \dfrac{n + 1}{2}\right)
\end{cases} \Bigg\} \text{ for } n \text{ odd.}
$$

Mowshowitz shows that the orbits of $L_{n,n}$ are the sets $O_i \times O_i$ for $1 \le i \le n/2$ when n is even and $1 \le i \le (n + 1)/2$ when n is odd and also $(O_i \times O_j) \cup (O_j \times O_i)$ for $i \ne j$ with $1 \le i, j \le n/2$ when n is even and $1 \le i, j \le (n + 1)/2$ when n is odd. Calculation of the information content of $L_{n,n}$ requires a discussion of the combination of graphs as developed in [95] lying beyond the scope of our treatment here. However we conclude this tyopic by simply stating Mowshowitz's result for the square lattice $L_{n,n}$:

$$I(L_{n,n}) = \begin{cases} 2\log\dfrac{n}{2} - \dfrac{n-2}{n}, \text{ if } n \text{ is even} \\ 2\left(\dfrac{n-1}{n}\log\dfrac{n}{2} + \dfrac{1}{n}\log n\right) - \left(\dfrac{n-1}{n}\right)^2 \text{ if } n \text{ is odd.} \end{cases}$$

If we use the chromatic graph method for determining the partition of the vertices we find the information content measure for $L_{n,n}$ is as follows:

$$I_c(L_{n,n}) = \begin{cases} \log 2 & \text{if } n \quad \text{ is even} \\ -\dfrac{m-1}{2m}\log\dfrac{m-1}{2m} - \dfrac{m+1}{2}\log\dfrac{m+1}{2m} \end{cases}$$

if n is odd, where $m = n^2$ is the number of lattice vertices.

This follows from the fact that a plane square lattice graph is bipartite for any value of n.

OPTIMIZATION ON GRAPHS

(This topic was brought to the author's attention by E. Schroeppel.)

In the previous section we saw that the information content of a graph can be defined by means of partitions of its vertex set. Indeed the concept of vertex set partitions arises in a number of theoretical and applied aspects of graph theory. Here we wish to bring to the reader's attention a class of optimization problems that may be formulated in terms of vertex set partitions. Though the material discussed arose out of engineering design concepts we place it here under the general heading of operations research because the ideas and solution algorithms seem to have wide applicability.

An electrical network and in particular a printed circuit may be thought of as based on a graph where the vertices represent logical (or actual) elements and the edges represent connections. If a signal is to pass through such a network with prescribed direction the appropriate model is a digraph. Various design problems ([97]) consider the grouping of basic elements into sets which together with the connections between sets constitute the complete network. In those problems one is interested in the time required for a signal to traverse the complete network. It is supposed that the time required to pass between elements of the same set is negligible while a unit time is required to pass between elements of adjacent sets. Thus for a given network the goal is to collect the elements into sets in such a way as to minimize the signal passage time. If the connections are thought of as representing a unit distance in the interset case and zero distance in the intraset case the problem is to partition the elements into sets so that the maximum distance arising in the network is minimal. Other optimization problems may also be formulated by employing similar notions but we shall consider the minimal "distance" problem as a typical illustration.

Representation of these problems in terms of graphs is achieved by letting the vertices represent the basic elements of the network and the edges of the graph represent the connecting branches of the network. The problem is to partition the vertex set so that the maximum length path in the resulting network is minimum. Edges connecting vertices in the same set have zero length, edges between sets have unit length. Of course such a problem is trivial unless there exists some forms of constraints on the number of elements (vertices) allowed in any one set. In the cases studied here we shall require that any set have at most M vertices. Problems such as these are in general rather difficult to solve for any but small size networks where complete testing of (essentially) all cases is feasible. Lawler, Levitt, and Turner have developed an algorithm for the minimum distance problem in [97] and we shall discuss some aspects of their work here. The complete algorithm becomes somewhat involved and we refer the interested reader to [97]. As a reference note we should mention that this class of problems is sometimes spoken of as graph decomposition or logic partitioning, particularly in the (electrical) engineering literature.

One may consider the graph decomposition problem on various graphs and we shall consider directed trees and directed rooted trees. Since all graphs are directed in this discussion we shall assume directedness without specific reference in each case. A rooted tree is called a sink tree if the root vertex is the only vertex with zero out degree and is called a source tree if the root is the only vertex with zero in degree. Vertex v_i is a predecessor of vertex v_j if there exists a (directed) path from v_i to v_j. In this case v_j is a successor of v_i.

In developing algorithms for the graph decomposition problem [97] takes the following lemma as its starting point.

Lemma. There exists an optimal partition of the vertices such that the graph formed by the vertices lying in any part of the partition and their incident edges is a connected graph (in the undirected sense).

This lemma greatly simplifies the search for optimal partitions since it states that one need not consider partitions with any part being a non-connected graph. That is to say that the vertices in any part tend to be near to each other rather than widely separated on the graph.

It is convenient to attach a weight to each vertex which we denote by w_i for vertex v_i. The solution algorithms produce labels on the vertices which indicate the optimal partition. If the largest label assigned to any predecessor of vertex i is the number k we say all predecessors having label k are k-predecessors of vertex i in a particular labeling. The total weight of all

k-predecessors of vertex i is denoted by $w_i(k)$. It is the sum of all weights assigned to the k-predecessors of v_i.

Consider a sink tree for which the total weight in any part cannot exceed M. An optimal partition is to provide clustering of vertices so that the maximal path length employing intercluster edges is minimal. The solution algorithm from [97] is very simple in this case:

STEP 0. Label all source vertices zero.

STEP 1. Find any unlabeled vertex v_i, all of whose predecessors have labels. Let k be the largest label given to any predecessor of v_i. If $w_i + w_i(k) \leq M$, apply label k to v_i; otherwise use label $k + 1$.

Repeat Step 1 until all vertices are labeled.

Cluster all vertices having the same label that are connected subgraphs. This yields an optimal partition.

It is important to note that only connected subgraphs form clusters and every vertex in such a cluster has the same label. As an example of this algorithm Figure 131 shows an optimal clustering with minimal greatest distance of 2. Labels produced by the algorithm are shown at each vertex. In this example $M = 3$ is the maximum number of vertices allowed in any cluster.

One may also be interested in finding optimal partitions in which a minimal number of clusters (parts) are employed. The algorithm does not necessarily yield such a minimal part optimal partition. Reference [97] mentions that algorithmic techniques exist for reducing the number of parts. Often observation can be employed to reduce the number and the reader may consider Figure 131 with a view toward producing a new partition in which the $M = 3$ condition is maintained but the number of parts is less (there are several).

Reference [97] points out that a source tree can be optimally partitioned by an algorithm similar to the one given above by changing sink to source and

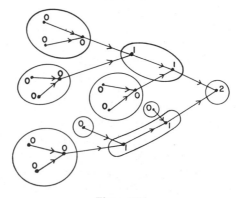

Figure 131

predecessor to successor. Nonrooted trees are dealt with by a more involved algorithmic method. This requires the division of the tree into rooted trees, labeling of the resulting rooted trees by means of the sinktree and source tree algorithms given above, and testing for an optimal partition of the total. In case the result is not optimal one carries out modification of the labeling. The algorithm is based on the following lemma.

Lemma. For a nonrooted tree, there exists a set of edges E such that removal of those edges separates the tree into rooted trees, and each source-sink path in the tree contains at most one edge of E.

The nonrooted tree is separated by means of the following algorithm.

STEP 0. Select any source vertex and all of its successor vertices as an initial source tree. Mark all vertices of this tree.

STEP 1. Find an edge, exactly one end vertex of which is marked and put that edge in the set E. If the marked vertex has the edge as an inward edge (outward edge) the vertex at the other end of the edge and all of its predecessors (successors) is selected as a sink (source) tree. Mark each vertex of that tree.

Repeat Step 1 until every vertex is marked. The set E will consist of all edges put into it by this process. Removal of E will result in separation of the original tree into rooted source and sink trees.

The complete optimal selection algorithm is not difficult, but is more involved than we wish to present here for the general reader. Rather than continue the details of that algorithm we turn to a rather different type of optimal partitioning due to Gorinshteyn and discussed in [98].

Gorinshteyn considers the minimization of connectedness between the parts of a partition. That is the number of interpart edges is to be minimized. It is assumed that the number of vertices in each part is fixed. This is done in terms of a specific partition of n, the number of vertices, into r parts m_1, $m_2, \ldots, m_r$. Thus it is desired to find, from all partitions of the n vertices into r parts having the prescribed sizes (number of vertices), a partition with minimal number of interpart edges. This problem differs from the previous graph decomposition in several ways. The reader should note in particular that the number of vertices in a part is not bounded from above as before but is specified exactly. Of course the objective being minimized is different in each case as well and the present problem is considered on undirected graphs.

Some form of efficient algorithmic procedure is clearly required in a problem of this kind since the total number of possible partitions is (with $m_0 \equiv 0$)

$$\prod_{k=1}^{r=1} \binom{n - m_1 - m_2 - \cdots - m_{k-1}}{m_k},$$

which is extremely large for any non-trivial values of n and r.

The procedure presented in [98] begins with an initial partition and develops new partitions in an efficient way until an optimal partition is reached. The procedure employs bounds on the number of interblock edges so that large classes of partitions may be implicitly tested as candidates for optimality. This is the idea behind what are commonly referred to as branch and bound methods.

Let V_i denote the set of vertices in the ith part of the initial partition for $i = 1, \ldots, r$. Any other partition can be formed by transferring vertices from parts of the initial partition into other parts. Thus if V_i^* denotes the i part composition of another partition

$$V_i^* = \bigcup_{j=1}^{r} V_{ji}$$

where V_{ji} is the set of vertices transferred from V_j to the new ith part V_i^*. Let $v_{ji} = |V_{ji}|$ then

$$0 \leq v_{ji} \leq \min (m_j, m_i), \tag{22}$$

$$\sum_{j=1}^{r} v_{ji} = m_i, \tag{23}$$

$$\sum_{i=1}^{r} v_{ji} = m_j. \tag{24}$$

The numbers v_{ji} constitute elements of an $r \times r$ matrix whose jth row indicates how the vertices of V_j transform into new sets. The ith column shows how the vertices arrive into the new V_i^* set.

All partitions of the vertex set V can be divided into equivalence classes in which all partitions in a particular class have the same matrix $K = (v_{ji})$. The number of classes is determined by the number of different sets of solutions to the above conditions (22, 23, and 24) for the elements v_{ji} of K. The reader should note that we are considering every set of m vertices as distinct whereas it might be necessary, due to the incidence structure of a graph, to consider several sets of m vertices the same. Such consideration complicates matters a good deal, particularly in a general discussion, and also in solution techniques, so we shall only consider the case where all sets are distinct. In case some of the partition numbers m_i have the same value symmetry makes some cases the same so that not all solutions to the system of constraints on the v_{ij} yield meaningful transformation matrices. For example Gorinshteyn discusses an example where $n = 9$, $r = 3$, and $m_i = 3$ for $i = 1, 2, 3$. In that case he shows nine distinct classes with characterizing transformation matrices. The matrix

$$K' = \begin{pmatrix} 3 & 0 & 0 \\ 0 & 0 & 3 \\ 0 & 3 & 0 \end{pmatrix}$$

does not represent a transformation but merely a rearrangement of the parts and illustrates our remark about symmetry. The elements of K' satisfy the v_{ji} conditions but K' is not a meaningful transformation; it does not yield a different partition. Three of the nine matrices that do give significant transformations are:

$$\begin{pmatrix} 2 & 1 & 0 \\ 1 & 2 & 0 \\ 0 & 0 & 3 \end{pmatrix}, \quad \begin{pmatrix} 2 & 1 & 0 \\ 0 & 2 & 1 \\ 1 & 0 & 2 \end{pmatrix}, \quad \begin{pmatrix} 2 & 1 & 0 \\ 1 & 1 & 1 \\ 0 & 1 & 2 \end{pmatrix}$$

and the reader may wish to construct the remaining six matrices. As a further very simple example consider the case where $n = 4$, $r = 2$, $m_1 = m_2 = 2$ for which there is a total of six permutations if all are distinct, shown by the three cases in Figure 132 and three more in which V_1 and V_2 are interchanged. If we take the first case in Figure 132 as the initial partition, $V_1 = (1, 3)$, $V_2 = (2, 4)$ then the transformation matrices are

$$\begin{bmatrix} 1 & 1 \\ 1 & 1 \end{bmatrix} \quad \text{and} \quad \begin{bmatrix} 2 & 0 \\ 0 & 2 \end{bmatrix}$$

but the latter (giving the three cases omitted from the Figure) does not correspond to significantly different partitions, that is, the connectivity cannot be changed in those cases.

The algorithmic method developed by Gorinshteyn employs a lower bound estimate c_p for the number of interpart edges b_p at the stage p. Let $S(A, B)$ denote the number of edges with one end incident on a vertex of set A and the other end incident on a vertex of set B. Let b_i denote the number of interpart edges (connectedness) of the initial partition. Then

$$b_p = b_i + \sum_{\substack{i, j \\ i \neq j}} s(V_{ji}, V_j - V_{ji}) - \sum_{\substack{i, j, k \\ j \neq k, j \neq i}} s(V_{ji}, V_{ki})$$

where the terms count edges added or lost as interpart edges when vertices are transformed into different parts due to the modification of the initial partition to form the pth partition. This expression can be transformed into the following more useful form:

$$b_p = b_i + \sum_{\substack{i \neq j}} s(V_{ji}, V_j - V_{ji}) - \sum_{\substack{i, j, k \\ i \neq j \\ j \neq k}} s(V_{ji}, V_k) + \sum_{\substack{i, j, k, l \\ i \neq j, j \neq k \\ l \neq i}} s(V_{ji}, V_{kl}).$$

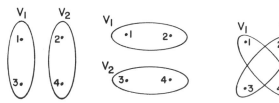

Figure 132

If the last term is removed from this expression we obtain the lower bound $c_p \leq b_p$ on connectedness which can be put in the form:

$$c_p = b_i + \sum_j g_j{}^v$$

where

$$g_j = \sum_{i,\, i \neq j} s(V_{ji}, V_j - V_{ji}) - \sum_{\substack{i,\, k, \\ i \neq j,\, j \neq k}} s(V_{ji}, V_k)$$

and the superscript v indexes the different variants of partition within a class. The range of v for each class x and for each part i is specified by the quantity:

$$\tau(x, i) = \prod_{k=1}^{r=1} \binom{m_i - v_{i1} - \cdots - v_{i(k-1)}}{v_{ik}}$$

where $v_{i0} \equiv 0$.

For the ith part, $\tau(x, i)$ part components $g_j{}^v$ for $v = 1, 2, \ldots, \tau(x, i)$ are computed. Thus for any class of partitions x we have the quantities:

$$G_i{}^x = (g_i{}^1, g_i{}^2, \ldots, g_i{}^{\tau(x, i)}) \qquad \text{for} \quad i = 1, \ldots, r.$$

From these we can calculate all estimates c_p of connectedness of partitions of the class x by using one element from each set $G_i{}^x$. These are obtained in increasing order of size by sorting the sets $G_i{}^x$ and then sorting some estimates of connectedness.

The branch and bound algorithm proceed as follows.

Formulate all transformation matrices for the partition classes. Calculate the $G_i{}^x$ sets for each class and also the minimal estimate c^x for each class. Order the classes according to the minimal estimates c^x of connectedness.

Start with an initial partition having connectedness b_p (try and select a partition with small connectedness to reduce the work) and set the running value d of minimal connectedness equal to the initial value b_p.

Select the class with minimal c^x. Find the partition of that class that corresponds to the estimate c^x. This is done by noting the index values of the $g_i{}^x$ used to calculate c^x. For that partition calculate b_p and test its value against the current d value. If b_p is less we have found a specific partition that improves the connectedness so we change d to the new (smaller) value. Continue to examine the partitions of the same class for which the lower bound estimates are less than the current d value. Once a lower bound is reached, in this sorting process, which equals or exceeds d we drop further considerations of the class and move to another class. Of course at each step one computes b_p and updates d as necessary.

The next class is selected as the next highest c^x value. If that value is equal to or greater than the current d then no further class must be examined and one already has the minimum partition corresponding to the partition that specified d.

The method is involved and requires considerable background calculations for its utilization. However it is a formal method for solving a difficult problem. Gorinshteyn gives an example with $n = 9$, $r = 3$ and $m_i = 3$ for $i = 1, 2, 3$ which has a total of 280 possible partitions. His algorithmic method produces a solution after an examination of only 29 partitions which of course is a considerable improvement over complete examination.

It is important to note that a major step in the method is to list the lower bound values for all combinations of the g_i^x within a class that has been selected for examination. One then sees which partitions of the class need to be studied since they have bounds that are less than the current d value. One also sees the order in which to study those partitions.

As an example of Gorinshteyn's method for minimal connectedness partitioning consider the graph G in Figure 133. In this example $n = 6$ and we require $r = 3$ with $m_1 = 3$, $m_2 = 2$, and $m_3 = 1$. It is simple to see that the minimum connectedness is 5 for graph G under the given partition structure. Many partitions yield that minimum and we will carry out the algorithm until it terminates to illustrate the several aspects of the method.

The partition $(1, 6, 5)$ $(2, 3)$ (4) is taken as the initial partition, for which $b_p = 7$. There are five classes of partitions, specified by the following transformation matrices:

$$K_1 = \begin{pmatrix} 3 & 0 & 0 \\ 0 & 1 & 1 \\ 0 & 1 & 0 \end{pmatrix}, \qquad K_2 = \begin{pmatrix} 2 & 0 & 1 \\ 0 & 2 & 0 \\ 1 & 0 & 0 \end{pmatrix}, \qquad K_3 = \begin{pmatrix} 2 & 1 & 0 \\ 1 & 1 & 0 \\ 0 & 0 & 1 \end{pmatrix},$$

$$K_4 = \begin{pmatrix} 2 & 1 & 0 \\ 1 & 0 & 1 \\ 0 & 1 & 0 \end{pmatrix}, \qquad K_5 = \begin{pmatrix} 1 & 1 & 1 \\ 1 & 1 & 0 \\ 1 & 0 & 0 \end{pmatrix}.$$

The necessary background data are presented in Table 11.

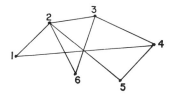

Figure 133

Table 11

Partition Part i	Class x	Row of Matrix $v_{i1}\ v_{i2}\ v_{i3}$	Partition of sets V_i				Part Components $g_i{}^v$
			v	V_{i1}	V_{i2}	V_{i3}	
1	3, 4	2 1 0	1	1, 6	5		-1
			2	1, 5	6		-2 (min)
			3	5, 6	1		-1
	2	2 0 1	1	1, 6		5	-1 (min)
			2	1, 5		6	0
			3	5, 6		1	-1
	5	1 1 1	1	1	5	6	-1
			2	1	6	5	-3 (min)
			3	5	1	6	-1
			4	5	6	1	-3
			5	6	1	5	-2
			6	6	5	1	-2
	1	3 0 0	1	1, 6, 5			0 (min)
2	3, 5	1 1 0	1	2	3		-2 (min)
			2	3	2		0
	4	1 0 1	1	2		3	-3 (min)
			2	3		2	-1
	2	0 2 0	1		2, 3		0 (min)
	1	0 1 1	1		2	3	0 (min)
			2		3	2	1
3	1, 4	0 1 0	1		4		-1 (min)
	2, 5	1 0 0	1	4			-1 (min)
	3	0 0 1	1			4	0 (min)

The minimal lower bounds are computed from the minimal $g_i{}^v$ values for each class, for example, for class K_2 the bound is $c^2 = b_p + g_1{}^2 + g_2{}^2 + g_3{}^2$ where the $g_i{}^v$ are the minimal values. Thus $c^2 = 7 - 1 + 0 - 1 = 5$ and similarly we find $c^1 = 6$, $c^3 = 3$, $c^4 = 1$, and $c^5 = 1$. We take K_4 as our starting class since it is the lowest number class (an arbitrary tie breaking device) with minimal c^x.

$$G_1^4 = (-1, -2, -1), G_2^4 = (-3, -1), G_3^4 = (-1)$$

which leads to the following cases in increasing order of the lower bound estimate:

$$c = 1 \text{ yields } g_1^2 + g_2^1 + g_3^1$$

and we see that g_1^2 yields (1, 5) (6) ( ), g_2^1 yields (2) ( ) (3) and g_3^1 yields ( ) (4) ( ) as the contributions to the parts. This case, therefore, corresponds to the partition (1, 2, 5) (4, 6) (3).

Each entry in the algorithm table (Table 12) is obtained in a similar way.

Table 12

p	Partitions			c_p	b_p	d_p	
	V_1	V_2	V_3				
0	1, 6, 5	2, 3	4		7	7	Initial partition
1	1, 2, 5	4, 6	3	1	6	6	
2	1, 2, 6	4, 5	3	2	5	5	← Optimal partition
3	3, 5, 6	1, 4	2	2	6	5	Class K_4
4	1, 3, 5	4, 6	2	3	8	5	
5	1, 3, 6	4, 5	2	4	6	5	
6	3, 5, 6	1, 4	2	4	6	5	
7	1, 2, 4	3, 6	5	1	5	5	
8	2, 4, 5	3, 6	1	1	5	5	
9	2, 4, 6	1, 3	5	2	7	5	
10	2, 4, 6	3, 5	1	2	7	5	Class K_5
11	1, 2, 4	3, 5	6	3	5	5	
12	1, 3, 4	2, 6	5	3	5	5	
13	2, 4, 5	1, 3	6	3	6	5	
14	3, 4, 5	2, 6	1	3	5	5	
15	3, 4, 6	1, 2	5	4	5	5	
16	3, 4, 6	2, 5	1	4	5	5	
17	1, 2, 5	3, 6	4	3	5	5	
18	1, 6, 2	5, 3	4	4	6	5	Class K_3
19	5, 6, 2	1, 3	4	4	6	5	
							Algorithm ends

We search in one class until the lower bound is one unit below the present d value (smallest connectivity so far) or until the class has no more entries. Then we go to the next highest lower bound estimate class and so forth. We terminate when there are no classes remaining with estimates that are less than (by one unit) the best value obtained for d.

Because of the simple form of the graph G the reader will not be impressed by the algorithm as a solution method for this example! However, it is intended to make the method clear so that it may be applied in cases where the solution is not at all as obvious as in the example. Certainly the method is not simple to use. In fact it is difficult to use; but it is intended for the solution of very difficult problems. It is interesting to note the occurrence of several optimal partitions in the table. We must take the algorithm to termination even though we have obtained optimal partitions (in this example our second case was optimal) since before termination we do not know the partitions are optimal. Of course in this simple problem we do in fact know optimal partitions but in general we would not.

EXERCISES

1. For the following PERT network calculate expected time t_e for each activity, Calculated Expected time T_E and Latest Allowable time T_L for each event. Find the slack for each event. Find a critical path through the network.

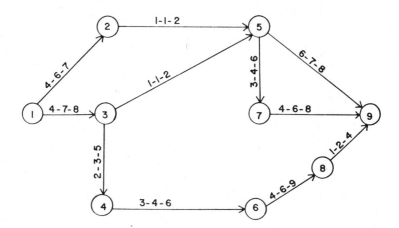

2. Using the Ford-Fulkerson algorithm find the maximum flow in the following network with arc capacities as shown [use the initial flow: $X_{12} = 3$, $X_{25} = 3$, $X_{57} = 3$, $X_{13} = 2$, $X_{34} = 2$, $X_{46} = 2$, $X_{67} = 2$, all other $X_{ij} = 0$]. Find a

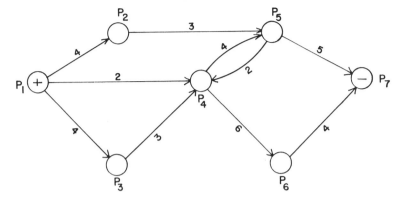

cut having value equal to the maximum flow. Can you find a cut that does not disconnect source and sink? Do such cuts always exist in standard flow networks?

3. Use the cascade algorithm to compute the path lengths between vertices of the 5-graph and the 10-graph with (adjacent) distance matrices given below. A dash as a matrix element denotes non-adjacent vertices (used to designate a value of infinity):

$$
\begin{pmatrix}
0 & 2 & - & - & 1 \\
2 & 0 & 8 & - & 2 \\
- & 8 & 0 & 2 & - \\
- & - & 2 & 0 & 3 \\
1 & 2 & - & 3 & 0
\end{pmatrix},
\qquad
\begin{pmatrix}
0 & 6 & - & 8 & - & 1 & 2 & - & 10 & - \\
6 & 0 & 1 & - & 1 & - & - & - & - & - \\
- & 1 & 0 & - & - & - & 1 & - & - & - \\
8 & - & - & 0 & 2 & - & - & - & - & 1 \\
- & 1 & - & 2 & 0 & - & - & - & - & - \\
1 & - & - & - & - & 0 & - & 10 & - & 10 \\
2 & - & 1 & - & - & - & 0 & - & - & - \\
- & - & - & - & - & 10 & - & 0 & 10 & - \\
10 & - & - & - & - & - & - & 10 & 0 & - \\
- & - & - & 1 & - & 10 & - & - & - & 0
\end{pmatrix}
$$

4. Find the information content in the wheel on 5 vertices. Use coloring to define the vertex partition. Use the transitive sets of the automorphism group to define the vertex partition. Compare the two results.

5. Apply the optimal partitioning algorithm to the sink tree shown below. Find an optimal partitioning for $M = 3$ and one for $M = 4$. Compare the results.

6. Apply Gorinshteyn's method with $n = 7, r = 3, m_1 = 2, m_2 = 3$, and $m_3 = 2$ to find a minimal interpart edge partition for the graph with adjacency matrix A.

$$A = \begin{pmatrix} 0 & 1 & 0 & 0 & 1 & 1 & 0 & 3 \\ 1 & 0 & 1 & 0 & 1 & 0 & 0 & 3 \\ 0 & 1 & 0 & 1 & 0 & 0 & 1 & 3 \\ 0 & 0 & 1 & 0 & 1 & 1 & 1 & 4 \\ 1 & 1 & 0 & 1 & 0 & 0 & 0 & 3 \\ 1 & 0 & 0 & 1 & 0 & 0 & 1 & 3 \\ 0 & 0 & 1 & 1 & 0 & 1 & 0 & 3 \end{pmatrix}$$

Chapter 9

Applications in Social Science and Psychology

Graph theory and related topics are providing important mathematical tools for analytic studies in fields where other branches of mathematics, though highly successful in Physics and related sciences, have not been particularly useful. As these applications are more fully developed there is an indication that the models constructed will not only be extremely useful to the science for which they are designed but allow the application of many other types of mathematics to be applied in these areas as well. The powerful methods of mathematical analysis have proven useful in the physical sciences due to the physical structure already present in these subjects upon which such methods could operate. In less structured sciences such as the Social Sciences, Psychology and Biology the same methods of analysis could not be applied. Graph theoretical models may provide sufficient structure in these subjects upon which the analytic methods can operate.

The use of graph models alone provides interesting and useful applications in many subjects. We wish to indicate the type and range of these applications by a few examples in this chapter.

Before turning specifically to examples in the Social Sciences we mention an application of graph theory to genetics. Applications of graph theory in Biology are related to applications in the Social Sciences because of their role in providing structural models in both fields. The applications in Physics and Engineering are more well structured situations.

One problem in genetics is the testing of various models of gene linkage. One can consider a linear linkage model and test such a model by means of experiments. For microorganisms having a standard form and mutant forms the mutants arise due to alterations in the connections of the gene structure. In the experiments one determines whether or not the modified parts of two mutant genes intersect in the (gene) structure or not. An undirected graph can be formed from such data. The vertices represent mutant genes and any pair is connected if their modified parts are found to intersect. Thus a test for linear structure corresponds to testing the resulting graph to see if it is an interval graph as discussed by Fulkerson and Gross in [16]. We have discussed interval graphs, following [16] in Chapter 1. An alternative model for the gene structure is the circular-arc graph as discussed by Tucker [17]. These applications indicate an interesting intersection of experimental and theoretical genetics, statistical analysis of data, and graph theory.

ORGANIZATIONAL STRUCTURE

The members of an organization can be represented by vertices and ranking within the organization indicated by directed edges of a digraph. This yields an abstraction of the familiar organization chart used in corporation studies. If an edge is directed from vertex a to vertex b this means that individual a is the superior of individual b within the organization. Such a representation is shown in Figure 134.

There are many uses of such a representation. One of the most interesting is

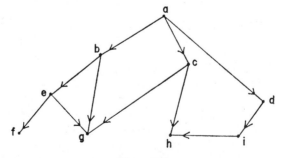

Figure 134

the definition of Social Status within an organization. This illustrates how a concept such as status can be quantified with the help of graphs.

An organization graph should contain no (directed) cycles for when this occurs a person could be his own superior and be subordinate to another to whom he is superior, etc. Thus we shall usually assume that an organization graph is a cycle free digraph. Although this assumption is reasonable, it may in fact fail to hold in some organization structures. For example among birds it is not uncommon to find cyclic pecking order (e.g., see Collias [99]).

The level of subordinacy is the length of the shortest path from the superior to the subordinate. In Figure 134, i is two levels below a, h is one level below i and g is two levels below a (note a path of length three exists between a and g but two is the shortest length).

Let $S(X)$ denote the status of individual X within the organization. We first formulate reasonable properties that might be desired of a quantitative measure of status; then one such measure will be defined.

1. $S(X)$ shall be integral valued.
2. If X is the superior of no individual then $S(X) = 0$.
3. If a new subordinate is added to X without other organizational changes, then $S(X)$ increases.
4. If a subordinate of X is moved to a lower level relative to X without other organizational changes, then $S(X)$ increases.

A measure of status satisfying these four properties is $h(x)$ defined by

$$h(X) = \sum_k k n_k$$

where individual X has n_k subordinates at level k. This measure has an important property. Among all possible measures satisfying the four desirable properties there are none having lower values. Thus if $m(X)$ is a measure satisfying the four desirable properties we have $m(X) \geq h(X)$ for all X.

A fifth desirable property for a status measure is that if a is the superior of b then $S(a) > S(b)$. This may not be the case for $h(X)$ defined above. Consider the graph in Figure 135.

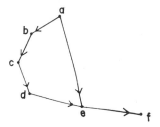

Figure 135

$$h(a) = 1(2) + 2(2) + 3(1) = 9$$
$$h(b) = 1(1) + 2(1) + 3(1) + 4(1) = 10$$

hence though a is superior to b we have $h(a) < h(b)$. Property 5 may not hold for the measure $h(x)$.

If the definition of level is charged to the length of the longest path from superior to subordinate the measure $h(x)$ will satisfy condition 5 as well as the other four.

It is characteristic of these applications that measures are seldom unique and selection from among several appropriate measures is usually difficult.

ENUMERATION OF WAYS

In the study of organizational structure and also in many other applications of graph theory, it is important to be able to give the number of (distinct) simple paths or ways of various lengths between any pair of vertices. If A is the adjacency matrix of a graph G we have seen that the elements of A^k give all edge sequences of length k between pairs of vertices. That simple result is of little use since the enumeration includes cases with edges repreated, cycles and so forth. The enumeration of distinct ways of length k between each pair of vertices of an n-graph G suggests itself as a natural problem in graph theory. However, it is interesting that the main results in this area are to be found in three papers all of which are from the journal Psychometrika which deals with mathematical techniques in the social psychological sciences. This is a good illustration of how an essentially mathematical question finds its solution as a result of requirements arising from applied research. There are two distinct approaches to the enumeration of ways. One method is to formulate a direct expression for the number of ways; the other method utilizes an iterative approach giving ways of length k in terms of the number of $k - 1$ ways. Neither result is simple; the former seems useful for ways of length four or less. The latter is probably better for larger ways. We shall describe both procedures.

A procedure for the direct evaluation of the number of ways (called paths in the references) is given by Ross and Harary in [100]. It employs a formal framework upon which one creates specific results by detailed analysis of each length of way. The formal framework utilizes special partitions of the integer k, the way length, (i.e., number of edges in the way). One considers special partitions of k into three parts

$$k = k_1 + k_2 + k_3$$

where k_1 and k_3 are nonnegative but not both zero and $k_2 > 1$. Moreover, as the subscripts indicate, the order of the parts is significant. Let us call paths that are not ways, redundant paths. Such paths may be related to the special three part partitions. A redundant path from v_i to v_j satisfies the partition (k_1, k_2, k_3) if the number of edges from the initial vertex v_i to any vertex which is repeated other than that vertex's last appearance, is k_1; the number of edges from this appearance of the repeated vertex to a later appearance is k_2; and the number of edges from this second appearance to the terminal vertex v_j is k_3. In determining the number of redundant paths, we must note that some such paths satisfy more than one partition. However, there can be two partitions of k (of the special type which we shall refer to merely as partitions in the remainder of this section) such that no redundant path satisfies both.

Two facts are very useful. The maximum number of partitions which are satisfied by a redundant k-length path is: $u = 2 \binom{\lceil \frac{k+1}{2} \rceil}{2}$. The number of partitions of k is $q = \binom{k}{2} - 1$. The proof of the first result may be considered an exercise; the proof of the second result is as follows. The number of partitions of k may be considered the number of partitions of k or less into two parts. The first part is non-negative and the second part is greater than one but less than k, order of parts being significant. When the second part is $k - 1$ the first part may be 0 or 1 (note that the third part is fixed by our selection of the first two parts). As the second part is decreased by one the number of possible first parts increases by one. This leads to a total number of $2 + 3 + \cdots + (k - 1) = \binom{k}{2} - 1$ partitions.

Let r_{ij} denote the number of redundant paths from v_i to v_j. Let $p_1, p_2, \ldots, p_q$ denote the q partitions of k and let a_x be the number of redundant paths satisfying p_x. Let $a_{x_1 x_2}$ be the number of redundant paths satisfying both p_{x_1} and p_{x_2} and in general $a_{x_1 x_2 \cdots x_t}$ be the number of paths satisfying all t of the partitions $p_{x_1}, \ldots, p_{x_t}$. Further, let $A_t = \sum a_{x_1} \cdots a_{x_t}$ where the sum is over all subscripts satisfying $x_1 < x_2 < \cdots < x_t$. So that for example

$$A_1 = \sum_{x=1}^{q} a_x, \quad A_2 = \sum_{\substack{x, y-1 \\ x < y}}^{q} a_{xy},$$

and so forth.

The principle of inclusion and exclusion then yields the result:

$$r_{ij} = A_1 - A_2 + A_3 - \cdots + (-1)^{1+u} A_u.$$

In this expression the a_x values depend on the particular graph under investigation as expressed by its adjacency matrix A.

If we let $R_k^{(t)}$ have as its (i, j) entry the sum A_t then the matrix $R_k = (r_{ij})$ is given by the formula:

$$R_k = \sum_{t=1}^{u} (-1)^{t+1} R_k^{(t)},$$

which is the matrix of k length redundant paths in a graph G.

The matrix of k-length ways, denoted by W_k, is then given by $W_k = (A^k - R_k) - d(A^k - R_k)$, where $d(H)$ is a matrix formed from a matrix H by setting all off diagonal elements to zero and retaining the diagonal elements unchanged. The second term arises because the diagonal elements of A^k and R_k will not in general both enumerate all redundant cycles; however, we want the diagonal elements of W_k to be zero since we are enumerating non-cyclic ways.

Elementwise matrix multiplication is useful in expressing some of the results. It is written $A \times B$ and defined by $A \times B = (a_{ij} b_{ij})$.

Application of the general structural formula above is possible for any k in principle but gets extremely involved for larger values of k. However, for $k = 3$ and $k = 4$ the results are not too complicated.

For $k = 3$ we have $R_3 = R_3^{(1)} - R_3^{(2)}$. The partitions of 3 are $p_1: 1 + 2 + 0$, and $p_2: 0 + 2 + 1$. The matrix of all three edge paths satisfying p_1 is $A \cdot d(A^2)$ since one first goes from one vertex to a different vertex v_x in one step then returns to the same (repeated) vertex v_x in two steps. Similarly the 3 edge paths satisfying p_2 are enumerated by $d(A^2) \cdot A$. Thus $R_3^{(1)} = A \cdot d(A^2) + d(A^2) \cdot A$.

The matrix of all 3 edge paths satisfying both partitions is $A \times A^T$ (where A^T is A transpose). Therefore $R_3 = (A \cdot d(A^2) + d(A^2) \cdot A) - A \times A^T$. From this we can obtain $W_3 = (A^3 - R_3) - d(A^3 - R_3)$.

Higher values of k become involved; we shall merely state the result of R_4, formulas for R_5 and R_6 are given in Ross and Harary [100].

$$R_4 = (A \cdot d(A^3) + d(A^3) \cdot A) + (A^2 \cdot d(A^2) + d(A^2) \cdot A^2) + A \cdot d(A^2) \cdot A$$
$$- A \times (A^2)^T - 2(A \times A^T) \times A^2 - (A \cdot (A \times A^T) + (A \times A^T) \cdot A).$$

We shall illustrate the above results for ways of length $k = 3$ in the graph G of Figure 136. Of course, the ways of length one are given by the elements of

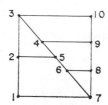

Figure 136

A and paths (including redundant paths) of length two are given by A^2. For the graph G we find matrices (1) and (2).

$$A^3 = \begin{pmatrix}
0 & 4 & 0 & 2 & 1 & 2 & 4 & 1 & 1 & 1 \\
4 & 0 & 6 & 0 & 6 & 1 & 1 & 2 & 2 & 0 \\
0 & 6 & 0 & 7 & 0 & 2 & 1 & 2 & 0 & 5 \\
2 & 0 & 7 & 0 & 6 & 1 & 2 & 1 & 6 & 0 \\
1 & 6 & 0 & 6 & 0 & 5 & 2 & 2 & 1 & 3 \\
2 & 1 & 2 & 1 & 5 & 2 & 5 & 5 & 2 & 1 \\
4 & 1 & 1 & 2 & 2 & 5 & 2 & 5 & 1 & 1 \\
1 & 2 & 2 & 1 & 2 & 5 & 5 & 2 & 5 & 0 \\
1 & 2 & 0 & 6 & 1 & 2 & 1 & 5 & 0 & 5 \\
1 & 0 & 5 & 0 & 3 & 1 & 1 & 0 & 5 & 0
\end{pmatrix},$$

(1)

$$R_3 = \begin{pmatrix}
0 & 4 & 0 & 0 & 0 & 0 & 4 & 0 & 0 & 0 \\
4 & 0 & 5 & 0 & 5 & 0 & 0 & 0 & 0 & 0 \\
0 & 5 & 0 & 5 & 0 & 0 & 0 & 0 & 0 & 4 \\
0 & 0 & 5 & 0 & 5 & 0 & 0 & 0 & 5 & 0 \\
0 & 5 & 0 & 5 & 0 & 5 & 0 & 0 & 0 & 0 \\
0 & 0 & 0 & 0 & 5 & 0 & 5 & 5 & 0 & 0 \\
4 & 0 & 0 & 0 & 0 & 5 & 0 & 5 & 0 & 0 \\
0 & 0 & 0 & 0 & 0 & 5 & 5 & 0 & 5 & 0 \\
0 & 0 & 0 & 5 & 0 & 0 & 0 & 5 & 0 & 4 \\
0 & 0 & 4 & 0 & 0 & 0 & 0 & 0 & 4 & 0
\end{pmatrix}$$

$$W_3 = \begin{pmatrix}
0 & 0 & 0 & 2 & 1 & 2 & 0 & 1 & 1 & 1 \\
0 & 0 & 1 & 0 & 1 & 1 & 1 & 2 & 2 & 0 \\
0 & 1 & 0 & 2 & 0 & 2 & 1 & 2 & 0 & 1 \\
2 & 0 & 2 & 0 & 1 & 1 & 2 & 1 & 1 & 0 \\
1 & 1 & 0 & 1 & 0 & 0 & 2 & 2 & 1 & 3 \\
2 & 1 & 2 & 1 & 0 & 0 & 0 & 0 & 2 & 1 \\
0 & 1 & 1 & 2 & 2 & 0 & 0 & 0 & 1 & 1 \\
1 & 2 & 2 & 1 & 2 & 0 & 0 & 0 & 0 & 0 \\
1 & 2 & 0 & 1 & 1 & 2 & 1 & 0 & 0 & 1 \\
1 & 0 & 1 & 0 & 3 & 1 & 1 & 0 & 1 & 0
\end{pmatrix}.$$

(2)

Therefore for example we find there are three distinct ways of length three between v_5 and v_{10}, namely (v_5, v_4, v_3, v_{10}), (v_5, v_4, v_9, v_{10}), and (v_5, v_2, v_3, v_{10}).

Now we shall describe an iterative procedure for enumerating ways of length k for a graph G in terms of enumerations of ways of length $k-1$ on certain subgraphs of G.

Let G_j denote the subgraph of an n-graph G obtained by removing all edges incident on vertex v_j. Thus G_j retains v_j as an isolated vertex. This allows a simple representation of the adjacency matrix of G_j, denoted by B_j. The matrix B_j is obtained from A by setting all elements in the jth row and jth column of A to zero, that is, $B_j = (b_{xy})$ where $b_{xy} = a_{xy}$ for $x \neq j$ and $y \neq j$, and $b_{jy} = 0$ for all y, and $b_{xj} = 0$ for all x. As before we let W_k denote the matrix of all ways of length k in G. Let $W_k(G_j)$ denote the matrix of all ways of length k in the subgraph G_j. Parthasarathy [101] gives the following iterative method based on these concepts.

The jth column of W_k is equal to the jth column of the matrix $W_{k-1}(G_j)$ $\cdot A$ for $j = 1, \ldots, n$. The result follows directly from the construction of G_j with isolated vertex v_j.

Though this result may seem simpler than the previous method, it requires the computation of all the $W_{k-1}(G_j)$ matrices and can in fact involve about the same amount of work. It is useful when all values of k are required up to a high number since one can build them up by iteration.

It should be remarked that either method applies to graphs or to digraphs. In the case of digraphs the enumeration is for directed ways.

As an example of the iterative method, let us consider W_4 for the graph G in Figure 136. There are at least two procedures one might use. One is to develop the $W_3(G_j)$ directly and the other is to use the iterative method to compute each $W_3(G_j)$ from $W_2(H)$ matrices where H designates the various subgraphs of G_j appropriate to the calculation of $W_3(G_j)$. The later procedure indicates the full computational scheme that underlies the iterative procedure. It involves a number of cases and formulations. In some instances the former procedure can be employed to advantage, decreasing the extent of the formal calculations.

For this illustration we will compute the first column of W_4 by obtaining $W_3(G_1)$. We will obtain $W_3(G_1)$ from the terms $W_2(G_{1j})$ and from the adjacency matrix B_1 of G_1, where G_{1j} denotes the subgraph of G_1 obtained by isolating vertex v_j, $j = 2, 3, \ldots, 10$. This involves nine subgraphs of G_1 and the nine matrices $W_2(G_{1j})$. From those matrices we obtain matrix (3).

$$
W_3(G_1) = \begin{pmatrix}
0 & 0 & 0 & 0 & 0 & 0 & 0 & 0 & 0 & 0 \\
0 & 0 & 1 & 0 & 1 & 0 & 1 & 1 & 3 & 0 \\
0 & 1 & 0 & 2 & 0 & 2 & 0 & 2 & 0 & 1 \\
0 & 0 & 2 & 0 & 1 & 1 & 2 & 1 & 1 & 0 \\
0 & 1 & 0 & 1 & 0 & 0 & 1 & 2 & 1 & 3 \\
0 & 0 & 2 & 1 & 0 & 0 & 0 & 0 & 2 & 1 \\
0 & 1 & 0 & 2 & 1 & 0 & 0 & 0 & 1 & 1 \\
0 & 1 & 2 & 0 & 2 & 0 & 0 & 0 & 0 & 0 \\
0 & 3 & 0 & 1 & 1 & 2 & 1 & 0 & 0 & 1 \\
0 & 0 & 1 & 0 & 3 & 1 & 1 & 0 & 1 & 0
\end{pmatrix} \tag{3}
$$

Matrix (3) gives the number of distinct ways of length three in the subgraph G_1 of graph G. From it we obtain the elements $w_{i1}{}^4$ in the first column of W_4 as the first column of the matrix $W_3(G_1) \cdot A$. The result is shown below. Number of distinct ways of length 4 in the graph G (Figure 136) between vertex v_1 and each other vertex:

Vertex number	1	2	3	4	5	6	7	8	9	10
Number of ways	0	1	1	2	2	0	1	1	4	1

The four ways between v_1 and v_9 are $(v_1, v_7, v_6, v_8, v_9)$, $(v_1, v_2, v_5, v_4, v_9)$, $(v_1, v_2, v_3, v_{10}, v_9)$, and $(v_1, v_2, v_3, v_4, v_9)$. It will be clear to the reader in working on these examples that a computer must be utilized if results are to be obtained in applying the above enumeration methods in all but simple cases (such as the illustration).

CONSISTENCY OF CHOICE

When a person is required to make a choice among several objects it is difficult to measure the degree of internal consistency between his choices. One method of studying this problem is to let the vertices of a digraph represent the choices which may be made. By presenting the choices to a person in pairs he need only choose between two alternatives. If item a is preferred to item b the directed edge (a, b) is introduced into the digraph. In this way by formulating all binary choices we construct a complete antisymmetric graph.

Now consider what happens when more than two choices are presented to the person. For example, we may consider the 3 choice cases. When the 3 vertices in question fail to lie on a circuit (directed) all is well since the selection follows the previous pair wise selection without difficulty. However, when the vertices are on a circuit the selection is no longer clear and any selection will be inconsistent with at least one of the previous binary selections. In Chapter 6 we have discussed the maximum number of 3 circuits in a complete antisymmetric graph on n vertices and denoted this number by $C_3(n)$. If R is the number of 3 circuits in our particular preference graph then the measure of inconsistency I may be defined as $R/C_3(n)$.

The most consistent selection situation would yield an I of zero (corresponding to no 3 circuits) and the least consistent yields a value of $I = 1$.

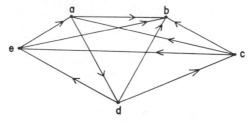

Figure 137

As an example consider Figure 137.

For this graph the adjacency matrix is (4).

$$A = \begin{pmatrix} 0 & 1 & 0 & 1 & 0 \\ 0 & 0 & 0 & 0 & 0 \\ 1 & 1 & 0 & 0 & 1 \\ 0 & 1 & 1 & 0 & 1 \\ 1 & 1 & 0 & 0 & 0 \end{pmatrix} \begin{matrix} 2 \\ 0 \\ 3 \\ 3 \\ 2 \end{matrix} \quad \text{row sums.} \qquad (4)$$

Using the formulas given in Chapter 6:

$$C_3(n) = \frac{5(25-1)}{24} = 5$$

and

$$R = \frac{5.\,4.\,9}{12} - \tfrac{1}{2}(4+9+9+4) = 2.$$

Hence for this graph the measure of inconsistency is $I = 0.4$ which is not particularly bad.

Many aspects of graph theory seem to have application to psychology and applications in that field are increasing all the time. The reader will find more of this material in Harary, Norman, and Cartwright [4] and he may also consult with recent issues of Psychometrika. The writings of K. Lewin play an important role in the initial attempts at modeling psychological studies, [102].

We shall give one more illustration from the field of psychology that is closely related to concepts of consistency and was one of the first situations to be treated by graph theory models and methods. It is the study of attitude balance initially introduced by F. Heider and discussed in Berger, et al. [103].

Heider's study utilizes *attitudes* (such as like, love, distrust, and so forth) denoted by L and *causal units* (such as build, give a speech, introduce a theory, and so forth) denoted by U. Compounds of attitudes and causal units are combined to form a behavior structure among persons, denoted by p and o; involving "things" denoted by x. Both persons and things are called entities by Heider. Compounding is denoted by $+$, the negative attitude is denoted by $\sim L$, the negative causal unit is denoted $\sim U$, and an attitude or causal relation is denoted by juxtaposition. A typical behavior structure is

$$pLo + oUx + pLx.$$

Heider's theory of balance consists of a number of rules stating which types of behavior structures are balanced and which are not. Using those rules the above structure is balanced while $pLo + oUx + p \sim Lx$ is not balanced. It should be remarked that a behavior structure is not judged as balanced or not on the basis of ones "feelings about it" but according to the theory, with extensive sets of rules. Harary and others [4], have converted much of Heider's involved theory into a digraph model representation. In that model the balance of a behavior structure is directly calculated from the properties of the digraph. Generalizations and extensions have been developed with the help of the graph theory model.

Vertices and directed edges are not sufficient to produce a complete model of a behavior structure. In addition it is necessary to assign a plus ($+$) or minus ($-$) sign to each edge. The resulting object is called a signed digraph and corresponds to a behavior structure according to the following interpretations. Vertices are entities (persons or things), edges are relations, an assignment of ($+$) means the relation holds, and assignment of a ($-$) means the negative of the relation holds. A cycle has the sign resulting from the algebraic product of the signs of its component edges. A signed digraph is balanced if all its cycles are of positive signs. If a signed digraph has no cycles it is (vacuously) balanced. For example the graph in Figure 138 is balanced for any attitude, say L. It represents the behavior structure $ALC + A \sim LB$. Note the attitude or causal unit applies to the graph in an indirect way; it is not placed directly on the graph structure.

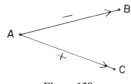

Figure 138

The concept of balance on a graph model can apply to various applications and involve either digraphs or graphs. We shall discuss the questions of balance in graphs in the remainder of this section. The results apply to undirected paths and cycles in a graph or digraph, (i.e., the directions are ignored).

It is in general difficult to determine balance by obtaining the sign of every cycle. The following theorem provides additional criteria (a proof is given in Harary [4]).

Theorem. In any signed graph S the following statements are equivalent.

1. S is balanced (every cycle has a positive sign).

2. All paths joining a pair of vertices have the same sign, for every pair of vertices.

3. The vertices of S can be divided into two disjoint sets (one of which may be null) such that each positive edge joins vertices of the same set and each negative edge joins vertices of different sets.

It may be that not every cycle of a signed graph S is positive. Though S is not balanced in such a case it does have some degree of balance related to the excess of positive cycles over negative cycles. In practice such incomplete balance may be more important than the strict (and hence hard to achieve) concept of balance itself. One measure of the degree of balance, denoted $B(S)$, is the ratio of the number of positive cycles c^+ to the total number of cycles c, that is,

$$B(S) = \frac{c^+}{c}$$

and clearly $0 \le B(S) \le 1$ where $B(S) = 1$ if and only if S is balanced. We have seen that c is often difficult to evaluate and that is also true of c^+ except for small (or simple structure) graphs. However, $B(S)$ is a useful concept and there are some results that can be helpful in its evaluation. One of the most useful such results reduces the calculation to smaller parts of the graph S. The parts used are biconnected subgraphs having maximal number of vertices, called blocks. Thus a block is a connected subgraph without articulation points such that it is a proper subgraph of no other biconnected subgraph. Since any cycle must lie completely in a block, it follows that if S consists of r blocks:

$$B(S) = \frac{c_1^+ + \cdots + c_r^+}{c_1 + \cdots + c_r}$$

where c_i and c_i^+ denote the number of cycles and positive cycles in the ith block respectively. This relation is useful for graphs composed of small blocks; for example, a cactus like graph where each block has simple cycle structure. For the graph shown in Figure 139 we have three blocks numbered 1, 2, and 3. In this graph $c_1 = 6, c_1^+ = 2, c_2 = 1, c_2^+ = 1, c_3 = 3, c_3^+ = 1$

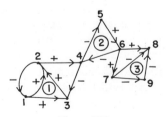

Figure 139

and the degree of balance is 0.4. Note that in this example S is a digraph though the direction of edges plays no part in the degree of balance.

For larger structures it may become unimportant for the larger cycles to be positive. This leads to the concept of limited balance or H-balance. A signed graph S is called H-balanced if every cycle of length less than or equal to H is positive. This concept is investigated with the help of the valency array $V_a = (v_{ij})$ of the signed graph S. Let $A_s(S)$ denote the signed adjacency matrix of the signed graph S; it is defined as follows $A_s = (a_{ij}^s)$ where the adjacency matrix is $A = (a_{ij})$ and $a_{ij}^s = \text{sig}(v_i, v_j) a_{ij}$. Here $\text{sig}(v_i, v_j)$ is the sign of the edge (v_i, v_j). Note that in this theory we use simple graphs or digraphs with at most one directed edge between two vertices. More general graphs are not needed for the applications in these areas and in addition would not lead to the results given because of the more involved form of their adjacency matrices.

Two vertices may have no common edge, one common edge, or two common edges. If two vertices have no common edges they are called o-adjacent; if all their common edges are positive they are called p-adjacent; if all their common edges are negative they are called n-adjacent, and if their common edges have one negative and one positive sign they are a-adjacent (the a denotes ambivalence). The valency array is defined in terms of these concepts. It consists of letters as follows:

$$
\begin{array}{llll}
v_{ij} = o & \text{if} \quad a_{ij}^s = a_{ji}^s = 0, & v_{ij} = p & \text{if} \quad a_{ij}^s + a_{ji}^s > 0, \\
v_{ij} = n & \text{if} \quad a_{ij}^s + a_{ji}^s < 0, \quad \text{and} & v_{ij} = a & \text{otherwise.}
\end{array}
$$

For blocks 1 and 2 from the graph of Figure 139 the valency array is (5).

$$
\begin{pmatrix}
o & a & p & o & o & o \\
a & o & p & p & o & o \\
p & p & o & n & o & o \\
o & p & n & o & n & n \\
o & o & o & n & o & p \\
o & o & o & n & p & o
\end{pmatrix}
\tag{5}
$$

Let $V_a^k = (v_{ij}^k)$ denote the kth array produced by operating on V_a as though it were a matrix using the arithmetic set forth in Tables 13 and 14 for the elements of V_a.

Table 13. Addition Table   **Table 14. Multiplication Table**

+	o	p	n	a
o	o	p	n	a
p	p	p	a	a
n	n	a	n	a
a	a	a	a	a

·	o	p	n	a
o	o	o	o	o
p	o	p	n	a
n	o	n	p	a
a	o	a	a	a

If $V_a(S)$ is the valency matrix of a signed graph S then $v_{ij}{}^k$ is o, p, n or a when the (vertex-edge) sequences of length k between vertices v_i and v_j are nonexistent, all positive, all negative, or have both signs, respectively.

The H-balance concept can be studied in terms of sequence lengths of various types since the cycles will be positive or negative depending on the number and sign of the sequences of which the cycles are composed. Moreover, the valency is of some direct interest as are the kinds of sequences between vertices (given by the arrays $V_a{}^k$).

GRAPH THEORY MODELS FOR URBAN STUDIES

Urban studies seem to provide an area in which graph theoretic techniques might prove extremely useful.

The mathematical formalism of the physical sciences provides basic structures upon which standard engineering studies can be developed. Such structures are not available for a large part of urban study. Though many urban problems relate to engineering, (e.g. traffic, pollution, etc.), important areas of social-economic-political investigation cannot draw upon normal physical structures for their analyses. Those areas may be studied in essentially three related ways:

1. On the basis of non-quantitative, social-economic-political theory utilizing observation, analogy, opinion, and expert judgment by trained and experienced investigators;

2. By means of statistical data involving the collection, analysis, and interpretation of those data in both semi-quantitative and qualitative terms;

3. In largely quantitative terms by introducing structure and defining indices and measures upon that structure which yield meaningful and useful interpretations in terms of social-economic-political analyses.

Certainly methods one and two must be employed, both prior to and concurrently with any utilization of method three if the structural models are to reflect reality. However, one and two type investigations alone may often provide only incomplete results in that the possibility for uncovering hidden elements, extrapolating and interpolating from specific findings, and formulating broad theoretical results, is hardly present without analyses based on some kind of quantifiable structure.

As illustrated in the previous sections of this Chapter linear graph theory has provided a means of introducing quantifiable structure in a number of recent sociological and psychological studies. In those applications graphs are employed as representational models which in some cases are quantified and yield explicational or theoretical-construct type information. As we have seen the vertices of the graphs usually represent individuals and edges denote relations between those individuals. For the most part specific graphs are employed and the analyses conducted on them are limited.

The non-engineering aspects of urban studies have much in common with the social and psychological sciences as regards problems of formulation of quantitative models. In both cases an initial structure must be imposed which models those aspects of a population under study. Quantification follows by imposing numerical indices and measures upon that structure.

This section illustrates the application of linear graphs to non-engineering aspects of urban studies by presenting a structural model of group membership within a community. (Of course many aspects of graph theory and network flow analyses are also useful in the more well structured investigations of what we are calling engineering aspects of urban studies, (e.g., location of utility and services, pollution control, etc.).) That model is quantified in several ways. Each quantification study will be discussed from the viewpoints of model analysis and interpretation. It will be seen that in addition to specific graphs, populations of random graphs may be utilized in studies based on population samples.

A collection of individuals in a community is considered to be a group when there is an unambiguous criteria for membership together with a reasonably well defined description of common activity, goals, or socioeconomic factors associated with that collection. Examples of such community groups are: community members who are policemen, teachers in community schools, members of the League of Women Voters and so forth. In forming a model the number of groups should not be extremely large but should be sufficient to include all significant groups relative to the intended analyses. Discussions as to what groups should be considered and how those groups should be defined require a combination of methods one and two described above. That part of the model formulation may be motivated by model structure but does not otherwise depend on the structure method.

In the model being described groups are represented by vertices of a linear graph. There is exactly one vertex corresponding to each group. An edge between two vertices denotes joint membership of some individual in each of the two groups to which those vertices correspond. We shall call this the simple model and its structure is a simple graph. A more general model results when each joint member is represented by an edge so that if, for example, three teachers were also members of the League there would be three edges between the two vertices corresponding to those two groups.

A number of studies are possible based on the graph model introduced above. Discussion of such studies is facilitated by introducing formal notation for the model. A formal definition of the graph model will now be given and will be followed by brief discussions of a number of special studies.

Let G_i denote the set of individuals in the ith group for $i = 1, 2, \ldots, N$ where N is the number of groups utilized in constructing the model. The graph structure will be denoted by $S = (V, E)$ where V is the set of vertices and E is the set of edges of the graph S. Set V will contain N elements v_i which correspond to the N groups G_i. Denoting an edge of S by the pair of vertices on which it is incident one observes that edge (v_i, v_j) is in E if and only if $G_i \cap G_j$ is not empty. In the simple model a single edge would be employed while in the general model a k-sample of k_{ij} such edges would be incorporated into the model, where $k_{ij} = |G_i \cap G_j|$, that is, the number of common elements. The edges may be assigned labels such as e_r but in simple graphs it is more useful to specify an edge by means of its corresponding vertex pair. In the case of general graphs, when multiple edges occur some index method may be required and we shall utilize the notation $e_{ij}(r)$, $r = 1, \ldots, k_{ij}$ to denote the rth edge between vertices v_i and v_j.

The simple structure model in which at most one edge is incident on any pair of vertices may be modified to yield a model of group membership reflecting statistical data about the population. Rather than construct the graph S on the basis of a complete investigation of all members of each group one can take the complete linear graph on N vertices, S_N as a basis for the structural model. In the graph S_N there is a single edge between each pair of vertices. A number p_{ij} is assigned to edge (v_i, v_j) which denotes the probability that in any particular construction of S (from a completely investigated population) the edge (v_i, v_j) would occur (i.e., $G_i \cap G_j$ would not be empty). The resulting graph model is denoted $S_N(P)$ and is a network based on the graph S_N and quantified by the edge numbers p_{ij} ($0 \leq p_{ij} \leq 1$ for all i, j). It is convenient to consider P as an $m \times m$ matrix with entries p_{ij} where $m = N(N - 1)/2$ (the number of edges in S_N).

A population may be sampled to obtain an estimate $P' = (p'_{ij})$ for the numbers p_{ij}. The resulting membership graph $S_N(P')$ is an estimate of specific graphs based on particular (complete) populations. The model $S_N(P')$ can be thought of as representing a larger population on the basis of a (restricted) sample of that population. Alternatively it may be used to estimate the structure for similar populations which did not form part of the sample used to estimate P. For example a model $S_N(P')$ might be developed based on population samples taken in Syracuse, N.Y. That model could then be used to represent the group membership structural model for similar cities. Here is another instance in which expertise and statistical information must be employed to implement the formation and use of the structural model. The word "similar" in the present context relates to the group definitions and to the goals of that study for which the model is developed.

The use of statistical sample graphs $S_N(P')$ seems to hold particular promise as a unifying method of analysis in model cities projects since the goals are similar in such projects. In addition a properly flexible definition of groups can introduce similarity in the structural models for such projects.

Estimation of the p_{ij} is described in terms of the following notation.

Let g_i be the number of individuals sampled that are found to belong to group G_i. Let g_{ij} be the number of individuals sampled that are found to belong to both G_i and G_j. Then let

$$P'_{ij} = \frac{g_{ij}}{g_i + g_j - g_{ij}}$$

be used as the estimate of p_{ij}.

Analyses carried out on the structural model S can also be carried out on $S_N(P)$ (the theoretical statistical structure) or on $S_N(P')$ (the sampled structure) in various ways as discussed under some of the following special studies.

The reader will recall that a path between two vertices v_i and v_j of a graph S is any sequence of edges and vertices of the form $(v_{i_1}, v_{k_1}), (v_{k_1}, v_{k_2}), (v_{k_2}, v_{k_3})$, $\ldots, (v_{k_r}, v_j)$. In general an edge or a vertex may occur several times in such a sequence. However, we shall direct our attention to simple paths in which an edge occurs exactly once and a vertex occurs exactly once (when they occur at all), we have called such a path a way. The length of a way is the number of edges it contains ($r + 1$ in the above example). The distance between vertices v_i and v_j was defined in Chapter 2, it is denoted by d_{ij}, the length of a way of minimum length between v_i and v_j (there may be several ways of the same length between two vertices).

Two vertices v_i and v_j are said to be adjacent or edge connected if the edge (v_i, v_j) is in the graph S. If v_i and v_j are adjacent the distance between them is one.

For a simple graph S the adjacency matrix A shows distances of length one in S.

A graph is connected if there is at least one path between any two vertices. Otherwise, a graph is not connected but consists of a collection of subgraphs that we have called components, each of which is itself a connected graph. In a connected graph all distances are finite. The distance between vertices lying in distinct components of S is taken to be infinity (∞).

Denote the nth power of the adjacency matrix by A^n and its elements by a_{ij}^n. Then the distance matrix D is obtained as follows, where d_{ij} denotes the distance between vertices v_i and v_j:

$$d_{ii} = 0 \qquad \text{for } i = 1, \ldots, N$$
$$d_{ij} = \infty \qquad \text{if } v_i \text{ and } v_j \text{ are in distinct components of } S.$$

Otherwise, d_{ij} is the smallest integer power n such that $a_{ij}^n > 0$. Thus D can be obtained from A by (somewhat involved) calculation.

In the group membership model distance represents a measure of separation between groups in terms of common membership between groups. Such membership provides information channels, leadership functions, etc. and may therefore lead to very useful analyses. For example, one community may have a graph structure with no distances greater than three while another community may have several distances greater than three. The later community is less closely connected than is the former. Studies of each could indicate what that phenomenon meant while the structural models which brought it to light might also indicate how useful changes could be made.

The concept of distance in a graph may be used to provide useful information and studies. However, since the distance concept is treated as an $N \times N$ matrix it may not be as useful for purposes of quantification as some scalar quantities which are based on the distance concept. Some of those quantities will now be reviewed and extended to random (sample) graphs.

As in Chapter 2 we let I denote the index set $\{1, 2, \ldots, N\}$. The radius R of a graph S is defined as follows:

$$R = \min_{i \in I} \max_{j \in I} d_{ij}.$$

A vertex v_r of S is said to be a center (of S) if

$$R = \max_{j \in I} d_{rj}.$$

The diameter H of a graph S is defined as follows:

$$H = \max_{i, j \in I} d_{ij}$$

that is, H is the greatest element in D, the maximum distance between any pair of vertices of S.

As we have discussed previously the radius R indicates the degree to which some vertex or vertices of S (centers) are close to the other vertices. On the other hand H indicates separation of vertices.

In terms of the group membership model a center is a particularly important group for it is closest to the other groups in terms of direct chains of individuals. The radius R measures the closeness of a community to the extent that the groups comprising the model represent the community. Separation of community groups is indicated by H.

These concepts and measures are defined on a particular graph structure S. However, they can be interpreted on statistical graphs of the form $S(P)$ (or $S(P')$ in which the values p_{ij} are estimated by sample values p'_{ij}). In $S(P)$ each edge occurs with a probability value as previously described. In such a model one can define the expected distance matrix $\bar{D} = (\bar{d}_{ij})$ in which $\bar{d}_{ij}$ is the expected distance between vertices v_i and v_j. The $\bar{d}_{ij}$ do not depend on particular groups but only on the number N of groups comprising the model and on the values p_{ij} (or the estimates of these quantities, p'_{ij}). However, the $\bar{d}_{ij}$ are difficult to calculate for graphs having more than a few (e.g., five or six) vertices. One method of arriving at the $\bar{d}_{ij}$ values is by monte-carlo simulation techniques. However, there is much to be gained from more explicit probabilistic formulations.

From the matrix $\bar{D}$ the expected radius $\bar{R}$, the expected diameter $\bar{H}$, and expected centers can be determined in the same way the corresponding deterministic quantities are determined from the distance matrix D.

Such expected measures can be used to extrapolate or interpolate statistical findings from limited data as previously described under the discussion of $S(P')$. They have another interpretation as well. A structure S defined for a complete population (or interpreted as so defined) may change over time, members dropping from some groups and joining others. The structure $S(P)$ and measures based on it can be used to interpret the likelihood and the significance of such individual changes.

The quantities D, R, and H may be considered random variables defined on the structure $S(P)$. In addition to the expected values $\bar{D}$, $\bar{R}$, and $\bar{H}$ described above other useful statistical quantities may be defined for those random variables. The variance is one such quantity. They can be computed for small graphs but become difficult to calculate for larger graphs. However, such measures often provide useful information and for the models under consideration here they might be obtained with reasonable effort since the number of groups might not be too great.

As defined in Chapter 2 the coefficient of internal stability or independence number of a graph S is a measure of the extent to which the vertices are nonadjacent. A set of vertices in S is internally stable (independent) if no two vertices of the set are adjacent. Let J^* denote the class of all internally stable sets of S with J denoting a typical member of J^*. The coefficient of internal stability is denoted by $s_i(S)$ and defined by:

$$s_i(S) = \max_{J \in J^*} |J|$$

where $|J|$ is the number of vertices in J.

One can extend this concept to statistical graphs $S(P)$ by considering expected adjacency.

Interpretations of sets J and the index $s_i(S)$ in the context of the group membership model arise in both the deterministic case S and the statistical case $S(P)$. For a collection of N completely independent groups $s_i(S) = N$ whereas the case of complete intergroup membership yields $s_i(S) = 0$. It may be taken as axiomatic that for many purposes of community development the zero case represents an ideal situation.

A set of vertices J is externally stable or dominating if every vertex of S that is not in J is adjacent to at least one vertex of J. Denote the class of all externally stable sets by J^{**}. The coefficient of external stability or domination number of a graph S is denoted by $s_e(S)$ and defined by:

$$s_e(S) = \min_{J \in J^{**}} |J|$$

This quantity measures the extent to which a number of vertices has access to the other vertices by means of a single edge.

Extensions and interpretations of this concept can be made along lines previously discussed.

A graph S can be thought of as representing a structure to some particular degree of complexity. If one wishes to obtain a more detailed representation that will allow finer grained analyses, the vertices of S may be expanded into graphs themselves (which then form subgraphs of S). Such an expanded graph is denoted by S^+. These types of graphs have been discussed in Chapter 7 as hierarchy graphs.

The group membership model lends itself to the kind of magnification introduced above. If it becomes desirable to study the detailed structure of some group, that group can be represented as a subgraph in S^+ (to some level of complexity) rather than as a vertex in S. The procedure for constructing S^+ is to divide the group to be studied in detail into as many subgroups as needed, determine common subgroup membership and formulate the resulting subgraph of S^+.

To employ the models described in this section, one might well start with a relatively simple case having only a few groups (each representing a rather large and in detail, complex element of the community). Then as one becomes more familiar with that model and acquires additional data more detailed models could be built up by the expansion of vertices into subgraphs (thus including more groups, each being less complex than the macrogroups from which the initial model was constructed).

The measures and concepts introduced may be applied at various levels of detail. Their dependence on detail would provide useful information in addition to the particular values they assume in each case.

EXERCISES

1. Select (or invent) a small organization with which you are familiar and construct the digraph representing ranking of individuals within that organization. Calculate the status measure for individuals. See if the four desirable aspects of status hold for that measure in the case you are considering.

2. Construct a preference digraph for 5 or 6 objects you own or persons you know. Formulate the measure of inconsistency for your preferences. Construct a signed digraph to represent a group of 5 people to which you belong. Use a common relation, such as friendship, to determine the directed edges and signs. Is the structure balanced? What is the measure of balance $B(S)$?

3. Select any vertex of the Petersen graph and label it v_1. Enumerate the number of distinct ways of length 4 between v_1 and each of the other 9 vertices.

4. Represent the rooms (including halls and stairways) in your home by vertices of a graph. Connect any pair of vertices that represent adjacent, connected rooms. Enumerate the distinct ways of length 3 from the kitchen (vertex) to all other rooms.

5. Consider how you might utilize the urban group membership model given in the text within your community. What data are required to build the model? How would you gather those data? In what ways might the formulation and analysis of your model be used in local decision making?

6. Compute the measure of inconsistency for the following complete antisymmetric graph. If the graph represents the binary preference selections of an individual for 5 objects, how consistent do his choices seem?

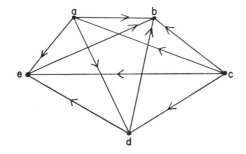

Chapter 10

Application in Physics

PERCOLATION PROCESSES

Many physical processes can be interpreted as though they represented the passage of particles through a medium. Indeed the diffusion of one gas through another is such a process. A less obvious example is spread of a contagious disease over a population. Also problems of local changes of state resulting in energy changes to a body can be put into forms similar to these problems.

There are two basically different ways to formulate probabilistic models of such processes. In one case the random behavior is assigned to the particle; such models are called diffusion processes. Many physical problems are well represented by such a model (random walk methods are often employed in their study) and a great deal of work has been done on them. Important as diffusion processes are, they are only indirectly related to linear graphs. However, the other model possible for such processes is closely connected with random graphs. In these models the probabilistic character of the process is assigned to the media rather than the particle. Such models are called percolation processes and find wide application, particularly to examples where diffusion models are inappropriate or extremely forced.

The random character assigned to the medium is most often interpreted as being the blocking of a fluid which is attempting to flow along the edges and through the vertices of the medium. In some applications such as the penetration of water into rock this model is directly related to the natural process, in other applications the fluid is purely hypothetical as in the spread of local energy charges in solid state media.

One can define bond percolation processes in which the probability of fluid passage is assigned to the edges or bonds of the medium or atom percolation processes where the random character is assigned to the vertices or atoms. Of course there can be a combined random medium but this can usually be avoided by proper definitions.

We shall illustrate the type of studies carried out in this subject by talking a little about bond percolation process. For each such process there is an equivalent atom percolation process and sometimes one is more easily treated than the other. Several major advances of the theory have been made by transforming between the two types of percolation processes.

Consider the case of a fluid wetting a rock. A single atom called the source atom is filled with fluid and following a bond percolation process this fluid spreads out over the medium passing through all unblocked bonds that it can reach. The problem is to study the extent of wetting in the medium. We let p be the probability that a bond is unblocked. All bonds have the same probability p of being unblocked independently of the state of the other bonds. This is a minor restriction for if we wanted to study bonds having different blocking probabilities we could introduce chains of bonds and atoms for such bonds.

Let $P_N(p)$ denote the probability that at least N atoms (in addition to the source atom) are wet in the medium by the fluid. The percolation probability is defined by

$$P(p) = \lim_{N \to \infty} P_N(p).$$

This is the probability that the fluid will wet infinitely many atoms. The evaluation of $P(p)$ is extremely difficult and one must usually carry out approximations to it by monte carlo simulation as was the case in the multiple correspondence statistics problems and other problems we have discussed in previous chapters.

Another important quantity studied in percolation theory is the critical probability p_0 defined as the least upper bound of the set of all p values for which $P(p) = 0$. If $p < p_0$ there is only local wetting of the medium but when $p > p_0$ the fluid can spread to infinite extent within the medium. Computation of p_0 is also very difficult and is most often approximated. However, since 1960 theoretical values for p_0 have been obtained for " simple " plane lattices. Though the examples are on simple structures the results are not at all simple to obtain and their evaluation is a considerable achievement due to M. Sykes and J. Essam, [104]. The results are shown in Table 15.

Table 15

Structure	p_0
Triangular lattice	$2 \sin \pi/18$
Square lattice	$1/2$
Hexagonal lattice	$1-2 \sin \pi/18$

Such results apply to the study of random graphs quite apart from the particular formulation given above. This illustrates an interesting interplay between purely abstract studies on graphs and studies motivated by the desire to model and understand physical problems. Of course the material above is still rather theoretical, dealing with idealized lattices of simple structure but it does help in speculation concerning more realistic situations. Moreover, for practical purposes many solids are so large relative to their fundamental point structure that they act like infinite lattices except on their boundaries or near points of imperfection. We might remark that many studies are devoted to those special points and boundaries for a number of practical reasons. In terms of linear graphs these situations correspond to the hierarchy graph concept in which the special region is made a subgraph of appropriate structure. That graph is thought of as having replaced a single vertex in the uniform structure graph. An example is shown in Figure 140, where the imperfection occurs at vertex v.

Employing such a model with an appropriate representation for the imperfection (possibly derivable from structure studies) allows its effect on properties of the lattice to be studied. For example, the effect on p_0 might be of interest. Unfortunately, it is unlikely that any direct analyses could be applied and simulation studies are about the only way to approach these complicated structures.

STRUCTURAL STATISTICAL PHYSICS—DIMERS

Phenomena which result from the combined action of many parts (particles or "systems") can be studied by statistical methods. Deterministic relations

Uniform triangular
lattice

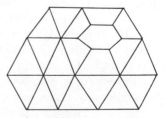

Imperfection at vertex v

Figure 140

often represent the average behavior of such aggregates. Thus the laws of Thermodynamics reflect the gross cumulative effect of many parts contributing to energy alterations which are sensed and classified as temperature (due to "heat"). The detailed study of such systems requires that the methods of statistics be applied to the particles comprising the system. Alternatively a system may have many states which it can occupy. The relative frequency of its being in a various state can be studied by substituting for the system a collection (ensemble) of like systems; the numbers which are in a given state being the relative frequency of single system occupancy (the ensemble concept of Gibbs).

In physics of many particle or many state systems the laws of statistics are well satisfied and the very large numbers produce a form of structure in such geometrically structureless media as gases. If in addition we study many state systems which have considerable structure, such as crystals, the statistical mechanics becomes related to linear graph enumerations in many ways, some direct, and others much less obvious.

A major technique in statistical mechanics is to formulate the partition function for a system under study. This is usually denoted by Z and is obtained as a sum or integral over the energy states of the system. The individual terms contributing to Z are usually of exponential type in the state energy. From the partition function Z the free energy F of the system may be obtained as

$$Z = e^{-cF}$$

where c is a positive constant (in many cases we may interpret c as $1kT$ where k is Boltzmann's constant and T is the absolute temperature). Once these quantities have been obtained for a system its aggregate behavior can be described by thermodynamic quantities which are specified by F and Z. Examples of such quantities are the specific heat per molecule, the pressure and the entropy.

Here we are particularly interested in the fact that when the statistical mechanics of structured media is considered the computation of the partition function is strongly dependent upon graph enumeration problems. If the individual states are represented by appropriate configurations throughout the vertex sites of the medium and site energy changes can be represented in an elementary way the major task in forming Z is the enumeration of configuration types. Many problems in physical science have such formulation, excluded volume problems in chemistry, models of solid state, order-disorder statistics in physics and condensation theory in physical chemistry being some typical areas where these methods have been applied. The success of such applications have been somewhat limited due to the extreme difficulty of these problems.

Those areas of study mentioned above in which the construction of the partition function largely depends on combinatorial aspects of graphs or other structures (such as lattices in one, two or three dimensions) can be considered to belong to structural (combinatorial) statistical mechanics. The transition from the physical system to an abstract mathematical form appropriate to enumerative formulation and the corresponding transformation of any solution to the enumerative problems back into physical terms requires considerable knowledge of applied statistical mechanics and the physics of the problem. Our interests are in the enumerative problems themselves as they relate to graphs and similar structures.

Probably the first problem of the type we are discussing was the Ising model of Ferromagnetism (as discussed by, for example, Montroll [105]) initially discussed in 1925, though treatment of the model by graph enumeration methods is rather recent starting about 1945. The study of Ferromagnet models continued to develop and at the same time it was realized that many other problems could be treated by similar methods. Many people became interested in the possibility of studying physical problems of a structural statistical mechanical nature by converting to enumeration problems for linear graphs. The resulting enumeration problems are in general extremely difficult to solve and often one must resort to simulation techniques for approximate numerical answers. Even so these problems are so interesting that people who become familiar with them are apt to contract what is often called the " Ising Disease ". A major symptom of this " disease " is an almost continual sketching of lattice diagrams; a consequence of the disease is a state of frustration at being unable to solve any problem to one's satisfaction.

We shall illustrate one type of enumeration arising from physical problems of the structural statistical type. These are known as dimer covering problems. These problems have their origin in the possible arrangements of diatomic molecules on various lattices such as rectangular lattices. The term dimer comes from this arrangement problem. Many other problems can be studied in terms of the dimer coverings (e.g., the Ising models can be looked at from this point of view) so that they form a somewhat unifying formulation and method in structural statistical mechanics.

The dimer has a bond which is the same length as an edge of the lattice structure. There are then two types of problems concerned with dimer covering. For a given lattice we ask if a dimer covering is possible (this is an existence problem). When such coverings exist we wish to enumerate them.

Two important types of plane lattices upon which dimer covering studies may be discussed are shown in Figure 141. A is the well known square lattice and B is often called a bathroom-tile or wall-tile lattice. By changing the dimensions, problems with covering by two different length dimers can be considered and so forth. In Figure 142 we show two dimer coverings of a

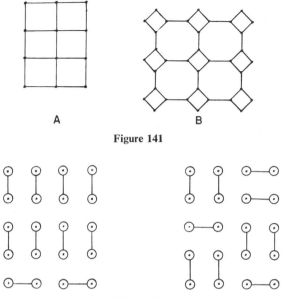

Figure 141

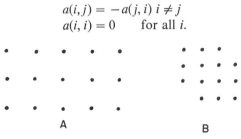

Figure 142

rectangular lattice. In Figure 143 rectangular lattices for which there are no dimer coverings are shown. Certainly any lattice with an odd number of vertices (sites) such as the one in A can never be covered by dimers. However, as we have seen the 8×8 chessboard with the two squares lying at opposite ends of a major diagonal removed cannot be covered by dominoes. Covering by dominoes is equivalent to dimer covering on a lattice. A similar example, shown in B, illustrates the existence of cases with an even number of vertices that cannot be covered by dimers.

A useful concept for studying the number of dimer coverings is the Pfaffian associated with a square antisymmetric matrix A. Let the elements of A be defined by means of ordered pairs of the numbers $1, 2, \ldots, 2k$ so that $a_{i,j} = a(i, j)$. Then A is $2k \times 2k$ and

$$a(i, j) = -a(j, i) \; i \neq j$$
$$a(i, i) = 0 \quad \text{for all } i.$$

A   **B**

Figure 143

The order of the matrix is even since we deal with dimer coverings and there are none when the number of vertices is odd. The Pfaffian of A, written $P_f(A)$ is a combinatorial sum based on the elements of A. Other combinatorial sums of this type are the Permanent and the Determinant of A. The Pfaffian is similar to the Determinant. A standard symbolic notation for the Pfaffian is:

$$P_f(A) = |a_{ij}| = |a(1, 2)\ a(1, 3) \cdots a(1, 2k)$$
$$a(2, 3) \cdots a(2, 2k)$$
$$\cdots$$
$$\cdot\ \cdot$$
$$a(2k - 1, 2k)|$$

The Pfaffian is defined as the combinatorial sum:

$$P_f(A) = \sum_P s_p\, a(p_1, p_2)\, a(p_3, p_4) \cdots a(p_{2k-1}, p_{2k})$$

where the sum is over terms corresponding to special permutations of the integers $1, \ldots, 2k$. A typical permutation is denoted by the letters $p_1, \ldots, p_{2k}$ and s_p is the sign of the permutation (as in the case of Determinants), $s_p = 1$ if P is an even permutation (results from an even number of interchanges of pairs of numbers) and $s_p = -1$ if P is an odd permutation. The sum is restricted to those permutations for which:

$$p_1 < p_2,\, p_3 < p_4, \ldots, p_{2k-1} < p_{2k}$$

and

$$p_1 < p_3 < \ldots < p_{2k-1}.$$

For example

$$P_f(A) = |5 \quad 3 \quad 2| = 35 - 12 + 2 = 25.$$
$$1 \quad 4$$
$$7|$$

For the matrix A defined by this example it is also true that the determinant of A gives $\det(A) = (25)^2 = P_f^2(A)$. In fact this illustrates a general property of the Pfaffian by which it can be evaluated in many cases; namely, for a square antisymmetric matrix A: $P_f^2(A) = \det(A)$.

Now let us consider the dimer covering problem. Suppose we wish to enumerate the distinct coverings of a lattice having $2k$ points. The points may be denoted by $p_1, p_2, \ldots, p_{2k}$ where they are numbered in such a way that any pair (p_{2i-1}, p_{2i}) for $i = 1, \ldots, k$ represents the end points of a dimer. Distinctness of different dimer coverings is specified if we require that $p_1 < p_2,\, p_3 < p_4, \ldots, p_{2k-1} < p_{2k}$ and that $p_1 < p_3, \ldots, < p_{2k-1}.$

Let N_x and N_y denote the number of horizontal and vertical dimers respectively in any covering. Then an enumerative generating function for dimer coverings may be written in which each term specifies the number of coverings having a given value of N_x and of N_y. We shall only deal with a rectangular shaped square lattice having m rows and n columns. For such structures the generating function may be written

$$Z_{mn}(x, y) = \sum_{N_x, N_y} g_{m,n}(N_x, N_y) \, x^{N_x} \, y^{N_y}$$

following Fisher's notation (see [106]). In the generating function $Z_{mn}(x, y)$, $g_{m,n}(N_x, N_y)$ denotes the number of dimer coverings of an mxn rectangular square lattice, using N_x horizontal and N_y vertical dimers. A value of zero for any coefficient indicates no coverings are possible of the specific type indicated.

Because of the similarity in the restrictions on the point orderings for dimer pairs and for permutations in the Pfaffian expansion there is seen to be a one to one correspondence between the terms of the dimer enumerating function and the Pfaffian based on the same letter pairs. In $Z_{m,n}(x, y)$ the symbols x and y are used to specify the type of covering by means of the powers N_x and N_y those symbols assume in a given term. Thus the total number of coverings is given by $Z_{m,n}(1, 1)$. However in a more general way the x and y may denote much more than combinatorial separation. In fact, we are free to interpret those variables in any useful way. That is important to us because the terms of the generating function are non-negative (since they count) but the terms of the Pfaffian alternate in sign. Thus, though there is a one to one correspondence between the terms of the two sums they are not equal sums unless the elements of the Pfaffian can be selected properly. It is necessary to formulate those elements in such a way that the odd terms in the Pfaffian will contain a negative sign to overcome the intrinsic negative sign carried by those odd terms. In so doing the even terms must not be affected. At least two methods have been used to achieve the desired result; we shall follow the method of Fisher and Temperley [107]. An alternative method due to Kasteleyn [108] is discussed by Montroll [105].

In terms of statistical mechanics if x and y represent the activities of horizontal and of vertical dimers respectively then $Z_{m,n}(x, y)$ is the configurational grand partition function of the process related to those activities. Thus the coefficients $g_{m,n}(N_x, N_y)$ are directly related to the formulation of physical processes on the lattice.

The development of $Z_{m,n}(x, y)$ may be based on h operators A_{ij} where h is considered greater than or equal to the number of bonds in the lattice. These operators satisfy the conditions:

$$A_{ij} A_{rs} + A_{rs} A_{ij} = 2\delta_{rs}{}^{ij}$$

and $A_{ij}^2 = I$ where I is the identity and $\delta_{rs}^{ij} = 0$ if $(i,j) \neq (r,s)$ and is 1 if $(i,j) = (r,s)$. For the formal part of the development there is no need to give specific representations for these operators. They serve to set up a formal generating function and then they are represented in a direct way so as to yield the desired results. The trace of these operators satisfies:

$$Tr\{A_{ij}\} = 0, \quad Tr\{A_{ij}A_{rs}\} = 0 \quad \text{if } (i,j) \neq (r,s)$$
$$Tr\{A_{ij}A_{rs}A_{pq}\} = 0 \quad \text{if } (i,j) \neq (r,s) \neq (p,q), \dots \text{ etc.}$$

and in a representation in t dimensions $Tr\{A_{ij}A_{ij}\} = t$.

The vertices of the lattice are numbered by $k = 1, 2, \dots, mn$. The operator A_{ij} is associated with the edge (bound) joining vertices i and j. Consider the form V_k specified by the operators associated with the four nearest neighbor vertices to vertex k (recall we are dealing with a square lattice). Let p_i, $i = 1, \dots, 4$ denote the nearest neighbor sites then $V_k = x_{kp_1}^{1/2} A_{kp_1} + x_{kp_2}^{1/2} A_{kp_2} + y_{kp_3}^{1/2} A_{kp_3} + y_{kp_4}^{1/2} A_{kp_4}$. We observe that $T^* = Tr\{\pi_{k=1}^{mn} V_k\}$ will have terms such that an operator A_{kp_i} will occur twice in a term or not at all, otherwise the terms will be zero. Any nonzero term will be of the form:

$$(-1)^p \, x_{k_r p_r} x_{k_s p_s} \cdots y_{k_u p_u} y_{k_v p_v} \cdots y_{k_w p_w} t$$

which corresponds to horizontal dimers at vertices (k_r, p_r), (k_s, p_s), ... and vertical dimers at vertices (k_u, p_u), (k_v, p_v), ... and so forth. At this point the reader will observe the reason for defining V_k with one half powers of the x and y variables. By so doing we now have them as linear factors. Thus to each distinct dimer covering of the rectangular lattice there corresponds a term in T^*. As in our introductory presentation we again encounter an indicated variation of sign built into the expression. The exponent p is called the parity of the number of interchanges of adjacent operators in the term that are required to bring like operators together in pairs. Because of the one to one correspondence of the terms and the dimer coverings it is desirable to arrange the operators so that p is always even. When that is done we set $x_{kp_i} = x$ and $y_{kp_i} = y$ for all cases and obtain the basic formulation:

$$Z_{m,n}(x,y) = t^{-1} Tr\{\pi_{k=1}^{mn} V_k(x,y)\}.$$

It is interesting that something as basic as the way vertices are assigned numerical indices yields success or failure of the above development. One way to make the parity p come out even in every case is to number the vertices as follows. Start with the lower left corner vertex and number it one, move to the right numbering successive vertices in increasing order until the lower right corner has been reached and numbered n. Then move up one level, number the next vertex $n + 1$ and proceed to the left with increased numbering. This

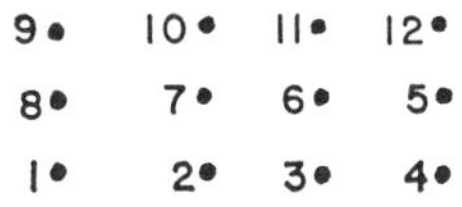

9• 10• 11• 12•

8• 7• 6• 5•

1• 2• 3• 4•

Figure 144

zig zag system is illustrated for a 3×4 lattice in Figure 144. It is by no means obvious that this sytem will yield the correct parity; a proof is given in Fisher, [106].

The next basic step is to express the trace form as a Pfaffian. When operators A_i satisfy the anticommutation properties specified for the A_{ij} above, it can be shown that:

$$t^{-1} Tr\{\pi V_r\} = |(v_{ro} v_{so} - \sum_{i=1}^{4h} (-1)^{r+s} v_{ri} v_{si})|$$

where $V_r = v_{ro} I + \sum_{i=1}^{4h} v_{ri} A_i$. The limit $2b$ merely indicates an even integer and we have mn even for our dimer covering problems. This concludes the basic aspects of the development and one may now formulate the generating function specifically. That formulation utilizes the fact that $Z_{m,n}(x, y)$ is a Pfaffian and hence its square can be expressed as a determinant. The Pfaffian itself takes the form:

$$Z_{m,n}(x, y) = |(k, p)|$$

where $(k, p) = xv_{kp} + y\eta_{kp}$ in which the quantities $v_{kp} = 1$ or 0 accordingly as the vertices k and p are or are not connected by a horizontal dimer and the η_{kp} are defined similarly for vertical dimers. Thus the operator formalism becomes very specific. The numbering of vertices is of course zig zag in form to insure proper parity. The result is:

$$Z_{m,n}^2(x, y) = (-1)^{m[n/2]} 2^{mn} \prod_{s=1}^{m} \prod_{r=1}^{n} \left(x \cos \frac{\pi r}{n+1} + i y \cos \frac{\pi s}{m+1} \right)$$

where $i = \sqrt{-1}$.

Assuming n is even and using the relation:

$$\frac{\sin h \gamma}{\sin \gamma} = 2^{h-1} \prod_{r=1}^{n-1} \left[\cos \gamma + \cos \left(\frac{\pi r}{h} \right) \right], \quad \text{for } h - 1 \text{ even,}$$

one obtains:

$$Z_{m,n}(x, y) = 2^{n/2[m/2]} \prod_{s=1}^{[m/2]} \prod_{r=1}^{n/2} \left(x^2 + y^2 + x^2 \cos \frac{2\pi r}{n+1} + y^2 \cos \frac{2\pi s}{m+1} \right) J_m(x)$$

where $J_m(x)$ is 1 for m even and is $x^{n/2}$ for m odd. This expression may be used to enumerate dimer coverings on rectangular square lattices.

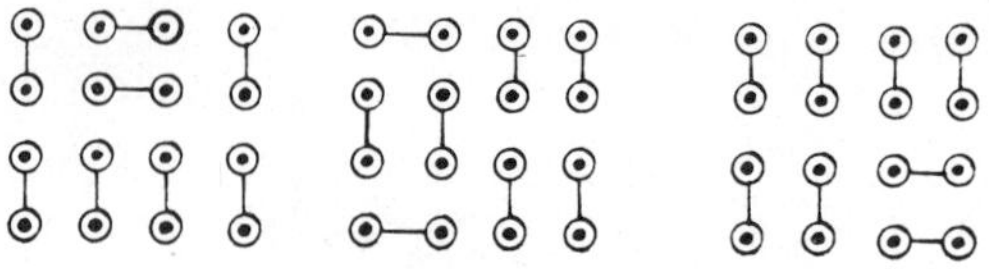

Figure 145. Dimer coverings.

For example when $m = n = 4$ we have the expression:

$$Z_{4,4}(x, y) = 2^4 \prod_{s=1}^{2} \prod_{r=1}^{2} \left(x^2 + y^2 + x^2 \cos \frac{2\pi r}{5} + y^2 \cos \frac{2\pi s}{5}\right)$$

$$= x^8 + 9x^6 y^2 + 16x^4 y^4 + 9x^2 y^6 + y^8.$$

Thus there is a total of $Z_{4,4}(1, 1) = 36$ distinct dimer coverings of the 4×4 lattice. Of these there is one with only horizontal dimers and one with only vertical dimers. There are 9 coverings using 6 horizontal and 2 vertical dimers and there are also 9 using 2 horizontal and 6 vertical; three of these are shown in Figure 145. There are 16 coverings using 4 dimers of each type.

The problem of enumerating the number of coverings of an 8×8 chess-board with dominos may be obtained from the generating function developed in this section. For that problem we have:

$$Z_{8,8}(1, 1) = 2^{16} \prod_{s=1}^{4} \left\{\left(2 + \cos \frac{2\pi s}{9} + \cos \frac{2\pi}{9}\right) \left(2 + \cos \frac{2\pi s}{9} + \cos \frac{4\pi}{9}\right)\right.$$

$$\left. \cdot \left(2 + \cos \frac{2\pi s}{9} + \cos \frac{6\pi}{9}\right) \left(2 + \cos \frac{2\pi s}{9} + \cos \frac{8\pi}{9}\right)\right\},$$

which yields

$$Z_{8,8}(1, 1) = 2^4 (901)^2, \qquad \text{a rather large number of coverings.}$$

It is interesting to study the behavior of $Z_{m,n}(x, y)$ as the lattice becomes large. For that purpose one defines $Z(x, y)$ by the equation: $\ln Z(x, y) = \lim_{m,n \to \infty} (mn)^{-1} \ln Z_{m,n}(x, y)$. Using the expression given above for $Z_{m,n}(x, y)$ one obtains:

$$\ln Z(x, y) = \frac{1}{\pi} \sum_{r=0}^{\infty} (-1)^r \frac{(x/y)^{2r+1}}{(2r + 1)^2} + \frac{1}{2} \ln y.$$

When $x = y = 1$ we get the limiting expression for the number of dimer coverings as: $\ln Z(1, 1) = G/\pi$ where

$$G = 1 - 3^{-2} + 5^{-2} - 7^{-2} + \cdots = 0.915966 \cdots \text{ is Catalan's constant.}$$

This gives the approximate result:

$$Z_{m,n}(1, 1) \sim e^{G/\pi mn} = (1.791)^{mn/2}.$$

For example $ln\ Z_{8,8}(1, 1) = 17.380$ and $ln\ (1.791)^{32} = 18.624$ so that for the relatively small 8×8 lattice the limit approximation is only about 7 percent in error.

GRAPH ENUMERATION IN PHYSICS

A number of studies in theoretical Physics can be put into forms in which the enumeration of certain types of linear graphs plays a fundamental part. One of the earliest instances of such considerations occurred in the work of Uhlenbeck and Ford in the theory of gas condensation [109]. Other areas of application are the Ising and Heisenberg models of a ferromagnet and antiferromagnet, entropy of ice, percolation processes, and generally to statistical mechanical studies on structured media. As we have discussed previously a major technique of such investigations is to express the statistical mechanical quantity of interest in terms of a series. In such a framework the coefficients of the series can often be made to depend on the number of graphs of a certain type. The types of graph usually depend on the structure of the medium such as lattice properties in the case of models on a crystal. It is beyond the scope of this book to enter into the physical background or the statistical mechanical aspects of these developments. We wish to discuss some of the graph theory aspects that grow out of those studies and we will do so in a general way, without relating to the physical aspects. Sykes, Essam, Heap, and Hilery give a discussion of some of the related physical developments in [110]. Here our only connection with physics is to observe how a topic in (pure) graph theory grew out of and plays an important role in several types of physical problems. The general topic of graph enumeration is a broad one. We have considered various enumerations elsewhere in the book (e.g., trees) and there are many unsolved enumeration problems in graph theory. The physical problems which motivate the present enumeration studies consider restricted types of graphs. Even so general solutions are not available and the research deals with specific counting techniques. Much of the published results consist of dictionaries of graphs (picture and data tables) and discussion of how the results were obtained with indications of applications.

This section is based on a paper of Heap [111] and on a paper of Nagle [112]. A basic aspect of the graph enumeration problem considered here is the numbering of graphs in a sequential ordering according to one or more features of the graph. There are various ways to create such graph orderings for specific types of graphs. Nagle deals primarily with the question of developing such orderings. Heap also addresses the ordering question but also deals with the enumeration problem and presents some enumerative results. We will mainly follow Heap's presentation, using material from Nagle to supplement the ordering discussions.

The type of graph under discussion is an undirected multiply-connected graph with multiple edges allowed but without self loops. Thus the graphs are connected and have no articulation vertices. This type of graph is called a star by the people who work with them in the context of physical problems. A star in this sense is to be distinguished from a star graph as we have defined it in Chapter 1. We shall follow their example and call these graphs stars in this section. The applications do not require the enumeration of all stars on a given set of vertices. We shall now consider the special types of stars that are to be enumerated.

Two graphs that differ only in the number of vertices of degree two that they contain are homomorphic to each other. A contraction homomorphism (as discussed in Chapter 1) consists in the removal of a vertix of degree two and its replacement by a single edge. One may also place a vertex of degree two on any edge of a graph to obtain a new graph, homomorphic to the original. Heap deals with the enumeration of stars without vertices of degree two, which are called homomorphically irreducible stars (or star graphs). Such graphs are the most important for the physical applications mentioned above.

Homomorphic graphs, defined in terms of the vertices of degree two relation, all have the same cyclomatic number $c(G) = m - n + 1$ (for connected n-graphs with m edges, see Chapter 1). The enumeration of homomorphically irreducible stars is carried out for the various values of the cyclomatic number. It is convenient to use HI to denote " homomorphically irreducible." The reader can observe that there is only one such graph for $c(G) = 2$ and there are four for $c(G) = 3$. Essam and Sykes have shown that there are 17 HI stars with $c(G) = 4$. In [111] Heap reports that there are 118 HI stars with $c(G) = 5$ and he tabulates all the HI stars with cyclomatic number ≤ 5. The results reported were developed by a computer search method that utilized an ordering of the graphs in question.

Linear graphs can be ordered in various ways, the most appropriate way often depends on the application (and also the type of graph). Nagle discusses the graph ordering problem in [112] and it is also treated in [110] and [111]. One way to order graphs is to order their adjacency matrices but this requires a little consideration. Permutation of the vertex labels results in different adjacency matrices for the same graph (all such matrices are similar but this fact plays no role here). Two matrices are ordered by scanning across each row from left to right and from top row to bottom row if the first pair of unequal elements is such that $a_{ij} < b_{ij}$ then $A < B$. Nagle introduces this ordering to define the canonical (adjacency) matrix of a graph as the matrix that precedes all others for that graph. Once each of the stars is presented by its canonical matrix the individual stars are ordered by using the same method on the canonical matrices. One possible modification of this ordering method should be pointed out. If some specific condition is required such as the assignment of a specific label to some vertex then the formation of the canonical matrices is carried out under that restriction (rather than over the set of all possible adjacency matrices). In considering the specific enumeration of stars (or indeed any graph type) it is convenient to pre-order the graphs into subsets upon which a complete ordering, following the above procedure, can be imposed. Criteria for preordering are arbitrary to some extent. Nagle and Heap take the cyclomatic number as the first preordering criteria. Stars are classified by the $c(G)$ values and treated in more detail within each specific cyclomatic number class. As we have mentioned Heap presents results for the first five such sets.

A somewhat finer classification is by the degree-tuple of a graph. Let D_i denote the degree of vertex i and $(D_1, D_2, \ldots, D_n)$ be the degree tuple in which $D_i \leq D_j$ for $i > j$. The degree tuples are ordered by saying $D > D'$ if the first unequal component, say the kth, satisfies $D_k > D'_k$. In carrying out the full enumerations several sub-sets are defined and ordered by such means before the final detailed ordering by canonical matrices is imposed.

Heap shows that the distinct HI stars with cyclomatic number c (for $c \geq 3$) can be derived from those having cyclomatic number $c - 1$ by considering all possible ways of carrying out the following operations:

(a) Joining any two distinct vertices by an edge.

(b) Inserting a vertex of degree two on any edge and joining it to any other vertex by an edge,

(c) Inserting two vertices of degree two on any two (not necessarily distinct) edges and joining them by an edge.

Any of these procedures increases the cyclomatic number by exactly one as required and can be shown to generate all HI stars of cyclomatic number c.

Heap has employed computer techniques utilizing the above operations to generate new HI stars. He has enumerated the HI stars for $c = 5$ from the known values for $c \leq 4$. That enumeration includes graphs with up to 8 vertices and 12 edges. The number of planar stars are also enumerated since they are of particular interest for various applications. A dictionary (pictorial) of all HI stars for $c = 5$ is given in [111]. All HI stars with $c = 2$ and $c = 3$ are shown in Figure 146 three out of the total of 17 HI stars with $c - 4$ are shown

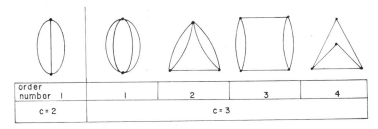

Figure 146

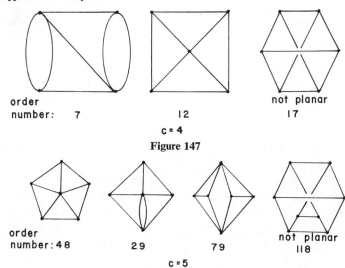

order
number: 7   I2   I7

c = 4

Figure 147

order
number: 4 8   2 9   7 9   not planar
                                                    II8

c = 5

Figure 148

in Figure 147 and four out of the total of 118 *HI* stars with $c = 5$ are shown in Figure 148.

Table 16.   Number of HI Stars

Cyclomatic Number	Number of vertices n	2	3	4	5	6	7	8	Total	
	Number of Edges									
2	3	1							1	1
3	4	1							1	
	5		1						1	
	6			2					2	4
4	5	1							1	
	6		2						2	
	7			5					5	
	8				4				4	
	9					5^p			5	17
5	6	1							1	
	7		3						3	
	8			13					13	
	9				24				24	
	10					38^{p^2}			38	
	11						23^{p^2}		23	
	12							16^{p^3}	16	118
		4	6	20	28	43	23	16		

Some statistics relative to the enumeration of *HI* stars for $c \leq 5$ are given in Table 16 which shows results taken from [111] but not stated there in the present form.

In Table 16 an expression p^k means that k of the stars in the classification are not planar. Heap also enumerates the number of *HI* stars for $n \leq 8$ and $m \leq 12$ some of which are included among the enumeration for $c \leq 5$ tabulated above. The largest set is for $n = 6$ and $m = 12$ which contains 914 *HI* stars.

An area of study that is somewhat related to the concepts discussed in this section, though distinct from it, is the application of special graph types in biology and chemistry. We shall not enter into any discussion of that topic here beyond pointing out [113] by Lederberg who has made a number of contributions in the field. In Chapter 4 we mentioned some results of Lederberg in connection with hamilton graphs. In fact those results [52] were motivated by the study of molecular structure.

EXERCISES

1. Find all dimer coverings of the following lattice:

$$
\begin{matrix}
\bullet & \bullet & \bullet & \bullet \\
\bullet & \bullet & \bullet & \bullet \\
\bullet & \bullet & \bullet & \bullet \\
\bullet & \bullet & \bullet & \bullet
\end{matrix}
$$

Consider the same problem in three dimensions where the lattice consists of two of the above planar lattices separated by one lattice space.

2. There are 17 homomorphically irreducible stars with cyclomatic number 4. A dictionary of these is given in [111] and 3 are shown in the text as Figure 147. Draw as many of these as you can and study their graph properties (such as chromatic number, cycle basis, coefficients of internal and external stability, and spectrum).

3. Consider the formulation of information distribution in a community as a percolation process. Interpret the meaning of the percolation process. Interpret the meaning of the percolation probability in this situation and the role played by the underlying (graph) structure. In particular, distinguish a triangular lattice community from a square lattice community.

4. Evaluate the Pfaffian:

$$
\begin{vmatrix}
3 & 6 & 1 \\
 & 2 & 5 \\
 & & 4
\end{vmatrix}
$$

Appendix

In this Appendix we present the basic ideas of Linear and Linear Integer Programming. Only the highlights are presented and this material is in no sense a substitute for a careful study of the subjects considered. There exists an extensive literature on these subjects but we wish to avoid a long list of references. The reader can refer with profit to the following:

Dantzig [114] and Ford and Fulkerson [90] for Linear Programming and closely related concepts, and Balinski [115] and Saaty, [40] for Integer Programming.

LINEAR PROGRAMMING

Let x and c denote n-vectors with elements x_i and c_i respectively. Let b denote an m-vector with elements b_j and A denote an $m \times n$ matrix with elements a_{ij}. The standard linear programming problem is defined to be:

$$\text{minimize } z = c^t x,$$
$$\text{subject to } Ax = b,$$
$$x \geq 0.$$

Almost no linear programming problem will be initially in the standard form but every linear programming problem can be put into this standard form by appropriate manipulation. In this statement we exclude problems which, though linear in their analytic aspects, contain more complex constraints such as x integral valued (meaning, of course, the elements of x are to be integers). Such can also be put into the standard form with the extra constraints still present as an additional requirement. Hence when we speak of the standard linear programming problem, it is to the above formulation that we refer.

The usual type of constraint is an inequality rather than an equality. A problem formulated with such constraints might originally have k (rather than n) variables. A typical constraint for such a problem is:

$$a_{11} x_1 + \cdots + a_{1k} x_k \le b_1$$

By adding an auxiliary variable x_{k+1} satisfying constraint $x_{k+1} \ge 0$ we obtain an equality constraint,

$$a_{11} x_1 + \cdots 3\, a_{1k} x_k + x_{k+1} = b.$$

without changing the original problem at all. Thus every solution to the new problem (in any algebraic sense of the term solution, a concept we shall discuss more fully below) will be a solution to the original problem. Had the inequality been in the opposite sense (as a greater than or equal to symbol) the auxiliary variable would have been subtracted. We therefore see that by the introduction of at most m auxiliary variables all the m constraints can be written as equalities, hence represented as $Ax = b$ as required by the standard form. In this form the original variables and auxiliary variables have been combined to form the components of the n-vector x.

Since the standard form is a minimization problem any linear maximization can be converted to standard form by multiplying the linear objective function $z = c^t x$ by minus one.

The constraints $x \ge 0$ are natural for programming problems in which amounts are assigned or not. In the more general application of linear programming techniques however it is important to be able to allow some (or all) of the original variables to be unrestricted in sign. To put such problems into the standard form suppose the variable x_i is to be unrestricted in sign; we introduce two new variables x_{i1} and x_{i2} and set $x_i = x_{i1} - x_{i2}$ throughout the problem, requiring that $x_{i1} \ge 0$, $x_{i2} \ge 0$. In this way general variables are introduced into the standard form.

Many typical features of linear programming problems can be illustrated by considering problems requiring two variables for their formulation. The value of such examples lies in the possibility for graphical representation of the region in which the variables satisfy the constraints and hence a pictorial or direct solution technique. We consider three of these two dimensional problems.

Example 1.   Maximize $2x_1 + x_2$

subject to $$x_1 + 4x_2 \le 4$$
$$4x_1 + 3x_2 \le 6$$
$$x_1 - x_2 \le 1$$
$$x_1, x_2 \ge 0.$$

The lines representing limits on allowed values of the variables are h_1: $x_1 + 4x_2 = 4$, h_2: $4x_1 + 3x_2 = 6$, h_3: $x_1 - x_2 = 1$ and the coordinate axes. By checking to see on which side of each line the zero lies (since zero satisfies every given constraint) we obtain the region of values to which x_1 and x_2 are constrained. By allowing the objective function $z = 2x_1 + x_2$ to take on increasing values we find the point of the constraint region that yields max z. This is illustrated by plotting some values of constant z together with the region of constraint in the x_1, x_2 plane as shown in Figure 149. The point $(9/7, 2/7)$ lying on the interaction of lines h_2 and h_3 yields the maximum value of z, z_{max} 20/7, consistent with the constraint conditions.

This example is characteristic of all linear programming problems. The constraint region is of the same general type as encountered here with boundary hyperplanes whose intersection forms vertices. Moreover an optimum value of the objective function z will always be found on one of these vertices (and possibly elsewhere). The fact that these results, as illustrated, hold in general allows the utilization of efficient algorithmic solutions to large linear programming problems. These two important properties seem obvious in the two dimensional case but require careful proof in higher dimensions.

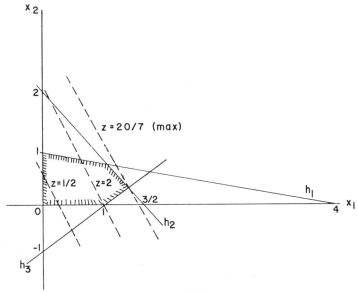

Figure 149

To conclude this example we may put it into standard form to obtain the problem:

minimize
$$-2x_1 - x_2$$

$$x_1 + 4x_2 + x_3 = 4$$

subject to
$$4x_1 + 3x_2 + x_4 = 6$$
$$x_1 - x_2 + x_5 = 1$$
$$x_i \geq 0, \; i = 1, \ldots, 5$$

Here three auxiliary variables have been introduced and the vector x is a five vector. In standard form a pictorial solution is no longer possible. The solution vector is, as determined above, equal to $(9/7, 2/7, 11/7, 0, 0)$.

Example 2. Here we consider the problem of Example 1 changing constraint h_1 to $21x_1 + 7x_2 \leq 29$. The picture is as shown in Figure 150.

The solution is the same as in Example 1; however in standard form the solution vector becomes: $(9/7, 2/7, 0, 0, 0)$ with an additional variable at the value zero. As before at least $n - m$ values are zero which, as we have remarked is the general case at optimal yielding (vertex) points. In this case, however, an additional variable is zero at the optimal since this vertex point represents a coalescence of two vertex points, the third constraint equation which would otherwise separate the point into two passes through the intersection of the other constraint equations. Such a situation is not uncommon in certain programming problems, though in this simple case it would seem that h_1 is

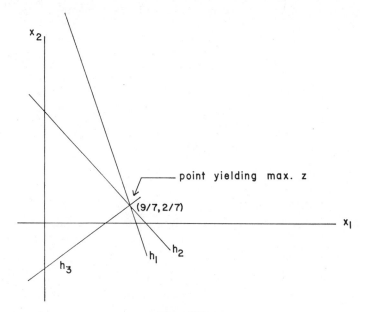

Figure 150

obviously redundant and should be dropped. These redundancies cannot be easily detected in large problems and can, moreover, come about in natural ways. The situation illustrated here in which more than $n - m$ of the variable values are zero is known as degeneracy. There is nothing particularly undesirable about the occurrence of such degeneracies except in certain theoretical situations and the problem may be dealt with by a modification of the algorithms for solution which removes degeneracy.

Example 3. The problem of Example 1 can be modified once again to illustrate one further basic aspect of linear programming. In this case we change the objective function to maximize $z = 4x_1 + 3x_2$.

It is clear that this linear form is parallel to h_2 and therefore will have the solution given by the point $(9/7, 2/7)$ as before. The optimum value is $z_{max} = 6$. The optimum is obtained on a vertex point as the general result we have indicated demands, however, a number of other points also yield the optimum. The set of points lying between $(12/13, 10/13)$ and $(9/7, 2/7)$ in h_2 all yield the optimal value for z and satisfy the constraints. We see, therefore, that although the optimum is always given by a vertex point other points may also yield the optimum.

Any n-vector x satisfying the constraint conditions $Ax = b$ is said to be a solution. If in addition a solution vector satisfies the non-negativity condition $x \geq 0$ it is called a feasible solution. We have seen in an example that an important class of solutions are those for which at least $n - m$ of the variables are at the zero value; these are called basic solutions to the constraints $Ax = b$. They are basic feasible solutions if their non-zero variables (components) are all positive.

The n-vectors x can be considered as points in euclidian n-space and special sets of points in such a space are of particular importance in mathematical programming; these are convex sets. Let $\{a_i\}$ and $\{x^i\}, i = 1, \ldots, k$ denote a set of k scalars and k n-vectors, respectively. The vector $y = \sum_{i=1}^{k} a_i x^i$ is said to be a convex combination of the vectors $\{x^i\}$ provided $\sum_{i=1}^{k} a_i = 1$ and $a_i \geq 0$ for $i = 1, \ldots, k$. In particular for two vectors x^1 and x^2 we can form their convex combination as $y = ax^1 + (1 - a)x^2$ where $0 \leq a \leq 1$. Such vectors y lie on the straight line segment joining the points x^1 and x^2 in n-space.

The set of points yielding feasible solutions to the standard linear programming problem form a convex set in n-space. Moreover, that set is bounded by hyperplanes and contains vertices; these latter corresponding to basic feasible solutions. Thus every feasible solution can be represented in terms of the basic feasible solutions (hence, their name "basic").

Corresponding to any particular basic feasible solution there is a canonica form for the standard linear programming problem. This form has two major aspects.

1. The objective function is expressed entirely in terms of variables not in the basis (basic feasible solution). This is accomplished by utilizing the constraint equations as needed to solve for the basic variables and substituting the results in the objective function.

2. The constraint equations $Ax = b$ are transformed so that each basic variable occurs in only one equation and has coefficient unity and each equation has only one basic variable. In this form, the new coefficient matrix is partitioned into an $m \times m$ identity submatrix and an $mx(n - m)$ sub-matrix so that one obtains the constraints in the form: $(I_m \ H)x = b'$ where the vector b' has the modified values of the original b vector elements. For convenience the expression is shown as though the first m variables $(x_1, \ldots, x_m)$ are the basic variables. Of course, any set of m variables may be the basic variables but by relabelling, if necessary, they can be expressed as shown.

A problem should be put into canonical form for use of the simplex method of solution.

Problem in standard form:

minimize $\qquad\qquad\qquad z = x_1 - 3x_2 + x_3$

subject to
$$2x_1 + x_2 - x_4 = 1$$
$$x_1 - 2x_2 + 8x_3 + x_5 = 3$$
$$5x_1 - 7x_3 + x_6 = 4$$
$$x_1 \geq 0, \ i = 1, \ldots, 6$$

A basic feasible solution is seen to be given by $x_1 = 4/5$, $x_4 = 3/5$ and $x_5 = 11/5$ with the non-basic variables x_2, x_3 and x_6 at zero. The cononical form for the basis (x_1, x_4, x_5) is as follows

$$z = \frac{4}{5} - 3x_2 + \frac{12}{5}x_3 - \frac{1}{5}x_6$$

$$x_1 - \frac{7}{5}x_3 \qquad\qquad + \frac{1}{5}x_6 = \frac{4}{5}$$

$$-2x_2 + \frac{47}{5}x_3 \qquad + x_5 - \frac{1}{5}x_6 = \frac{11}{5}$$

$$-x_2 - \frac{14}{5}x_3 + x_4 \qquad + \frac{2}{5}x_6 = \frac{3}{5}$$

Note that each column of A corresponds to the coefficients of a particular x variable in the constraint equations. To find a particular basic solution select those columns of A corresponding to the m variables which are to occur in the basic solution and use them as the columns of an $m \times m$ matrix B. Define x^m as an m-vector whose components are the m variables of the basis in the same order as their corresponding coefficient columns from A were placed in forming B. Then when the selected set forms a basic solution it is given as the solution of $Bx^m = b$ where one must be careful to interpret x^m properly as an m-vector.

We shall first assume a basic feasible solution is known and show how to proceed to the optimal solution. The procedure used is known as the simplex method. One starts in this case with a basic feasible solution and the standard linear programming problem in the canonical form associated with the given solution. The procedure will be given in short, four step outline and then the details of each step will be stated. A prime is always used to represent present values of a quantity which when unprimed is the value in the original standard formulation.

An outline of the procedure is as follows.

1. Select by some rule the coordinate x_s to be introduced into the basis to improve z if possible (certainly so as to not make z larger).

2. Select a variable x_r to be removed from the basis so as to cause no increase in z, and improve z as much as possible.

3. Obtain the canonical form for the new basic feasible solution.

4. Test for optimality. Include a test for no finite solution.

The details can be given various forms resulting in particular algorithms. The most basic is the simplex algorithm for which details follow.

1. $z = z_0 + c'_{m+1} x_{m+1} + \cdots + c'_n x_n$ is the present canonical form of the objective function, z_0 is the present value of z. If x_s is increased from zero the objective function has the value

$$z = z_0 + c'_s x_s.$$

If $c'_j \geq 0$ for $j = m + 1, \ldots, n$ then z cannot be improved by bringing in any such x_s and the present solution is optimal (major part of step 4).

If some $c'_j < 0$ for $j = m + 1, \ldots, n$ then increasing any of the non-basic variables corresponding to these will improve, (i.e., decrease) z. Let K denote the index set of those coefficients that are negative.

Here we select x_s such that $c'_s \leq c'_k$ where k ranges over the index set K.

2. We wish to see how large to make x_s while keeping the resulting solution feasible. This will determine x_r. In the canonical form transpose the x_s column to the right side of the equation. Then one obtains:

$$x_i = b'_i - a'_{is} x_s \qquad i = 1, \ldots, m$$
$$z = z_0 + c'_s x_s$$

and it is seen that x_r is determined by

$$\min \frac{b'_i}{a'_{is}} = \frac{b'_r}{a'_{rs}},$$

where minimization is over those values i such that $a'_{is} > 0$. The interchange of x_s for x_r has now been accomplished.

If r (the solution to the above condition) is not unique then the resulting new basis will be degenerate.

Note. In case $a'_{is} \leq 0$ for $i = 1, \ldots, m$; x_s can be increased without bound and no finite solution exists (completes step 4).

3. The new canonical form must now be obtained from the old form using a'_{rs} as pivot element. Denote new elements by an asterisk *. The new form is obtained by simple algebra, the total process being called a pivot. The necessary formulas are:

$$a'^{*}_{rs} = 1$$
$$a'^{*}_{is} = 0, i \neq r \quad \text{(sth column)}$$
$$a'^{*}_{ij} = a'_{ij} - a'_{is} a'^{*}_{rj} \quad \text{(all others)} \quad \begin{array}{l} i \neq r \\ i = 1, 2, \ldots, m \\ j = m + 1, \ldots, n \\ j \neq r, s \end{array}$$

$$a'^{*}_{rj} = \frac{a'_{rj}}{a'_{rs}} \quad j = m + 1, \ldots, n \quad \text{(rth row)}$$

$$c'^{*}_{j} = c'_{j} - c'_{s} a'^{*}_{rj} \quad j = m + 1, \ldots, n$$

The new basis variables are:

$$x_i = b'^{*}_i = b'_i - a'_{is} b'^{*}_r \quad i \neq r$$

$$x_s = b'^{*}_r = \frac{b'_r}{a'_{rs}}$$

$$z^{*}_0 = z_0 + c'_s b'^{*}_r$$

4. has been developed as part of steps 1 and 2.

The solution to a linear programming problem proceeds in two major stages. In Phase I an initial basic feasible solution is obtained or the existence of no such solution is indicated. In Phase II this initial solution is successively transformed until an optimal is obtained. Both phases can be carried out by the simplex algorithm described above.

The constraint conditions may be such that some variables have coefficient unity and the variables occur in only one equation each. If this number of variables is m and the b values are non-negative, those variables can be used directly as the original feasible basis.

In case the number of such variables is less than m one introduces artificial variables to take care of the unsatisfactory equations. Assume m such artificial variables are required. Denote them by $x_{n+1}, \ldots, x_{n+m}$, where x_{n+i} is added to equation i.

An initial feasible basic solution to the new problem is

$$x_{n+i} = b_i \qquad i = 1, \ldots, m$$
$$x_j \;\;\; = 0 \qquad j = 1, \ldots, n$$

where the constraint equations have been multiplied by -1 if necessary so that all $b_i \geq 0$.

We wish to remove all the artificial variables from the solution in order to obtain an initial solution to our original problem. Hence, we minimize the infeasibility form $w = x_{n+1} + \cdots + x_{n+m}$. The solution has $x_{n+i} = 0$ for $i = 1, \ldots, m$ and gives the desired initial solution.

If min w is not zero then the original problem has infeasible constraints and the problem has no (feasible) solution.

Many aspects of linear programming utilize the concept of the Dual program. Any given linear program may act as a Primal for which a specific Dual program is specified according to a strictly formalized definition. One usual definition is the following:

If a problem has the primal form (Min) $z = c^T x$ subject to $Ax \geq b$, $x \geq 0$ then the dual problem has the form (Max) $v = b^T y$ subject to $A^T y \leq c$, $y \geq 0$ in the dual variables y.

We shall not discuss duality in detail but simply state some of the most important results. Note that each dual variable corresponds to a constraint in the primal. If a primal constraint is an equality the corresponding dual variable is unrestricted in sign.

The duality theorem states that if feasible solutions exist for both problems then min z = max v.

If no feasible solution exists for the primal but a feasible solution exists for the dual the latter is an unbonded solution.

It is possible for neither problem to have a feasible solution.

The optimal solutions for each problem are related by means of the concept of simplex multipliers allowing one to obtain the full solution for one problem from the full solution of the other.

The values of the simplex multipliers associated with the optimal of the dual are the negative of the values of the variables for the optimal solution of the primal.

The optimal values of the dual variables are the values of the simplex multipliers associated with the optimal solution of the primal.

Any basis can be taken as the variables $x_1, \ldots, x_m$ by relabelling if necessary. We have seen that their values are found as the solution of

$$Bx^m = b.$$

To solve for the x variables for a general b value, that is, in terms of the components of b, we have

$$x^m = B^{-1} b.$$

Denote elements of B^{-1} by β_{ij}; this matrix is called the inverse of the basis.

In the process of forming the canonical representation of a system the basic variables must be eliminated from the objective function and corresponding new coefficients obtained. This may be done as follows:

Consider $Ax = b$. Multiply the ith equation by a real number $\pi_i, i = 1, \ldots, m$. Subtract all the resulting equations from the objective function $c^t x = z$. The result is

$$\sum_{j=1}^{n} (c_j - \sum_{i=1}^{m} a_{ij}\,\pi_i)\, x_j = z - \sum_{i=1}^{m} b_i\,\pi_i.$$

To obtain the canonical form of the objective function the first m coefficients must vanish. This yields the following m equations for the numbers π_i.

$$c_j - \sum_{i=1}^{m} a_{ij}\,_i\pi = 0 \qquad j = 1, \ldots, m.$$

The numbers π_i are called the simplex multipliers associated with the basis $x_1, \ldots, x_m$.

From them one finds:

$$c_j' = c_j - \sum_{i=1}^{m} a_{ij}\,\pi_i \qquad j = m+1, \ldots, n$$

and $\qquad z_0 = \sum_{i=1}^{m} b_i\,\pi_i.$

Thus the canonical form of the objective function and its value for the basis are determined by the associated simplex multipliers.

The procedure described as the simplex method can be carried out in several ways other than by the simplex algorithm described above. We shall only present one of those alternatives here because of its close connection with Integer programming. That method, called the Dual Simplex method (or algorithm) is based on the concepts of duality in linear programming. In this summary we can not enter deeply into the important and interesting subject of duality defined above which forms a central focal point for much of programming theory. However, we can give a formal expression of the Dual Simplex method without discussing duality concepts themselves.

A key point of the simplex method is that by using the canonical form a solution to the programming problem is achieved when the basis variables are non-negative (feasible) and the optimality condition is satisfied (objective coefficients non-negative for minimization). The Simplex algorithm achieves that goal by maintaining feasibility at each step and directing the iterations in such a way as to arrive at the optimality condition. The Dual Simplex algorithm maintains the optimality condition and carries out iterations on non-feasible solutions that are directed toward arriving at a feasible solution and hence a problem solution. That method in fact is carrying out the Simplex algorithm on what is called the Dual problem as the reader will find completely discussed in the references given in the introduction.

To apply the Dual Simplex algorithm the problem must be in canonical form with all objective coefficients non-negative. If a basis is not clear one can describe a Phase I procedure. However, for our purpose here it is sufficient to assume we have a problem in the form stated and to give the subsequent steps of the algorithm.

If in the canonical form the basis values are non-negative, the solution is optimal. Otherwise, we must select a variable to go out of the basis. The values are b_i' and we select $b_r' = \min b_i' < 0$ and take x_r out of the basis (note the change in order of operations from the regular simplex algorithm). To select the variable to enter the basis we use the criteria: introduce x_s' where

$$\min|c_j'/a_{rj}'| = c_s'/ - a_{rs}'$$

where we minimize over all negative a_{rj}' values (recall that in this algorithm $c_j' \geq 0$ for all j). If $a_{rj}' \geq$ for all j then one can not arrive at a feasible solution to the problem. The next step is to pivot on a_{rs}' to obtain the next canonical form. The algorithm will achieve an optimal feasible solution or else indicate that no feasible solution exists for the problem. Any ties in the selection of variables may be broken in any desired way.

INTEGER PROGRAMMING

In this section we will outline some major points in integer programming. It may be that a given problem will have an integer solution by using the simplex method, such a problem is said to have integrity. A major result of integrity studies is as follows.

Consider the problem with constraint set $Ax \leq a_0$, $x \geq 0$. Denote this set by (I) and assume all constants are integers.

Theorem (Hoffman and Kruskal). The extreme points of (I) have all integer coordinates for every choice of all integer vector a_0 if and only if A is unimodular.

If integrity can not be established one can solve by the simplex method and round off non-integer results. There are many objections to such a procedure however. The result may not be feasible. If feasible several different methods may be possible and one must check them all. The true integer optimal may not be among the rounded off feasible results, and so forth. Thus direct methods for solving programs in integers are required. There are two basic concepts used to develop such methods. One concept is to introduce additional constraints (hyperplanes) that remove non-integer solutions but do not remove integer solutions until a form of simplex operation arrives at an integer valued optimal solution. These are called cutting plane methods and a pioneer work on the subject is Gomory [116], The other concept is to convert the problem to a zero-one type if it is not already in that form (each x_i can be only zero or one) and enumerate all cases so that the optimal results. Since there are in general a vast number of cases any effective algorithm carries out the enumeration implicitly rather than by explicit consideration of each case. Such methods are often called branch and bound methods since they operate on branches of the tree of all possible cases and one drops many cases by considering bounds available for the objective function on a given branch. Two basic contributions to such methods are Balas [117] and Glover [118].

The most basic form of Gomory's cutting plane method can be given in simple terms. We shall do so here without discussing the background theory or the valuable extensions of the concepts to much better algorithms. In general, application of the method as presented below may be expected to converge very slowly to the solution.

Initial Steps

Introduce auxiliary variables as needed and put the optimization in the form of a minimization problem. Thus the initial form is as follows:

$$a_{11} x_1 + a_{12} x_2 + \cdots + a_{1n} x_n = a_{10}$$
$$\vdots \qquad \vdots \qquad \qquad \vdots \qquad \vdots$$
$$a_{m1} x_1 + a_{m2} x_2 + \cdots + a_{mn} x_n = a_{n0}$$
$$a_{01} x_1 + a_{02} x_2 + \cdots + a_{0n} x_n = a_{00}$$

where we suppose an initial basic feasible solution to the Primal problem is at hand and the above is in canonical form for that basis. Thus the initial solution values are a_{j0}. If $a_{0i} \geq 0$ for $i = 1, \ldots, n$ we have an optimal primal solution a_{j0}. If these are integer the solution is obtained, if not one introduces an additional constraint.

If $a_{0i} < 0$ for some i the given problem is solved by means of the Primal algorithm until it is optimized, that is, until $a_{0i} \geq 0$ for all i. If the final solution of such an algorithm is integral we are done; otherwise, introduce a new constraint.

Next Steps

When a new constraint has been added, we see that the situation is as follows:

The new problem is not optimal due to the form of the added constraint (discussed below). Since $a_{0i} \geq 0$ for all i the problem is exactly in the form desired for using the Dual algorithm. One proceeds using this until the optimum is reached. If this optimal is integral it is final solution, otherwise a new constraint is added, and so forth.

The constraint to be added when such is called for is an element of the module M generated by the vectors of fractional parts of the given constraints, over the integers. Thus we write $a_{ij} = n_{ij} + f_{ij}$ where n_{ij} is an integer and f_{ij} a fraction between 0 and 1 and call f_{ij} the fractional part of a_{ij}.

For example if $a_{ij} = -3/7$ then $n_{ij} = -1, f_{ij} = 4/7$.

Then the constraints yield the m vectors:

$$\bar{e}_i = (f_{i1}, f_{i2}, \ldots, f_{in}, f_{i0}) \; i = 1, \ldots, m$$

and the module M consists of all vectors $h = \sum_{i=1}^{m} \alpha_i \, \bar{e}_i$ where α_i are integers.

Any members of M can be utilized as added constraints as follows: Say the vector to be used is $(b_1, \ldots, b_n, b_0)$ then the added constraint is $\sum_{j=1}^{n} b_j x_j \geq b_0$ which leads to the new equation $-\sum_{j=1}^{n} b_j x_j + s = -b_0$ with the new auxiliary variable s. We see that the value of s is negative (recall $0 < b_0 < 1$) as mentioned above.

To Determine New Constraint

There is no fixed rule for this. One which can be used is to select one of the values $\bar{e}_i$ itself. (It is very hard to decide how more complex elements of M might be selected).

The following rule for selecting $\bar{e}_i$ is said to take a deep cut into the feasible region: (recall $a_{0j} \geq 0$ after the initial stage). Find minimum nonzero a_{0j}, then for that value of j, say j^* find minimum $a_{ij}{}^*$ (note this is the algebraic minimum) the row index i yielding this minimum is the index for $\bar{e}_i$.

Of course ties are broken by the usual methods.

Other rules for adding constraints can be used. Special forms for computation in the array can be used (this is often done in papers on the subject). These have nothing to do with the ideas of the method.

Unfortunately the combinatorial methods that may be applied to zero-one programming problems involve extensive definitions and detailed algorithms that do not allow an abbreviated summary. We can only formulate the class of problems treated and indicate the concepts used in developing the methods. For details the reader is advised to consult the references given above.

COMBINATORIAL METHODS FOR INTEGER PROGRAMMING

These methods treat problems in which the variables take only the values 0 or 1. Any bounded integer problem in which the upper bonds (on non-negative integer variables) are known can be converted to 0, 1 form by increasing the number of variables, for example, if $0 \leq y \leq N$ is an integer variable it is replaced by the 0, 1 variables x_i; $y = \sum_{i=1}^{N} x_i$, etc.

The standard 0, 1 problem is as follows.

$$\min z = c^t x$$
$$Ax \leq b$$
$$x = 0 \text{ or } 1$$

where $c \geq 0$.

Should the original problem not be in this form one first converts the inequalities and objective as needed by multiplying by -1. At that point the condition $c \geq 0$ may not prevail. To achieve $c \geq 0$ make the following change of variable:

$$x_j = x_j' \quad \text{for} \quad c_j \geq 0$$
$$x_j = 1 - x_j' \quad \text{for} \quad c_j < 0$$

This interchanges the values of x_j and x_j' (the new variable) and achieves the condition $c \geq 0$. It will in general introduce or modify existing constants in z but we drop such constants during the calculation.

Thus we must remember to *decode* the final result to recover original variable and objective function values. This is true for the several transformations that may be required to place the original problem into standard 0, 1 form.

Upon introduction of auxiliary (slack) variables y (which need *not* assume integer values) one has the standard 0, 1 problem [called problem P].

$$\min z = c^t x \tag{1}$$

$$Ax + y = b \tag{2}$$

$$x_j = 0 \text{ or } 1 \quad (j \in N) \tag{3}$$

$$y \geq 0 \tag{4}$$

where $c \geq 0$, A is $m \times n$ (as usual), etc.

Denote the jth column of A by a_j, and the index sets are $N = \{1, \ldots, n\}$, $M = \{1, \ldots, m\}$.

An $n + m$ vector $u = (x, y)$ is a solution (to P) if it satisfies (2) and (3), feasible solution if it satisfies (2), (3), and (4) and an optimal feasible solution if it satisfies all four conditions.

Problem P^s. A linear program satisfying (1), (2), (4) and

$$x_j \geq 0 \qquad (j \in N) \qquad\qquad (3a)$$

$$x_j = 1 \qquad (j \in J_s) \qquad\qquad (3b)$$

where $J_s \subseteq N$.

For $J_0 = \Phi$ (empty set) one gets problem P^0 the linear programming problem without any integer requirement on x.

The basic idea of algorithms is to try solutions formed by setting various x_j to 1 (forming problems P^s) and follow along the tree of trial solutions. Use the condition that $x_j = 1$ branches to the right and $x_j = 0$ branches to the left. Clearly the tree has 2^n roots (terminal points). If one went to the end and looked at z for every root one could select a minimizing solution directly.

The 0, 1 algorithms wish to drop out whole sets of roots and the branches leading to those roots by showing an optimal solution could not lie on any path leading from certain branches. This requires study of bounds on z for special branches and the introduction of efficient rules for dropping off branches.

These are called branch and bound methods and apply to a wide range of (optimization) problems.

In addition to the opportunity to suboptimize effectively there are two major values these methods have (over cutting planes, etc.): efficiency and simple arithmetic (addition only) leading to high accuracy.

Bibliography

In addition to the material cited as references throughout the book we wish to mention some items of a more general nature.

The reader may find a number of interesting topics in *A Seminar on Graph Theory*, F. Harary, Ed., Holt, Rinehart, and Winston, 1967.

Much has been done in the graph theory field by electrical engineers, particularly in certain applications. A classical text from that point of view is S. Seshu and M. Reed, *Linear Graphs and Electrical Networks*, Addison-Wesley, 1961.

In recent years a number of symposia have been held (under various titles) dealing with graph theory and combinatorial concepts. These have produced several important collections of works in book form. These collections continue to be produced at an increasing rate corresponding to the increased occurrence of such meetings and the expanding interest in the topic. These can be consulted with profit by the reader and some occur within our specific references. We list a few of the symposia collections without any attempt to give a complete list.

One of the earliest, hence influential, symposium volumes is *Theory of Graphs and Its Applications* (Smolenice Symposium, 1963), M. Fiedler, Ed., Academic Press, 1964. This volume is noteworthy in that it contains an extensive bibliography, rather complete to 1962, based on collections of Moon and Moser (their bibliography was informally produced).

Another extensive bibliography developed by J. Turner, designed to update and extend the Moon-Moser work is given in the symposium volume, *Proof Techniques in Graph Theory* (Proceedings of 2nd Ann Arbor Graph Theory Conference), F. Harary, Ed., Academic Press, 1969.

Some additional symposia volumes (not a complete list), in no particular order, are as follows.

Proceedings of IBM Scientific Computing Symposium on Combinatorial Problems (*1964*), IBM Corporation, 1966. In particular the section by V. Klee, "Convex Polytopes and Linear Programming."

Theory of Graphs—International Symposium (*Rome 1966*), P. Rosenstiehl, Ed., Gordon and Breach, 1967.

Graph Theory and Theoretical Physics (*Paris 1963*), F. Harary, Ed., Academic Press, 1968.

Theory of Graphs: Proceedings of the Tihany Colloquium (*Hungary 1966*), P. Erdos and G. Katona, Eds., Academic Press, 1969.

Recent Progress in Combinatorics (*Proceedings of 3rd Waterloo Conference on Combinatorics*), W. Tutte, Ed., Academic Press, 1969.

Combinatorial Structures and Their Applications (*Proceedings of Calgary Conference*), H. Guy, H. Hanani, N. Sauer, and J. Schonheim, Eds., Gordon and Breach, 1970.

International Conference on Combinatorial Mathematics (*1970*), A. Gewirtz and L. Quintas, Eds., New York Academy of Sciences, 1970.

An interesting survey of some researches in the field is to be found in: J. Turner and W. Kautz, *Graph Theory in the Soviet Union*, SIAM Review Supplement to Volume 12, 1970.

A basic article on Posa's theorem is found in: L. Pósa, "A Theorem Concerning Hamiltonian Lines," *Magyar Tud. Akad. Mat. Kutató Int. Közl*, **7** (1962), 225–226.

Matrices of zeros and ones are treated particularly well in: H. J. Ryser, "Matrices of Zeros and Ones," *Bull. Amer. Math. Soc.*, **66** (1960), 442.

Applications of graphs in the social sciences are developed in the following three books.

F. Harary and R. Norman, *Graph Theory as a Mathematical Model in Social Science,* University of Michigan Press, 1953.

C. Flament, *Applications of Graph Theory to Group Structure,* Prentice-Hall, 1963.

J. Kemeny and J. Snell, *Mathematical Models in the Social Sciences*, Ginn and Co., 1962.

Articles on graph theory and related topics occur in a wide variety of journals; however, the reader should notice that a rather significant number of articles in graph theory appear in the following three journals.

SIAM Journal on Applied Mathematics
Journal of Combinatorial Theory
Canadian Journal of Mathematics

To conclude our discussion of general reference material we bring a particularly important article to the reader's attention:

B. Grunbaum, "*Polytopes, Graphs, and Complexes*," *Bull. Amer. Math. Soc.*, **76** (1970), 1131–1201. This article touches on a number of important topics related to graph theory.

References

1. Essam, J., and M. Fisher, "Some Basic Definitions in Graph Theory," *Rev. Mod. Phys.*, **42** (1970), 272–288.
2. Lipschutz, S., *Outline of Theory and Problems of Set Theory and Related Topics*, Schaum, 1964.
3. Stoll, R., *Set Theory and Logic*, Freeman, 1961.
4. Harary, F., R. A. Norman, and D. Cartwright, *Structural Models—An Introduction to the Theory of Directed Graphs*, Wiley, 1965.
5. Fraley, R., K. Cooke and P. Detrick, "Graphical Solution of Difficult Crossing Puzzles," *Math. Mag.*, **39** (1966), 151–157.
6. Berge, C., *The Theory of Graphs and Its Applications*, Wiley, 1962.
7. Heller, I., and C. B. Tompkins, "An Extension of a Theorem of Dantzig's," *Linear Inequalities and Related Systems*, Annals of Mathematical Studies (38), Princeton University Press, 1956.
8. Hoffman, A. J., and G. J. Kruskal, "Integral Boundary Points of Convex Polyhedra," *Linear Inequalities and Related Systems*, Annals of Mathematical Studies (38), Princeton University Press, 1956.
9. Ryser, H. J., *Combinatorial Mathematics*, Wiley, 1963.
10. Williamson, J., "Determinants Whose Elements Are 0 and 1," *Amer. Math. Monthly*, **53** (1946), 427.
11. Tutte, W. T., *The Connectivity of Graphs*, Toronto University Press, 1967.
12. Tolkien, J., and E. Gordon, Eds., *Sir Gawain and the Green Knight*, Oxford, 1925.
13. Riordan, J., *An Introduction to Combinatorial Analysis*, Wiley, 1958.
14. Stewart, B. M., "Magic Graphs," *Canad. J. Math.*, **18** (1966), 1031–1059.
15. Ore, O., *Theory of Graphs*, American Mathematical Society, 1962.
16. Fulkerson, D. R., and O. A. Gross, "Incidence Matrices and Interval Graphs," *Pacific J. Math.*, **15** (1965), 835–856.

17. Tucker, A., "Characterizing Circular-Arc Graphs," *Bull. Amer. Math. Soc.*, **76** (1970), 1257–1260.

18. Harary, F., and L. Moser, "The Theory of Round Robin Tournaments," *Amer. Math. Month*, **73** (1966), 231.

19. Moon, J., *Topics on Tournaments*, Holt, Rinehart, and Winston, 1968.

20. Erdös, P., T. Grünwald, and E. Vazsonyi, "Über Euler-Linien Unendlicher Graphen," *J. Math. Phys.*, **17** (1938), 59–75.

21. Aardenne-Ehrenfest, T., and N. deBruijn, "Circuits and Trees in Oriented Linear Graphs," *Simon Stevin*, **28** (1950–1951), 203–217.

22. Ore, O., *The Four Color Problem*, Academic Press, 1967.

23. Busacker, R. G., and T. L. Saaty, *Finite Graphs and Networks*, McGraw-Hill, 1965.

24. Kac, M., "Can One Hear the Shape of a Drum?" *Amer. Math. Month*, **73** (1966), 1–23.

25. Harary, F., "The Determinant of the Adjacency Matrix of a Graph," *SIAM Rev.*, **4** (1962), 202–210.

26. Fisher, M. E., "On Hearing the Shape of a Drum," *J. Comb. Theory*, **1** (1966), 105–125.

27. Baker, G. A., "Drum Shapes and Isospectral Graphs," *J. Math. Phys.*, **7** (1966), 2238–2242.

28. Householder, A. S., *Principles of Numerical Analysis*, McGraw-Hill, 1953.

29. Turner, J., "Generalized Matrix Functions and the Graph Isomorphism Problem," *SIAM J. Appl. Math.*, **16** (1968), 520–526.

30. Blackett, D., *Elementary Topology*, Academic Press, 1967.

31. Cayley, A., "On the Mathematical Theory of Isomers," *Phil. Mag.*, **67** (1874), 444–446.

32. König, D., *Theorie der Endlichen und Unendlichen Graphen*, Chelsea reprint, 1936.

33. Aleksandrov, P. L., *Combinatorial Topology*, Vol. 1, Graylock Press, 1956.

34. Whitney, H., "Congruent Graphs and the Connectivity of Graphs," *Amer. J. Math.*, **54** (1932), 150–168.

35. Harary, F., and G. Chartrand, "Graphs with Prescribed Connectivities," in *Theory of Graphs*, P. Erdos, and G. Katona, Eds., Akademiai Kiado, 1968, pp. 61–63.

36. Harary, F., *Graph Theory*, Addison-Wesley, 1969.

37. Nettleton, R. E., K. Goldberg, and M. S. Green, "Dense Subgraphs and Connectivity," *Canad. J. Math.*, **11** (1959), 262–268.

38. Nettleton, R. E., "Some Generalized Theorems on Connectivity," *Canad. J. Math.*, **12** (1960), 546–554.

39. Beineke, L. W., "The Decomposition of Complete Graphs into Planar Subgraphs," in *Graph Theory and Theoretical Physics*, F. Harary, Ed., Academic Press, 1967, pp. 139–154.

40. Saaty, T. L., *Optimization in Integers and Related Extremal Problems*, McGraw-Hill, 1970.

41. Guy, R. K., "The Decline and Fall of Zarankiewicz's Theorem," *Proof Techniques in Graph Theory*, F. Harary, Ed., Academic Press, 1969.

42. Tutte, W. T., "Toward a Theory of Crossing Numbers," *J. Comb. Theory*, **8** (1970), 45–53.

43. Fáry, I., "On Straight Line Representation of Planar Graphs," *Acta Sci.-Math.*, **11** (1948), 229–233. In series *Acta Univ. Szeged.* (Hungary).

44. Mac Lane, S., "A Structural Characterization of Planar Combinatorial Graphs," *Duke Math. J.*, **3** (1937), 460–472.

45. Tutte, W. T., "An Algorithm for Determining Whether a Given Binary Matroid is Graphic," *Proc. Amer. Math. Soc.*, **11** (1960), 905–917.

46. Weinberg, L., "A Simple and Efficient Algorithm for Determining Isomorphism of Planar Triply Connected Graphs," *IEEE Trans. Circuit Theory*, **CT-13** (1966), 142–148.
47. Nash-Williams, C. St. J. A., "On Hamiltonian Circuits in Finite Graphs," *Proc. Amer. Math. Soc.*, **17** (1966), 466–467.
48. Ball, W. W. R., and H. S. M. Coxeter, *Mathematical Recreations and Essays*, Macmillan, 1947.
49. Tutte, W. T., "On Hamiltonian Circuits," *J. London Math. Soc.*, **21** (1946), 98–101.
50. Grünbaum, B., *Convex Polytopes*, Wiley, 1967.
51. Tait, P. G., "Listings Topologie," *Phila. Mag.*, *(Series 5)*, **17** (1884).
52. Lederberg, J., "Hamilton Circuits of Convex Trivalent Polyhedra," *Amer. Math. Month*, **74** (1967), 522–526.
53. Grace, D. W., "Computer Search for Non-Isomorphic Convex Polyhedra," Stanford Computation Center Tech. Report CS15, 1965.
54. Walther, H., "Ein Kubischer, Planarer, Zyklisch Fünffach Zusammenhangender Graph, der Keinen Hamiltonkreis Besitzt," *Wiss. Z. TH Ilmenau*, **11** (1965), 163–166.
55. Sachs, H., "Construction of Non-Hamiltonian Planar Regular Graphs," *Theory of Graphs—Rome Symposium (1966)*, P. Rosenstiehl, Ed., Gordon and Breach, 1967, 373–382.
56. Nash-Williams, C. St. J. A., "Decomposition of the n-Dimensional Lattice-Graph into Hamiltonian Lines," *Proc. Edinburgh Math. Soc. (Series 2)*, **12** (1960–1961), 123–131.
57. Kotzig, A., "Hamilton Graphs and Hamilton Circuits," *Theory of Graphs and its Applications—Smolenice Symposium (1963)*, M. Fiedler, Ed., Academic Press, 1964, pp. 63–82.
58. Mycielski, J., "Sur le Coloriage des Graphes," *Colloq. Math.*, **3** (1955), 161–162.
59. Grünbaum, B., "A Problem in Graph Coloring," *Amer. Math. Month.*, **77** (1970), 1088–1092.
60. Franklin, P., "A Six Color Problem," *J. Math. Phys.*, **13**, (1934), 363–369.
61. Yamabe, H., and D. Pope, "A Computational Approach to the Four-Color Problem," *Math. Comp.*, **15** (1961), 250–253.
62. May, K. O., "The Origin of the Four-Color Conjecture," *Isis*, **56** (1965), 346–348.
63. Kemeny, J. G., J. L. Snell, and G. L. Thompson, *Introduction to Finite Mathematics*, Prentice-Hall, 1957.
64. Liu, C. L., *Introduction to Combinatorial Mathematics*, McGraw-Hill, 1968.
65. Hall, M., *Combinatorial Theory*, Blaisdell, 1967.
66. Hoffman, A. J., "Some Recent Applications of the Theory of Linear Inequlities to Extremal Combinatorial Analysis," *Combinatorial Analysis*, R. Bellman, and M. Hall, Eds., American Mathematical Society, 1960, pp. 113–128.
67. Ore, O., "Graphs and Matching Theorems," *Duke Math. J.*, **22** (1955), 625–639.
68. Harary, F., "Determinants, Permanents, and Bipartite Graphs," Math Mag., **42** (1969) 146–148.
69. Turán, P., "On the Theory of Graphs," *Colloq. Math.*, **3** (1954), 19–30.
70. de Bruijn, N. G., "Polya's Theory of Counting," *Applied Combinatorial Mathematics*, E. F. Beckenbach, Ed., Wiley, 1964, pp. 144–184.
71. Hall, M., "Block Designs," *Applied Combinatorial Mathematics*, E. Beckenbach, Ed., Wiley, 1964, pp. 369–405.
72. Moon, J. W., "On the Distribution of Crossings in Random Complete Graphs," *SIAM J.*, **13** (1965), 506–510.
73. Koopman, B. O., "Combinatorial Analysis of Operations," *Proc. 1st Internat. Conf. Operations Res.*, Operations Research Society of America, 1957.

74. Marshall, C. W., "Full Service Probability for Regular Arrangements," *Oper. Res.*, **9** (1961), 186–199.

75. Finch, P. D., "Notes on a Combinatorial Problem of B. O. Koopman," *Oper. Res. Quart.*, **9** (1958), 169–173.

76. Marshall, C. W., "Double Correspondence Graphs," *SIAM J.*, **10** (1962), 211–227.

77. Austin, T. L., R. E. Fagen, W. F. Penny, and J. Riordan, "The Number of Components in Random Linear Graphs," *Ann. Math. Stat.*, **30** (1959), 747–754.

78. Gilbert, E. N., "Enumeration of Labelled Graphs," *Canad. J. Math.*, **8** (1956), 405–411.

79. Gilbert, E. N., "Random Graphs," *Ann. Math. Stat.*, **30** (1959), 1141–1144.

80. Erdös, P., and A. Renyi, "On Random Graphs I," *Publ. Math. Debrecen*, **6** (1959), 290–297.

81. Erdös, P., "Graph Theory and Probability," *Canad. J. Math.*, **11** (1959), 34–38.

82. Erdös, P., "Graph Theory and Probability II," *Canad. J. Math.*, **13** (1961), 346–352.

83. Robbins, H. E., "A Theorem on Graphs, with an Application to a Problem of Traffic Control," *Amer. Math. Month.*, **46** (1939). 281–283.

84. Farbey, B. A., A. H. Land, and J. D. Murchland, "The Cascade Algorithm for Finding all Shortest Distances in a Directed Graph," *Manage. Sci.* **14** (1967), 19–28.

85. Hu, T. C., "Revised Matrix Algorithms for Shortest Paths," *SIAM J. Appl. Math.*, **15** (1967), 207–218 (with corrections on p. 1517).

86. Land, A. H., and S. W. Stairs, "The Extension of the Cascade Algorithm to Large Graphs," *Manage. Sci.*, **14** (1967), 29–33.

87. Miller, R., *Schedule, Cost and Profit Control with PERT*, McGraw-Hill, 1963.

88. Muth, J., and G. Thompson, *Industrial Scheduling*, Prentice-Hall, 1963.

89. Ford, L. R., and D. R. Fulkerson, "Maximal Flows through a Network," *Canad. J. Math.*, **8** (1956), 399–404.

90. Ford, L. R., and D. R. Fulkerson, *Flows in Networks*, Princeton University Press, 1962.

91. Karreman, G., "Topological Information Content and Chemical Reactions," *Bull. Math. Biophys.*, **17** (1955), 279–285.

92. Rashevsky, N., "Life, Information Theory, and Topology," *Bull. Math. Biophys.*, **17** (1955), 229–135.

93. Trucco, E., "A Note on the Information Content of Graphs," *Bull. Math. Biophys.*, **18** (1956), 129–135.

94. Sabidussi, G., "The Composition of Graphs," *Duke Math. J.* **26** (1959), 693–696.

95. Mowshowitz, A., "Entropy and the Complexity of Graphs: I," *Bull. Math. Biophys.* **30** (1968), 175–204.

96. Mowshowitz, A., "Entropy and the Complexity of Graphs: II," *Bull. Math. Biophys.*, **30** (1968), 225–240.

97. Lawler, E. L., K. N. Levitt, and J. Turner, "Module Clustering to Minimize Delay in Digital Networks," *IEEE Trans. Computers*, **C-18** (1969), 47–57.

98. Gorinshteyn, L. L., "The Partitioning of Graphs," *Eng. Cybern.*, No. 1 (1969), 76–82.

99. Collias, N. E., "Problems and Principles of Animal Sociology," *Comparative Psychology*, C. P. Stone, Ed., Prentice-Hall, 1951, pp. 388–422.

100. Ross, I. C., and F. Harary, "On the Determination of Redundancies in Sociometric Chains," *Psychometrika*, **17** (1952), 195–208.

101. Parthasarathy, K. R., "Enumeration of Paths in Digraphs," *Psychometrika*, **29** (1964), 153–165.

102. Lewin, K., *The Conceptual Representation and the Measurement of Psychological Forces*, Duke University Press, 1938.

103. Berger, J., B. P. Cohen, J. L. Snell, and M. Zelditch, *Types of Formalization in Small-Group Research*, Houghton-Mifflin, 1962.

104. Fisher, M., and J. Essam, "Some Cluster Size and Percolation Problems," *J. Math. Phys.*, **2** (1961), 609–619.

105. Montroll, E. W., "Lattice Statistics," in *Applied Combinatorial Mathematics*, E. Beckenbach, Ed., Wiley, 1964, 96–143.

106. Fisher, M. E., "Statistical Mechanics of Dimers on a Plane Lattice," *Phys. Rev.*, **124** (1961), 1664–1672.

107. Fisher, M. E., and H. N. V. Temperley, "Dimer Problem in Statistical Mechanics—An Exact Result," *Phil. Mag.* **6** (1961), 1061–1063.

108. Kasteleyn, P. W., "The Statistics of Dimers on a Lattice," *Physica*, **27** (1961), 1209–1225.

109. Uhlenbech, G. E., and G. W. Ford, *Studies in Statistical Mechanics*, J. De Boer and G. E. Uhlenbeck, Eds., North Holland, 1962.

110. Sykes, M. F., J. W. Essam, B. R. Heap, and B. J. Hilery, "Lattice Constant Systems and Graph Theory," *J. Math. Phys.*, **7** (1966), 1557–1572.

111. Heap, B. P., "The Enumeration of Homeomorphically Irreducible Star Graphs," *J. Math. Phys.*, **7** (1966), 1582–1587.

112. Nagle, J. F., "On Ordering and Identifying Undirected Linear Graphs," *J. Math. Phys.*, **7** (1966), 1588–1592.

113. Lederberg, J., "Topology of Molecules," *The Mathematical Sciences*, G. Boehm, Ed. M.I.T. Press, 1969.

114. Dantzig, G. B., *Linear programming and Extensions*, Princeton University Press, 1963.

115. Balinski, M. L., "Integer Programming: Methods, Uses, Computation," in *Mathematics of the Decision Sciences*, Part 1, G. B. Dantzig, and A. F. Veinott, Eds., American Mathematical Society, 1968, pp. 179–256.

116. Gomory, R. E., "An Algorithm for Integer Solutions to Linear Programms" in *Recent Advances in Mathematical Programming*, R. L. Graves, and P. Wolfe, Eds., McGraw-Hill, 1963, pp. 269–302.

117. Balas, E., "Linear Programs with Zero-One Variables," *Operations Res.*, **13** (1965), 517–545.

118. Glover, F., "A Multiphase-Dual Algorithm for the Zero-one Integer Programming Problem," *Operations Res.*, **13** (1965), 879–919.

Author Index

317

Subject Index

319